Zebrafish Models in Neurobehavioral Research

AF773818

Edited by

Allan V. Kalueff

and

Jonathan M. Cachat

Department of Pharmacology and Neuroscience Program,
Tulane University Medical School, New Orleans, LA, USA

Humana Press

Editors
Allan V. Kalueff
Department of Pharmacology
and Neuroscience Program
Tulane University Medical School
70112 New Orleans, LA, USA
avkalueff@gmail.com

Jonathan M. Cachat
Department of Pharmacology
and Neuroscience Program
Tulane University Medical School
70112 New Orleans, LA, USA
cachatj@gmail.com

ISSN 0893-2336
e-ISSN 1940-6045
ISBN 978-1-4939-5922-8
ISBN 978-1-60761-922-2 (eBook)
DOI 10.1007/978-1-60761-922-2
Springer New York Dordrecht Heidelberg London

© Springer Science+Business Media, LLC 2011
Softcover re-print of the Hardcover 1st edition 2011
All rights reserved. This work may not be translated or copied in whole or in part without the written permission of the publisher (Humana Press, c/o Springer Science+Business Media, LLC, 233 Spring Street, New York, NY 10013, USA), except for brief excerpts in connection with reviews or scholarly analysis. Use in connection with any form of information storage and retrieval, electronic adaptation, computer software, or by similar or dissimilar methodology now known or hereafter developed is forbidden.
The use in this publication of trade names, trademarks, service marks, and similar terms, even if they are not identified as such, is not to be taken as an expression of opinion as to whether or not they are subject to proprietary rights.
While the advice and information in this book are believed to be true and accurate at the date of going to press, neither the authors nor the editors nor the publisher can accept any legal responsibility for any errors or omissions that may be made. The publisher makes no warranty, express or implied, with respect to the material contained herein.

Printed on acid-free paper

Humana Press is part of Springer Science+Business Media (www.springer.com)

NEUROMETHODS

Series Editor
Wolfgang Walz
University of Saskatchewan
Saskatoon, SK, Canada

For other titles published in this series, go to
www.springer.com/series/7657

Series Preface

Under the guidance of its founders Alan Boulton and Glen Baker, the Neuromethods series by Humana Press has been very successful since the first volume appeared in 1985. In about 17 years, 37 volumes have been published. In 2006, Springer Science + Business Media made a renewed commitment to this series. The new program will focus on methods that are either unique to the nervous system and excitable cells or which need special consideration to be applied to the neurosciences. The program will strike a balance between recent and exciting developments like those concerning new animal models of disease, imaging, in vivo methods, and more established techniques. These include immunocytochemistry and electrophysiological technologies. New trainees in neurosciences still need a sound footing in these older methods in order to apply a critical approach to their results. The careful application of methods is probably the most important step in the process of scientific inquiry. In the past, new methodologies led the way in developing new disciplines in the biological and medical sciences. For example, physiology emerged out of anatomy in the nineteenth century by harnessing new methods based on the newly discovered phenomenon of electricity. Nowadays, the relationships between disciplines and methods are more complex. Methods are now widely shared between disciplines and research areas. New developments in electronic publishing also make it possible for scientists to download chapters or protocols selectively within a very short time of encountering them. This new approach has been taken into account in the design of individual volumes and chapters in this series.

Wolfgang Walz

Preface

The use of animal models has become increasingly important for biomedical research over the past decade, enabling a better understanding of pathogenic pathways involved in a variety of human disorders. Within the realm of neurobehavioral research, animal models have played a crucial role in the development of new insights and theories of brain pathogenesis. Animal models such as mice, hamsters, and rabbits have been utilized in a multitude of neurobehavioral studies, yielding experimental data that have lead researchers to a better comprehension of neurobiology. As scientific research progresses, investigators are attempting to identify more novel animal models to utilize in new avenues of neurobehavioral research.

Zebrafish (*Danio rerio*) have become increasingly popular in biomedical research. Research conducted on these aquatic vertebrates has generated considerable discoveries not only in the areas of genetics and embryology but also in fields such as cardiology, endocrinology, and neuroscience. Zebrafish are promising animal models because of their high genetic homology with humans and quantifiable behavioral and neuropathological phenotypes analogous to humans.

The use of zebrafish to investigate the pathological mechanisms underlying neuropsychiatric disorders and behavior quantification is explored in depth in this book. The opening **Chapter** 1 is a comprehensive review of zebrafish behavior, ecology, taxonomy, reproduction, and genetics. This chapter emphasizes the need for continued experimentation in cognition, behavior, and field-based studies, resulting in a more thorough understanding of the zebrafish model.

Critical to survival in a natural habitat and strongly influencing their behavior, the olfactory system in zebrafish is explored in **Chapter** 2. Zebrafish possess three distinct types of olfactory sensory neurons, which integrate with other areas of the brain to induce various physiological and behavioral effects in response to odors. Olfaction allows zebrafish to detect nearby food, predators, and mates, in addition to conveying information relating to spawning sites, reproduction, dangerous environments, and the distinction between self and kin. Advanced knowledge of the neurological basis of olfaction is key to a better understanding of zebrafish wild type and anxiety-related behavior.

Chapter 3 focuses on the emergence of zebrafish as an effective model to study stress and anxiety. This chapter presents a concise introduction to anxiety-induced endocrine and behavioral responses in zebrafish. Since zebrafish possess all the classical vertebrate transmitters, and their neuroendocrine system yields robust cortisol responses to stress, zebrafish models enable greater insight into neural mechanisms associated with anxiety-related disorders. Furthermore, this chapter illustrates the importance of behavioral assays, genetic manipulation, pharmacological treatment, and video tracking for analysis of the phenomena involved in anxiety-related phenotypes.

While zebrafish demonstrate promising potential in the field of anxiety and stress-related research, they have also emerged as valuable models in other areas of neurobehavioral research. **Chapter** 4 describes how the effects of nicotine on processes such as learning, memory, and stress are similar to those exhibited by humans and rodents. The authors' analysis suggests that zebrafish may present significant translational capabilities in research as a model for the behavioral effects of nicotine.

Based on the establishment of zebrafish as a suitable model for behavioral research, **Chapter** 5 details the process for quantitative trait loci (QTL) mapping and how it attempts to discover the specific causative genes responsible for variations in complex behavioral traits in zebrafish. Because of the strides taken recently in the study of zebrafish behavior, QTL mapping would not only lead to a greater understanding of zebrafish activity, but also strengthen its application as a genetic model.

Chapter 6 explores the effects of alcohol on several strains of zebrafish. Like anxiety, alcoholism is a serious brain disease for which the pathogenic mechanisms are not well understood. Alcohol abuse in the world is on the rise, making a genetic model for the development of alcoholism vital. Using a noninvasive evaluation technique, the acute and chronic effects of ethanol on zebrafish were observed, clarifying the genetic factors involved in alcoholism.

Along the same line, the authors of **Chapter** 7 explore the use of zebrafish as a model of drug dependence and relapse behaviors in humans. These robust reactions to nicotine and alcohol not only reinforce the use of zebrafish as a behavioral model of addiction but also strengthen the notion that zebrafish may be utilized to discover various genetic factors underlying drug dependence, withdrawal, and relapse.

As previously mentioned, many neuroscientists seek to gain a more concrete understanding of the pathogenic mechanisms that induce neurobiological disorders and behavior. However, in some cases, an error in the mechanism of the neural circuitry is not the only contributing cause of behaviors or diseases that are expressed. **Chapter** 8 examines the impact of neurotoxic chemicals on the nervous system and their potential to increase susceptibility to neurodegenerative disorders. In this chapter, the authors utilize the high sensitivity of zebrafish to environmental changes to investigate the correlation between the influence of environmental neurotoxins and neurodegenerative disorders. This research analyzes alterations in the biogenic amine system following exposure to pesticides, as well as the detrimental effect of neurotoxins on the nervous system.

Other experiments that examine the neural effects of environmental factors are explored in **Chapter** 9. This chapter analyzes predator-avoidance behavior exhibited by zebrafish, which is induced by external environmental factors such as alarm pheromone. The predator-avoidance behavior displayed by zebrafish is based upon learned recognition of external environmental cues. Exploration into the process of learned recognition in zebrafish will enable researchers to gain a more tangible understanding of the mechanisms that underlie cognitive processes of learning and memory.

In **Chapter** 10, the authors discuss avoidance behavior in zebrafish. Similar to the learned recognition phenomenon, inhibitory avoidance paradigms provide insight into the learning and memory capabilities of zebrafish. While the behavioral phenotypes of small teleost fish have frequently been considered to be dominated by reflex and instinct, recent studies have suggested a more complex phenotype influencing emotional, social, and reproductive behavior. The authors employ new experimental models with zebrafish, and area to investigate learning and memory, and area of research that will contribute to a more comprehensive understanding of the zebrafish brain and behavior.

Further exploring the zebrafish neurocognitive domain, **Chapter** 11 reviews previous studies on the spatial cognitive abilities of zebrafish. Mounting evidence, summarized in this chapter, demonstrates the capability of zebrafish to learn from visual cues that identify potential risk or reward. The application of these tests may serve as an insightful resource by which the spatial cognition of zebrafish can be illuminated.

Finally, **Chapter** 12 describes common larval zebrafish behaviors. While the behavioral phenotypes of adult zebrafish are important to study in detail, the functionality of zebrafish larvae must be equally well understood in relation to its anatomical size and development. This chapter explores the scope of larval behavior, from movement to stimuli response to more complex behaviors such as swim bladder inflation, sleep, and social behavior. While a general repertoire may be established, specific behavioral tendencies are influenced by environmental factors such as temperature or nearby predators. Future experimentation is necessary to correlate the synergistic aspects of behavior and neurobiological development in zebrafish larvae.

Overall, this book emphasizes the growing importance of zebrafish in neurobehavioral research. As a promising alternative to mammalian animal models, zebrafish yield robust physiological responses analogous to humans but do not possess the complex behavioral phenotypes exhibited by many other animal models. This book portrays an extensive, thorough perspective on the emergence of zebrafish as a robust animal model in neuroscience research. The contributors to this book are leading international scholars whose work spearheads innovative research projects in laboratories around the world. The themes discussed within this book, ranging from stress-related behaviors to learning and memory phenotypes, encompass a wide spectrum of the utility of zebrafish within neurobiological disciplines. This theoretical book, as vol. 52 of the Humana Press Neuromethods series, complements another book ("Zebrafish Neurobehavioral Protocols", vol. 51) of this series, which focuses on practical laboratory applications of these concepts. Together, these two volumes will serve as a useful source for scientists new to the field, as well as established researchers seeking valuable insight into the growing utility of zebrafish in behavioral neuroscience.

Allan V. Kalueff
Jonathan M. Cachat

Contents

Contributors

ARANI ANANTHATHEVAN • *Biological and Experimental Psychology Group, School of Biological and Chemical Sciences, Queen Mary University of London, London, UK*
BRETT K. BARTELS • *Department of Pharmacology and Neuroscience Program, Tulane University Medical School, New Orleans, LA, USA*
ESTHER BEESON • *Department of Pharmacology and Neuroscience Program, Tulane University Medical School, New Orleans, LA, USA*
CAROLINE H. BRENNAN • *Biological and Experimental Psychology Group, School of Biological and Chemical Sciences, Queen Mary University of London, London, UK*
HAROLD A. BURGESS • *Laboratory of Molecular Genetics, Eunice Kennedy Shriver National Institute of Child Health and Human Development, National Institutes of Health, Bethesda, MD, USA*
JONATHAN M. CACHAT • *Department of Pharmacology and Neuroscience Program, Tulane University Medical School, New Orleans, LA, USA*
PETER R. CANAVELLO • *Department of Pharmacology and Neuroscience Program, Tulane University Medical School, New Orleans, LA, USA*
LUCIANA COFIEL • *Laboratório de Neurociências, Departamento de Fisioterapia, Universidade Federal de São Carlos, São Carlos, SP, Brazil*
CYNTHIA A. DLUGOS • *Department of Pathology and Anatomical Sciences, School of Medicine and Biomedical Sciences, University of Buffalo/State University of New York, Buffalo, NY, USA*
ARTI DOSHI • *Biological and Experimental Psychology Group, School of Biological and Chemical Sciences, Queen Mary University of London, London, UK*
MARCO F. ELEGANTE • *Department of Pharmacology and Neuroscience Program, Tulane University Medical School, New Orleans, LA, USA*
SALEM I. ELKHAYAT • *Department of Pharmacology and Neuroscience Program, Tulane University Medical School, New Orleans, LA, USA*
KANDICE FERO • *Laboratory of Molecular Genetics, Eunice Kennedy Shriver National Institute of Child Health and Human Development, National Institutes of Health, Bethesda, MD, USA*
ADAM M. STEWART • *Department of Pharmacology and Neuroscience Program, Tulane University Medical School, New Orleans, LA, USA*
GEORGIANNA G. GOULD • *Department of Physiology, University of Texas at Health Science Center at San Antonio, San Antonio, TX, USA*
JOSHUA L. HAIGHT • *Behavioral Neuroscience Program, Connecticut College, New London, CT, USA*
PETER C. HART • *Department of Pharmacology and Neuroscience Program, Tulane University Medical School, New Orleans, LA, USA*
WHITLEE HAYMORE • *Department of Pharmacology and Neuroscience Program, Tulane University Medical School, New Orleans, LA, USA*
ALLAN V. KALUEFF • *Department of Pharmacology and Neuroscience Program, Tulane University Medical School, New Orleans, LA, USA*

LAYLA K.M. KILY • *Biological and Experimental Psychology Group, School of Biological and Chemical Sciences, Queen Mary University of London, London, UK*
AUTUMN L. LAFFOON • *Department of Pharmacology and Neuroscience Program, Tulane University Medical School, New Orleans, LA, USA*
EDWARD D. LEVIN • *Department of Psychiatry and Behavioral Sciences, Duke University Medical Center, Durham, NC, USA*
ROSANA MATTIOLI • *Laboratório de Neurociências, Departamento de Fisioterapia, Universidade Federal de São Carlos, São Carlos, SP, Brazil*
SOPAN MOHNOT • *Department of Pharmacology and Neuroscience Program, Tulane University Medical School, New Orleans, LA, USA*
AMIT PARMAR • *Biological and Experimental Psychology Group, School of Biological and Chemical Sciences, Queen Mary University of London, London, UK*
SALMA PATEL • *Biological and Experimental Psychology Group, School of Biological and Chemical Sciences, Queen Mary University of London, London, UK*
JOSEPH A. SCHROEDER • *Department of Psychology, Connecticut College, New London, CT, USA*
ROWENA SPENCE • *University of St Andrews, St. Andrews Fife, Scotland*
SIDDHARTH GAIKWAD • *Department of Pharmacology and Neuroscience Program, Tulane University Medical School, New Orleans, LA, USA*
ANNA K. TIEN • *Department of Pharmacology and Neuroscience Program, Tulane University Medical School, New Orleans, LA, USA*
DAVID H. TIEN • *Department of Pharmacology and Neuroscience Program, Tulane University Medical School, New Orleans, LA, USA*
KATHLEEN E. WHITLOCK • *Centro de Genomicas Celular, Centro de Neurociencia de Valparaíso (CNV), Universidad de Valparaíso, Valparaíso, Chile*
BRIAN D. WISENDEN • *Biosciences Department, Minnesota State University Moorhead, Moorhead, MN, USA*
KEITH WONG • *Department of Pharmacology and Neuroscience Program, Tulane University Medical School, New Orleans, LA, USA*
DOMINIC WRIGHT • *IFM-Biology, Linköping University, Linköping, Sweden*
TOHEI YOKOGAWA • *Laboratory of Molecular Genetics, Eunice Kennedy Shriver National Institute of Child Health and Human Development, National Institutes of Health, Bethesda, MD, USA*

Chapter 1

Zebrafish Ecology and Behaviour

Rowena Spence

Abstract

The zebrafish is an important model organism in developmental genetics, neurophysiology and biomedicine, but little is known about its natural ecology and behaviour. It is a small, shoaling cyprinid, native to the flood-plains of the Indian subcontinent, where it is found in shallow, slow-flowing waters. Zebrafish are group spawners and egg scatterers, although females are selective with respect to sites for oviposition and males are territorial around such sites. Laboratory studies of zebrafish behaviour have encompassed shoaling, foraging, reproduction, sensory perception and learning. This chapter reviews these studies in relation to the suitability of the zebrafish as a model for studies in behavioural ecology.

Key words: Model organism, social behaviour, morphology, ecology, reproduction, development (ontogeny), evolution (phylogeny), natural habitat, diet, social behaviour, reproductive behaviour, cognitive behaviour, genetics.

1. Introduction

1.1. The Zebrafish as a Model Organism

The zebrafish, *Danio rerio* (Hamilton), is one of the most important vertebrate model organisms in genetics, developmental biology, neurophysiology and biomedicine (1–4). It has a number of attributes that make it particularly tractable to experimental manipulation. It is a small, robust fish, so large numbers can be kept easily and cheaply in the laboratory, where it breeds all year round. Females can spawn every 2–3 days and a single clutch may contain several hundred eggs. Generation time is short, typically 3–4 months, making it suitable for selection experiments. Zebrafish eggs are large relative to other fish eggs (0.7 mm in diameter at fertilisation), and optically transparent, the yolk being sequestered into a separate cell. Furthermore, fertilisation

A.V. Kalueff, J.M. Cachat (eds.), *Zebrafish Models in Neurobehavioral Research*, Neuromethods 52,
DOI 10.1007/978-1-60761-922-2_1, © Springer Science+Business Media, LLC 2011

is external so live embryos are accessible to manipulation and can be monitored through all developmental stages under a dissecting microscope (5). Development is rapid, with precursors to all major organs developing within 36 h and larvae displaying food seeking and active avoidance behaviours within 5 days post fertilisation, i.e. 2–3 days after hatching (5).

As a popular aquarium species, the zebrafish has been used in developmental biology for many years (e.g. (6)). Its current prominence as a model organism stems from the work of Streisinger et al. (7), who pioneered its use to apply molecular genetics to the study of vertebrate embryology, and Kimmel (8–10), who published detailed descriptions of cell differentiation and nervous system organisation (for review see (2)). The zebrafish was the subject of the first large-scale random mutagenesis screens to be conducted in a vertebrate (11). These screens, conducted in 1996 in Boston (12) and Tübingen (13), generated over 4,000 mutations and led to the identification of over 400 genes controlling vertebrate development. Since then there have been numerous technological advances (for review see (14–22)), culminating in the zebrafish genome project, based at the Sanger Institute in Cambridge, which began in 2001 and will shortly be completed (http://www.sanger.ac.uk). The zebrafish is increasingly important in biomedical research (23–25), particularly as a model of human disease (26, 27) and for the screening of therapeutic drugs (3, 28). Its strength as a model organism is that as a vertebrate it is more comparable to humans than invertebrate model species such as *Drosophila* (29, 30), while being more tractable to genetic and embryological manipulation than mammalian model species such as mice, in which such procedures are both more complicated and costly.

Over 400 labs worldwide now routinely use the zebrafish in fundamental and applied research (http://www.zfin.org) and there is an increasing interest in its use as a model for understanding the genetic basis of behaviour (18, 31, 32). **Figure 1.1** shows the number of papers on zebrafish behaviour published each decade since the 1970s, based on a search of Web of Science using "zebrafish" and "behaviour/behavior" as keywords. Despite this interest, it has attracted little attention from the behavioural ecology community, possibly because little is known about its natural ecology and few studies have been conducted on wild populations. Most laboratory lines of zebrafish are the product of many generations in captivity, which is likely to have resulted in selection for reproductive capacity, while relaxing selection for other traits, such as predator avoidance (33, 34). Thus, it is not clear in what respect and to what extent domesticated strains may differ from wild fish, nor how much inter-population variation exists in nature. This chapter reviews the current state of knowledge of the ecology and behaviour of the zebrafish. The term behaviour is

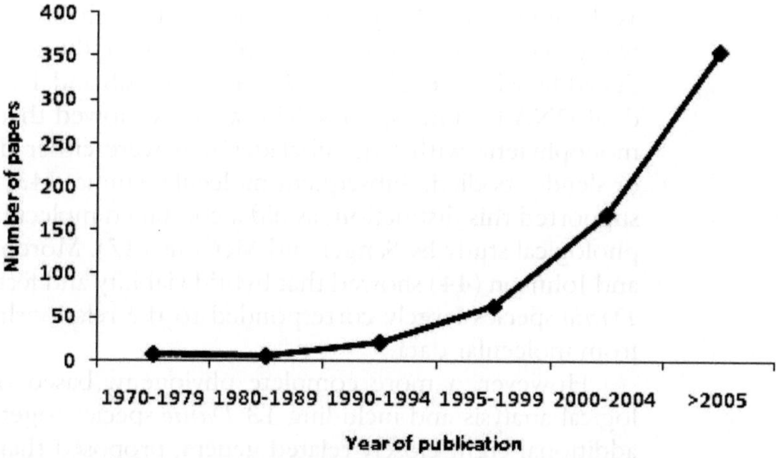

Fig. 1.1. The numbers of papers on zebrafish behaviour published since the 1970s, based on a keyword search in the Web of Science, up to mid 2009.

used not in the sense of a simple reflexive response to stimuli but rather for complex patterns of behaviour such as those involved in social and reproductive behaviour.

2. Taxonomy

2.1. Taxonomic Status

The zebrafish belongs to the family of freshwater fishes Cyprinidae, the most species-rich vertebrate family (35). There are currently approximately 44 danionin species (36), distributed throughout South and South East Asia. Their highest species diversity is in North East India, Bangladesh, and Myanmar (37). The name *Danio* derives from the Bengali name "*dhani*", meaning "of the rice field" (38). Danios are included in the subfamily Rasborinae (39). They are characterised by small size (<120 mm total length), the presence of a "danionin notch", in the ventromedial margin of the dentary, and a distinctive colour pattern based on alternating dark and light horizontal stripes, which may be broken up into blotches or bars.

Danio rerio was first described by Francis Hamilton, a surgeon with the British East India Company stationed principally in West Bengal at the beginning of the nineteenth century. He published *An Account of the Fishes Found in the River Ganges and its Branches* in 1822 that included 10 *Danio* species. *D. rerio* was later assigned to the subgenus *Brachydanio*, together with the other small *Danio* species with short dorsal fins and a reduced lateral line, *Danio* being reserved for the larger species of the group (40). *Danio* and *Brachydanio* were synonymised by Barman (37), as there were no diagnostic characters that reliably separated the two groups. The first molecular phylogeny of the group was produced by Meyer et al. (41, 42) based on 16S and 12S mitochondrial DNA for nine species. This analysis showed that *Danio* was monophyletic with two subclades that were either deep bodied or slender bodied. Subsequent molecular studies (43, 44, 45, 46) supported this distinction, as did a combined molecular and morphological study by Sanger and McCune (47). Moreover, Parichy and Johnson (44) showed that hybrid viability and fertility among *Danio* species largely corresponded to the relationships inferred from molecular data.

However, a more complete phylogeny, based on morphological analysis and including 13 *Danio* species together with an additional eight closely related genera, proposed that *Danio* was paraphyletic, the deep- and slender-bodied clades forming separate genera (48). The deep-bodied clade was thus assigned the distinct generic name of *Devario*, and includes most of the striped and barred danios (of which about 45 are considered valid), with *Danio sensu stricto*, (including *D. rerio*) restricted to nine species (48). A subsequent study using molecular data from a number of nuclear and mitochondrial genes and phylogenetic analysis confirmed this distinction, identifying *Danio* as monophyletic, being as closely related to *Chela, Microrasbora* and *Inlecypris* as to *Devario* (49). The closest relative of *D. rerio* is *D. kyathit* (49).

The two genera (*Devario* and *Danio*) cannot be reliably distinguished on the basis of proportional measurements alone, as there is considerable intra-species variation, mature females typically being deeper bodied than males or juveniles. Although *Devario* tend to be larger, one of the large species, *Danio dangila*, is included in *Danio* (36, 44, 45, 46). However, the two genera are ecologically quite distinct, *Devario* spp. occurring in hill streams with clear running water, while *Danio* spp. are confined to lowland areas, typically inhabiting slow-flowing, turbid rivers and pools (36).

2.2. Appearance and Morphology

Danio rerio rarely exceeds 40 mm body length (from the tip of the snout to the origin of the caudal fin (BL)). Its body shape is fusiform and laterally compressed, with a terminal oblique mouth directed upwards. The lower jaw protrudes further than the upper and the eyes are central and not visible from above. The diagnostic features for the species are an incomplete lateral line extending to the pelvic fin base, two pairs of barbels and five to seven dark blue longitudinal stripes extending from behind the operculum into the caudal fin (37). The anal fin is similarly striped, while the dorsal fin has a dark blue upper edge, bordered with white. The colour pattern comprises three types of pigment cell, dark blue melanophores, gold xanthophores and iridescent iridophores (50, 51). Developmentally, two stripes first form centrally with subsequent stripes being added sequentially above and below (43). As with many teleosts, the melanophores can be concentrated or dispersed in response to stimuli, which appear to function for both camouflage (in response to light intensity (18, 52)) and signalling (since fish typically darken during aggressive display (31, 53)). Colour change appears to be under some degree of cognitive control; fish which were subjected to cyclical alternations of black and white backgrounds over 20 days showed an increase in the speed and degree of aggregation and dispersal of melanophores (52). Males and females are of similar colouration, although males tend to have larger anal fins with more yellow colouration (54, 55). The sex of juveniles cannot be reliably distinguished without dissection and while gravid females have a more rounded body shape, the most reliable diagnostic feature is the presence of a small genital papilla in front of the anal fin origin (54).

2.3. Domestic Aquarium Strains

Zebrafish used for mutagenesis and screening are from lines bred in laboratories for many generations in order to maintain a stable genetic background. They are also "cleaned up"; i.e. bred selectively to remove embryonic lethal mutations. The main currently recognised wild-type lines from the Zebrafish International Resource Center are summarised in **Table 1.1**. For details of mutant lines see http://zfin.org

The "Leopard" danio, which displays a spotted colour pattern instead of stripes, was originally thought to be a separate species, described as *Brachydanio frankei* (56). However, neither molecular nor morphological analyses have differentiated between the two (41, 57) while hybrids were shown to produce fertile progeny (48). The Leopard danio is now known to be a spontaneous mutation of the wild-type *D. rerio* colour pattern (59), with homozygotes displaying a spotted pattern, while heterozygotes have a disrupted stripe pattern (60). Leopard danio mutants are primarily bred for the aquarium trade but also occur in nature (R. Spence, pers. obs.). Another aquarium variant is the "longfin" *D. rerio*, which is a dominant mutation resulting in elongated fins (61). The commonly used wild-type strain, TL or Tübingen Long-fin displays both the "leopard" and "longfin" mutations (www.zfin.org).

2.4. Pigment Patterns in *Danio spp.*

Comparison of pigment patterns among *Danio* species has provided insights into their evolutionary relationships. Larval danios of different species exhibit an identical pigment pattern, which only differentiates into the adult pattern in about the third week of development (43). Interestingly, several *D. rerio* pigment pattern mutations resemble other *Danio* species (44). This remarkable concurrence in appearance raises the possibility that the alleles expressed by zebrafish colour mutants are the same as those

Table 1.1
Wild-type zebrafish lines listed by the Zebrafish International Resource Center

Name	Description
AB	Derived from two lines purchased by George Streisinger from a pet shop in Albany, Oregon in the late 1970s. The currently used line *AB was derived from the original AB line in 1991–1992 by parthenogenesis
AB/Tübingen	An "official" line maintained as a cross but the term is also applied to crosses where the two parental lines are maintained separately
C32	Derived from laboratory strains at Oregon. The current C32bc9 stock is a derivative of Steve Johnson's inbred C32
Cologne	Isolated at the Reugels/Campos-Ortega Lab, University of Cologne
Darjeeling	Collected in Darjeeling in 1987 and sent to Monte Westerfield at Oregon. A much faster swimmer than other wild-type strains. Used extensively for mapping as it contains many polymorphic markers
Ekkwill (EKW)	From Ekkwill breeders in Florida and maintained in Grunwald lab, University of Utah
Hong Kong	Stock obtained from a Hong Kong fish dealer
HK/AB	Hybrid of Hong Kong and AB wild-type lines
HK/Sing	Hybrid of Hong Kong and Singapore wild-type lines
India	Stock obtained from expedition to Darjeeling (wild isolate)
Indonesia	Stock obtained from Indonesian fish dealer
Nadia	Wild-caught about 40 miles east of Calcutta. The fish were collected from stagnant ponds and flood plains. Imported in 1999 by a wholesaler in Oregon. Established in the Oregon laboratory from an initial breeding of about 10 individuals
Singapore	Stock obtained from Singapore fish dealer
SJA	SJA is an inbred line of *AB isolated at the Stephen L. Johnson Lab, Washington University Medical School. Unlike *AB, which is bred to retain polymorphisms, this subline is bred to reduce polymorphism and is at least 85% monomorphic
SJD	Isolated at the Stephen L. Johnson Lab
Tübingen	Wild-type short fins. Strain used by Sanger for the zebrafish sequencing project. It was cleaned up to remove embryonic lethal mutations from the background before being used for mutagenesis and sequencing
Tübingen long fin	Homozygous for *leo*^{t1}, a recessive mutation causing spotting in adult fish, and *lof*^{dt2} a dominant homozygous viable mutation causing long fins. This is not the line used in the Sanger zebrafish sequencing project. It is genetically different from TU because it was bred differently and not "cleaned up", and therefore retains a lot of polymorphisms
WIK	Derived from wild catch in India and used for genome mapping

expressed in related *Danio* species. Consequently, these alleles may have played a role in colour pattern diversification among species (44).

A spectacular array of adult pigment pattern mutants have been identified for zebrafish (59, 60). Many mutant colour patterns can be attributed to a single locus, and several pigment genes have been identified at the molecular level (63–65). In a study of colour pattern inheritance, Parichy and Johnson (44) showed that hybrids between zebrafish and four closely related *Danio* species all expressed pigment patterns resembling that of wild-type zebrafish. These findings imply that stripes may be ancestral in *Danio* spp. Thus the zebrafish may serve as a useful model for studying the genetic and developmental basis of colour pattern evolution as a mechanism for speciation (50, 51).

3. Ecology

3.1. Distribution and Habitat

The natural range of the zebrafish is centred around the Ganges and Brahmaputra river basins in north-eastern India, Bangladesh and Nepal, although in the past specimens have also been collected in the Indus, Cauvery, Pennar, Godavari and Mahanadi river basins (**Fig. 1.2**, **Table 1.2**). In addition, it has been reported as occurring in the Krishna river basin (38) and in the

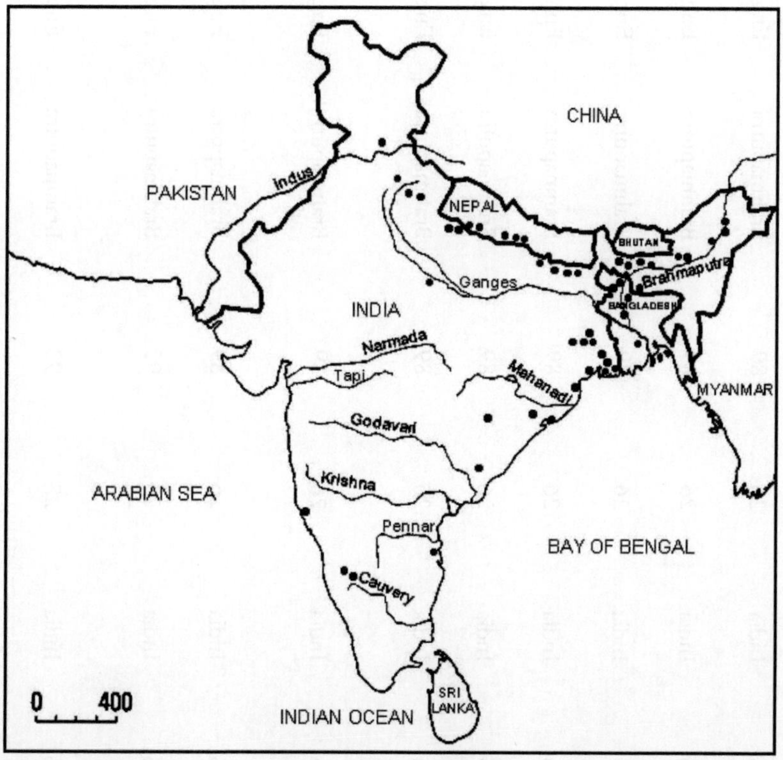

Fig. 1.2. The natural distribution of the zebrafish. Major river systems indicated. *Black dots* indicate recorded occurrences.

Table 1.2
Reported natural occurrences of *Danio rerio*

Year	Country	Latitude/longitude		Drainage	Source	Site description
2006	India	26	89	Brahmaputra	Engeszer et al. (73)	R. Jorai, slow flow, tea coloured, silt substrate, submerged vegetation
2006	India	26	89	Brahmaputra	Engeszer et al. (73)	R. Suthimari, medium flow, clear, silt substrate, abundant submerged vegetation
2006	India	26	89	Brahmaputra	Engeszer et al. (73)	R. Suthimari, clarity ~45 cm, medium flow, silt substrate, no vegetation
2006	India	26	89	Brahmaputra	Engeszer et al. (73)	R. Suthimari, clear, medium flow, silt substrate, abundant submerged and flooded vegetation
2006	India	26	89	Brahmaputra	Engeszer et al. (73)	R. Suthimari, clear, slow flow, rice paddy with yams, shaded
2006	India	26	89	Brahmaputra	Engeszer et al. (73)	Tributary of R. Rydak, clear, medium flow, gravel/cobble, silt substrate, abundant submerged vegetation
2006	India	26	89	Brahmaputra	Engeszer et al. (73)	Lefraguri swamp, clarity ~40 cm, no flow, silt substrate, abundant submerged and flooded vegetation
2006	India	26	89	Brahmaputra	Engeszer et al. (73)	R. Ghotamari, bottom visible, slow/medium flow, silt substrate, flooded and submerged vegetation
2006	India	25	92	Brahmaputra	Engeszer et al. (73)	Seinipoh stream, clarity ~100 cm, medium flow, gravel substrate, vegetation overhanging from bank
2006	India	25	92	Brahmaputra	Engeszer et al. (73)	Seinipoh stream, bottom visible, low flow, silt substrate, mature rice paddy
2006	India	25	92	Brahmaputra	Engeszer et al. (73)	Seinipoh stream, clarity ~50 cm, slow/no flow, silt substrate, flooded vegetation
2006	India	25	92	Brahmaputra	Engeszer et al. (73)	R. Dukan, bottom visible, slow flow, gravel/cobble substrate, vegetation overhanging from bank

2006	India	21	87	Ganges	Engeszer et al. (73)	Tarania village, clarity <3 cm, no flow, silt substrate, submerged and flooded vegetation
2006	India	21	87	Ganges	Engeszer et al. (73)	Tarania village, clarity <3 cm, no flow, silt substrate, submerged and flooded vegetation
2005	Bangladesh	22	90	Ganges	Spence et al. (66)	Ditch on campus of Khulna University, 3 m wide, <1 m deep, Secchi depth 50 cm, no vegetation. Grassy bank. Some shade
2005	Bangladesh	22	90	Ganges	Spence et al. (66)	Isolated channel of R. Golamari, near Khulna. Approx. area 200 × 1,500 m, 50 cm deep, Secchi depth 19 cm, vegetation at margins, silt substrate. Grassy bank. No shade
2005	Bangladesh	24	90	Brahmaputra	Spence et al. (66)	Isolated pond, Sutiakali, near Mymensingh. Approx. area 30 × 50 m, 15 cm deep, silt substrate, vegetated. Grassy bank. No shade
2005	Bangladesh	24	90	Brahmaputra	Spence et al. (66)	Isolated pond, Sutiakali, near Mymensingh. Approx. area 10 × 12 m, 40 cm deep, Secchi depth 15 cm, silt substrate, no vegetation. Grassy bank. Some shade
2005	Bangladesh	24	90	Brahmaputra	Spence et al. (66)	Isolated pond, Sutiakali, near Mymensingh. Approx. area 10 × 10 m, 1 m deep, Secchi depth 30 cm, silt substrate, no vegetation. Grassy bank. Some shade
2005	Bangladesh	24	90	Brahmaputra	Spence et al. (66)	Large semi-natural pond at Bangladesh Agricultural University field station, Mymensingh. Approx. area 105 m^2, 1 m deep, Secchi depth 30 cm, silt substrate, vegetation at margins. Vegetation on bank. No shade
2005	Bangladesh	24	90	Brahmaputra	Spence et al. (66)	Ditch connecting to paddy fields, Bangladesh Agricultural University field station, Mymensingh. 10 m wide, 50 cm deep, vegetation. Grassy bank. No shade

(continued)

Table 1.2 (continued)

Year	Country	Latitude/longitude		Drainage	Source	Site description
2005	Bangladesh	24	90	Brahmaputra	Spence et al. (66)	Small semi-natural pond at Bangladesh Agricultural University field station, Mymensingh. Approx. area 8 × 15 m, 65 cm deep, Secchi depth 15 cm, silt substrate, vegetation. Vegetation on bank. No shade
2005	Bangladesh	24	90	Brahmaputra	Spence et al. (66)	Channel adjacent to campus of Bangladesh Agricultural University, feeding into field station. 8 m wide, 75 cm deep, Secchi depth 15 cm, silt substrate, vegetation. Grassy bank. No shade
1998	India	22	88	Ganges	Fang & Roos, Swedish Museum of Natural History	About 65 km NNE of Calcutta, R. Tumapao close to Duma village, shore. Stream more than 100 m wide and >0.7 m deep with slow to moderately flowing, moderately turbid, brownish water. Adjacent to rice-field, no vegetation on bank, vegetation in water. Silt substrate
1998	India	24	87	Ganges	Fang & Roos, Swedish Museum of Natural History	Stream on Dumka-Rampurhat road, about 7 m wide and about 1 m deep, with slow-flowing, clear, greenish water. About 5% shade. Hilly area with grasses on land and on bank, some plants in water. Silt substrate
1998	India	24	87	Ganges	Fang & Roos, Swedish Museum of Natural History	Roadside stream about 62 km from Bhagalpur on Deughar-Bhagalpur road. Small stream 3 m wide and 0.5 m deep with fast running, clear, brownish water. No shade. Hilly area without vegetation on land, bank or in water. Sand substrate
1998	India	24	86	Ganges	Fang & Roos, Swedish Museum of Natural History	Stream on Jamtara – Deughar road, about 5–8 m wide and 0.4 m deep with moderate current and clear, uncoloured water. No shade. Hilly area with grass on land, no vegetation on bank or in water. Sandy substrate

1998	India	27	94	Brahmaputra	Fang & Roos, Swedish Museum of Natural History	Roadside ditch by the Sessa Tinali (Sessa crossing) on Dibrugarh – Jorhat road. About 15 m wide and 0.1–0.4 m deep with stagnant, brownish water. No shade. Plain with grass on land, no vegetation on bank, vegetation in water. Silt substrate
1998	India	27	95	Brahmaputra	Fang & Roos, Swedish Museum of Natural History	About 100 km SSE of Dibrugarh, small stream near R. Dilli. Stream about 2 m wide and 0.8 m deep with moderate current and yellow/brownish water. No shade. Plain with grass on land and bank, green algae in water. Silt substrate
1997	Bangladesh			Brahmaputra	Pritchard (67)	Small shallow pools in a dry river bed and an adjacent spring-fed pond in a village of the Santal tribal group near to the India-Bangladesh border. Some vegetation
1997	Bangladesh			Brahmaputra	Pritchard (67)	Artificial concrete channel at Northwest Fisheries, Saidpur, Bangladesh. Still, extremely turbid water. No vegetation
1997	Nepal			Brahmaputra	Pritchard (67)	Pond 10 km N of Tangail, next to Tangail–Madhupur highway. Clear water, silt substrate and vegetation. 1.5 m deep
1996	Nepal			Ganges	Pritchard (67)	Shallow ditches and pond on Ranpur campus of Royal Nepal Agricultural College, Chitwan. Clear, slow or still water, silt substrate, some vegetation
1996	Nepal	27	83	Ganges	Edds, Kansas University	Tribeni
1996	Nepal	28	80	Ganges	Edds, Kansas University	3 km W of Pipariya, Shuklaa Phataa Wildlife Reserve
1996	Nepal	28	80	Ganges	Edds, Kansas University	Confluence of 3 rivers (Chaudhar, Bahuni, Gobraiya) at Royal Shuklaa Phantaa Wildlife Reserve

(continued)

Table 1.2 (continued)

Year	Country	Latitude/longitude		Drainage	Source	Site description
1996	Nepal	28	80	Ganges	Edds, Kansas University	Raj-Marg highway, 9 km E of Mahendranagar
1996	Nepal	28	80	Ganges	Edds, Kansas University	Waters of Kailali district along Raj-Marg highway
1996	Nepal	27	84	Ganges	Edds, Kansas University	Narayangarh
1996	Nepal	26	86	Ganges	Edds, Kansas University	Just downstream from irrigation dam at Phattepur
1996	Nepal	26	86	Ganges	Edds, Kansas University	Just upstream from R. Koshi barrage
1996	Nepal	26	88	Ganges	Edds, Kansas University	Bhadrapur
1996	Nepal	26	87	Ganges	Edds, Kansas University	Belbari
1995	India			Ganges	McClure et al. (68)	Tributary of R. Song, Lachiwala, Dehra Dun, UP. 1–12 m wide, 16–57 cm deep, Secchi depth >35 cm. Substrate clay, silt, cobble, boulders. Shade 0–50%
1995	India			Ganges	McClure et al. (68)	Side channel of R. Pasuni, Janakikund, Banda, UP. Site characteristics as above
1995	India			Ganges	McClure et al. (68)	Rice paddy connected to R. Bhairab, near Bak Bungalow, Parganas, W. Bengal. Site characteristics as above
1993	India			Indus	Dutta (69)	Gadigarh stream, Jammu
1987	India			Cauvery	Roberts, California Academy of Sciences	NW/WNW of Mysore
1983	India	21	86	Mahanadi	Parshall, British Museum of Natural History	R. Salane
1975	Nepal			Ganges	Roberts, California Academy of Sciences	Chitawan Valley, 10 miles W of Narangar
1975	Nepal			Ganges	Roberts, California Academy of Sciences	Chitawan Valley, at Kasa Darbar or Dabar
1975	Nepal			Ganges	Roberts, California Academy of Sciences	Chitawan Valley, including Khagari Khola, 45 miles E and slightly N of Hetaura (Hitaura) and 11 miles SSE of Narangar

1975	Nepal			Ganges	Roberts, California Academy of Sciences	Chitawan Valley, low-lying mountain stream 1–2 miles S of Khoria Mohan in Someswar Hills (Hathimara Khola)
1975	Nepal			Ganges	Roberts, California Academy of Sciences	Chitawan Valley, R. Reu near confluence with R. Rapti
1975	Nepal			Ganges	Roberts, California Academy of Sciences	Farm pond 1–2 km east of Kalaiya or Khailaya
1972	India			Mahanadi	Rao, Zoological Survey of India	Koraput District, Orissa
1961	India	19	81	Godavari	Ross & Cavagnaro, California Academy of Sciences	9 miles north of Pharasgro, in pond among water plants
1961	India			Mahanadi	Lamba, Zoological Survey of India	Balaghat District, Madhya Pradesh
1957	India			Brahmaputra	v. Maydell, Zoological Museum of Hamburg	Raimona, R. Janali
1957	India	26	93	Brahmaputra	v. Maydell, Zoological Museum of Hamburg	Kaziranga, Mikir-Hills
1956	India	30	77	Brahmaputra	v. Maydell, Zoological Museum of Hamburg	Dharmawalla (Siwalik), R. Asan
1956	India	28	81	Brahmaputra	v. Maydell, Zoological Museum of Hamburg	Nishangara, Varei-Bach
1956	India	25	91	Brahmaputra	v. Maydell, Zoological Museum of Hamburg	Umsa, W. Assam, Khasi Hills
1956	India	25	91	Brahmaputra	v. Maydell, Zoological Museum of Hamburg	Garampani, Assam, R. Kopili
1955	India			Sharavathi	v. Maydell, Zoological Museum of Hamburg	Jog-Falls
1949	India	24	85	Ganges	Choata-Nagpur Survey, University of British Columbia	R. Barakar near Tillya dam

(continued)

Table 1.2 (continued)

Year	Country	Latitude/longitude		Drainage	Source	Site description
1940	India	21	81	Mahanadi	Herre, California Academy of Sciences	Bisrampur
1939	India			Brahmaputra	Hora, Zoological Survey of India	Darang District, Assam
1938	India	27	88	Brahmaputra	Hora, California Academy of Sciences	Kalimpong Duars and Siliguri Terai
1937	India	22	88	Ganges	Herre, California Academy of Sciences	R. Ganges delta at Pulta
1937	India			Cauvery	Rao, Zoological Survey of India	Stream on Kalurkatte Rd, Karnataka
1935	India			Ganges	Hora, Zoological Survey of India	Dehra Dun, Uttar Pradesh
1934	India			Cauvery	Sundberg, Swedish Museum of Natural History	Mysore
1932	India	31	75	Ganges	Khan, British Museum of Natural History	Phillaur (R. Sutley), Punjab
1929	India			Ganges	Mukerji, Zoological Survey of India	R. Ganges, Bhagalpur, Bihar
1926	Myanmar			Irrawady	Chopra, Zoological Survey of India	Mitkyina District, N. Myanmar
1917	India			Ganges	Southwell, Zoological Survey of India	Cooch Behar, W. Bengal
1911	India			Ganges	Annandale, Zoological Survey of India	Kalka hill stream, Haryana
1889	India	23	87	Ganges	Day, British Museum of Natural History	Bengal
1889	India	20	84	Mahanadi	Day, British Museum of Natural History	Orissa
1868	India	13	80	Pennar	Day, British Museum of Natural History	Madras

states of Rajasthan, Gujarat and Andhra Pradesh (river basins draining into the Arabian Sea) as well as northern Myanmar and Sri Lanka, although no location details are given (37). The reliability of some of the earlier records is questionable; either no specimens appear to have been collected (as in the case of records for Sri Lanka), or the specimen has been reclassified (as in the case of at least one species from Myanmar, now designated *Danio kyathit* (57)). Database records for this species should not be considered as complete. However, on the basis of confirmed occurrences, the zebrafish may be widely distributed over the Indian subcontinent; it may be overlooked in surveys on account of its small size and the fact that it has no value as a food fish, even to subsistence fishermen.

The Indian subcontinent has a monsoon climate with wide seasonal variation in the extent of freshwater habitats. Some of the major river systems, such as the Ganges, run through low-lying areas that flood extensively during the monsoon months. The floodplains are characterised by oxbow lakes and blind channels, which may have seasonal connections to the main river. In addition, these regions contain extensive areas of man-made lakes, ponds and irrigation channels constructed for fish and rice cultivation. There is a wide range of temperatures within the natural range of zebrafish, from as low as 6°C in winter to over 38°C in summer.

Zebrafish have typically been described as inhabiting slow-moving or standing water bodies, the edges of streams and ditches, particularly adjacent to rice fields (70, 38, 71). However, they are also reported as inhabiting rivers and hill streams (72). This inconsistency in habitat preference probably results from the taxonomic confusion between *Danio* and *Devario* (36). Three surveys have systematically described their habitat preferences; McClure et al. (68) captured zebrafish in three sites in the Ganges drainage in India, Spence et al. (66) captured them in nine sites in the Ganges and Brahmaputra drainages in Bangladesh, and Engeszer et al. (73) captured them in 14 sites in the Ganges and Brahmaputra drainages in India. In all three studies, zebrafish were found to occur in shallow water bodies with a visibility to a depth of ~30 cm, frequently in unshaded locations with aquatic vegetation and a silty substrate.

Zebrafish appear to be a floodplain rather than a true riverine species. They are most commonly encountered in shallow ponds and standing water bodies, often connected to rice cultivation. This association with rice cultivation may relate to the use of fertilisers that may promote the growth of zooplankton, a major component of the zebrafish diet (74). Rice paddies and shallow seasonal waters are also likely to be free from large predatory fish. Spence et al. (66) found no zebrafish in either rivers or temporary creeks that opened during the monsoon season. Where zebrafish are found in streams and rivers, these typically have a low flow regime and zebrafish were most often encountered at the margins (68, 73). Behavioural observations of their vertical distribution indicated that they occupy the whole of the water column and occur as frequently in open water as among aquatic vegetation (66).

3.2. Diet

Zebrafish are omnivorous, their natural diet consists primarily of zooplankton and insects, although phytoplankton, filamentous algae and vascular plant material, spores and invertebrate eggs, fish scales, arachnids, detritus, sand and mud have also been reported from gut content analysis (68, 69, 74). The majority of insects identified in these studies were aquatic species, or aquatic larval forms of terrestrial species, particularly dipterans. It has been suggested that zebrafish may have some value in mosquito control (69). The high proportion of planktonic items in their diet indicates that zebrafish feed primarily in the water column, however, terrestrial insects and arachnids are also consumed, suggesting surface feeding. The presence of inorganic elements and detritus suggests that zebrafish also feed from the substrate. In a study based on sampling over 12 months, dietary composition appeared to differ significantly among months although no clear seasonal pattern was apparent (74). Additional data are required to determine the extent to which food items in the gut of zebrafish reflect selectivity on the part of the fish as opposed to seasonal availability of different prey.

3.3. Growth and Mortality

Zebrafish growth is most rapid during the first 3 months following hatching; afterwards the growth rate starts to decrease to approximately zero by about 18 months (74). Growth rates of domesticated strains in the laboratory have been reported as higher than those for wild fish. Eaton and Farley (75) reported an annual growth rate of 183 mm y^{-1} during the first 45 days of development, compared to 72 mm y^{-1} during the first 2 months in nature (74). This difference in growth rates could result from inadvertent selection for rapid growth or as a consequence of higher food intake in captivity. The latter explanation is more likely, as F_2 offspring of wild-caught fish grow at an equivalent rate to domesticated strains under controlled conditions in the laboratory (C. Smith & R. Spence, unpublished data). Domesticated strains have also been reported to achieve a larger body size than some populations of wild fish (34). A length-frequency analysis based on sampling over 12 months from a lake population in Bangladesh showed the mean length of fish to be 25 mm after 1 year. The maximum BL observed was 35 mm (74), which is comparable to the typical range observed in laboratory strains. The size difference may be partly due to genetic factors (34, 76) with selection for fast growth and high fecundity among laboratory fish, but it may also reflect rearing conditions; in the laboratory, F_1 wild fish also achieve 35 mm BL after 18 months (R. Spence & C. Smith, unpublished data). Females tend to be larger than males both in domesticated and wild populations (74, 75, 77). The extent of variation in growth rates and body size among wild populations is unknown.

The zebrafish appears to be primarily an annual species in nature, the spawning season commencing just before the onset of the monsoon (74). Length-frequency analysis showed two distinct age classes during the summer months, representing reproductively mature 1+ year fish and a cohort of 0+ fish. Thus, the main period of rapid growth takes place during the monsoon months (June–September), a period of high temperatures (up to 34°C) and food availability (78).

Gerhard et al. (79) reported a mean life span of domesticated zebrafish of 42 months, with the oldest individual surviving for 66 months. However, instances of spinal curvature, a phenotype caused by muscle degeneration and commonly associated with senescence (79, 80), become apparent in domesticated and wild zebrafish after their second year in captivity (R. Spence, pers. obs.). Spinal curvature was not observed in a wild population (74) and it is likely that fish die in natural populations before this condition develops.

3.4. Assemblage

Where zebrafish are found, they tend to be among the most abundant species (66, 68, 73). Spence et al. (66) captured a total of 25 species from nine families that co-occurred with zebrafish over their range in Bangladesh, while Engeszer et al. (73) captured 36 species from 16 families. These were primarily small (<25 cm total length) indigenous species. Such species represent potential competitors of zebrafish. Zebrafish were often observed shoaling together with the flying barb *Esomus danricus* (Hamilton), another abundant cyprinid of similar size and appearance that is closely related to *Danio* (48). Other potential competitors are *Puntius* spp. and *Aplocheilus panchax* (Hamilton).

The other danionin species found with zebrafish were *Danio dangila* (Hamilton), *D. meghalayensis* (Sen & Dey), *Devario devario* (Hamilton), *Devario assamensis* (Barman) and *D. aequipinnatus* (McClelland). McClure et al. (68) reported significant differences in the characteristic temperature, pH and current speed of the habitats in which different danionin species occurred; the *Devario* species typically inhabited faster flowing water whereas zebrafish were captured in the margins of streams and rivers. This corresponds with Fang's (36) finding that the two genera occupy different microhabitats.

3.5. Predators

The commonest predatory taxa captured with zebrafish were snakeheads, *Channa* spp., and the freshwater garfish, *Xenentodon cancila* (Hamilton) (73, 64) although sampling protocols may have failed to capture other potential predators such as nocturnal catfish. Engeszer et al. (73) additionally captured the catfish *Mystus bleekeri* (Day) and the knifefish, *Notopterus notopterus* (Pallas). Mastacembelids, which also co-occur with zebrafish, are oophagous and may be predators of zebrafish eggs and embryos, while odonate larvae may be predators of larval and juvenile zebrafish (73). Adult zebrafish are also predators of zebrafish eggs and larvae. Avian predators such as the Indian pond heron, *Ardeola grayii* (Sykes), and the common kingfisher, *Alcedo atthis* L., are also ubiquitous in the floodplains of the Indian subcontinent and may feed on *D. rerio*.

Laboratory studies have shown that zebrafish display fright reactions in response to both visual and olfactory cues associated with predators. Dill (81, 82) used both living (largemouth bass, *Micropterus salmoides* (Lacepède)) and model predators to investigate zebrafish escape responses. The distance at which the response was elicited depended on the predator's size and its approach velocity. Reactive distance did not differ significantly between living and model predators, although escape velocity was higher with living predators. Over repeated trials on successive days, zebrafish responded earlier and flight distance increased. No decline in response was detected when zebrafish were retested after a 10-day break. This effect may be an example of secondary reinforcement; as the predator's approach was associated with a negative experience, the fish began to respond before the initial threshold was reached. Bass and Gerlai (83) compared the responses of zebrafish to a sympatric predator (the leaf fish, *Nandus nandus*), an allopatric predator (the compressed cichlid, *Nimbochromis compressiceps*), a sympatric harmless fish (the giant danio, *Devario malabaricus*) and an allopatric harmless fish (the swordtail, *Xiphophorus helleri*). The zebrafish, which were a domesticated line, showed an elevated fear response to the sympatric predator compared to the others. This would appear to indicate some kind of genetic involvement in anti-predator responses.

In common with other ostariophysian fishes, zebrafish show alarm behaviours in response to a pheromone that is released as a result of injury to the epidermal cells (84, 85). The strength of the response is proportional to the concentration of alarm substance in the water (86). Alarm behaviours include an increase in shoal cohesion and either agitated swimming or freezing on the substrate, a decrease in feeding rate and increase in aggression. These behaviours have been interpreted as having an antipredator function. Rehnberg and Smith (87) demonstrated that isolated zebrafish showed an alarm response to water containing alarm substance, so the response is independent of the presence of conspecifics.

3.6. Parasites

Little is known about the natural parasite fauna of zebrafish, or the role parasites play in their behaviour and ecology. In a preliminary study conducted in Bangladesh, based on an analysis of 120 specimens from seven sites, infection by 20 species of metazoan parasites and three protozoans was observed (R. Spence & C. Smith, unpublished data). The majority of parasites were larval stage digeneans, cestodes and acanthocephalans, while ectoparasite infection was rare. Infection by *Acanthostomum* sp., *Centrocestus* sp. and one diplostomoid species was observed in all the locations sampled, with 100% prevalence being observed for the metacercariae of *Acanthostomum* sp. in one site and *Centrocestus* sp. in two sites.

In laboratory stocks, infection by the microsporidian *Pseudoloma neurophilia* is common (88). It infects the central nervous system, cranial and spinal nerves, and skeletal muscle of zebrafish, causing emaciation, ataxia and spinal malformations. It is not clear whether vertical transmission of this parasite can occur in zebrafish. Captive zebrafish have also been subject to infection by the nematode *Pseudocapillaria tomentosa*, which infects the gut; symptoms include inflammation, emaciation and intestinal carcinomas (89). *P. tomentosa* can be transmitted directly and infects entire laboratory colonies. There are many possible explanations for this phenomenon but the finding that nematode infection appears to be rare in nature may indicate that zebrafish have not evolved natural immunity to the effects of parasitism by nematodes.

4. Reproductive Behaviour

4.1. Spawning Cycle

Much of the scientific literature on zebrafish reproduction has been concerned with how best to maximise the supply of eggs for research (reviewed by Laale (54)) and, until recently, almost nothing was known about the reproductive ecology of wild zebrafish. In zebrafish, all gonads initially develop as ovaries, which in males start to differentiate at approximately 5–7 weeks post hatching (10–15 mm TL) through an intersexual stage, finally developing into normal testes by approximately the third month of development (12–17 mm TL), depending on strain and rearing conditions (90, 91). The genetic mechanism of sex determination in zebrafish is unknown. However, there is evidence that food supply or growth rate affects sex determination, with faster growing individuals developing as females and slower growing individuals as males (92). Based on samples collected from a population in Bangladesh, sex ratios in nature appear to be 1:1 (74).

In the laboratory, domesticated zebrafish strains breed all year round whereas in nature spawning is more seasonal. However, larger females collected in January (outside the main spawning season) have been found to contain mature ova, indicating that reproduction may not be cued by season, but may instead be dependent on food availability, which is likely to co-vary with season (66). Furthermore, reproductive maturity appears to be related to size rather than age; wild and domesticated zebrafish appear to reach reproductive maturity at similar sizes, despite having different growth rates. Eaton and Farley (75) showed that domesticated zebrafish reared at 25.5°C reached maturity after 75 days, when females were 24.9 mm BL and males 23.1 mm. In laboratory conditions, F_1 wild zebrafish also reach reproductive maturity at approximately 23 mm BL (R. Spence, pers. obs.).

Pairs of zebrafish left together continuously spawn at frequent but irregular intervals (77) and a single female may produce clutches of several hundred eggs in a single spawning. In a study by Spence and Smith (93) inter-spawning intervals ranged from 1 to 6 days, with a mean of 1.5 days, producing clutches ranging from 1 to over 700 eggs, with a mean of 185 (± SD 149). Clutch size correlated positively with both female body size and inter-spawning interval. Eaton and Farley (77) reported that inter-spawning interval increased with age, from a mean of 1.9 days in 12-month-old fish to 2.7 days 3 months later. Clutch size also increased over this period from a mean of 158–195. No equivalent data are available for wild zebrafish, but inter-spawning intervals tend to be greater and clutch sizes smaller than domesticated strains (R. Spence, pers. obs.).

Ovulation is dependent on female exposure to male gonadal pheromones; male holding water, testis homogenates and testis fractions containing steroid glucuronides will induce ovulation but fail to do so in females rendered anosmic by cauterising the nasal epithelium (94, 95). Eaton and Farley (77) showed that exposure to a male for 7 h in the afternoon was sufficient to enable eggs to be stripped from females the following morning. However, eggs were never obtained from isolated females more than once in any 5-day period after exposure to a male. Thus it appears that all mature ova are released in a single spawning bout (77, 96).

The presence of a male is essential for females to spawn eggs. Females kept in isolation or older females can become "eggbound" (**Fig. 1.3a, b**) which can be lethal in severe cases. Dissections of eggbound females showed a 3 × 3 mm plug consisting of necrotic clumped eggs clogging the oviduct, preventing any further successful spawning (Gerlach unpublished results). Regular exposure to males and spawning dishes can prevent this development. Interestingly, despite the fact that egg production is non-continuous, females exposed to male pheromones for several

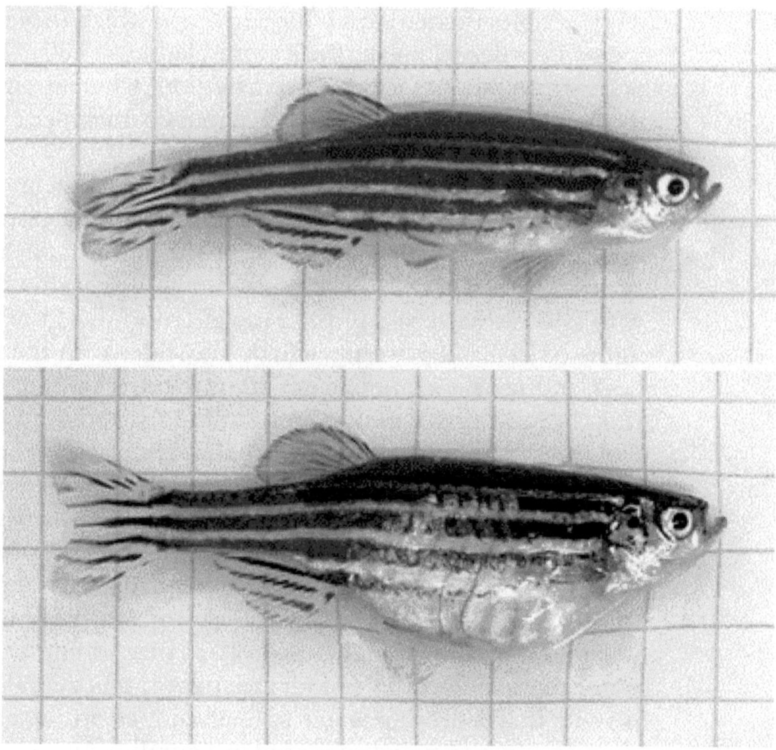

Fig. 1.3. Female zebrafish **a** before and **b** after being housed alone for 3 weeks. The belly of the females increased, on average, by 69 ± 24% ($n = 10$). Grid = 0.5 cm^2. (Reproduced by kind permission of Gabi Gerlach).

days prior to spawning produce more eggs of higher quality than females isolated for several days (97). This effect could be a consequence of the concentration of pheromones to which they are exposed. Bloom and Perlmutter (98) showed that both sexes produce pheromones that function as inter- and intra-sexual attractants, and have different effects at different concentrations. For both sexes, the intra-sexual response was elicited at a lower concentration than the inter-sexual response.

Eggs are non-adhesive and demersal, with a diameter of approximately 0.7 mm. They are released directly over the substrate with no preparation of the substrate by either sex and there is no parental care. Eggs become activated on contact with water and even in the absence of sperm, undergo a series of programmed developmental steps. Unfertilised eggs develop a perivitilline space but fail to develop beyond the first few cleavages (99). Hatching takes place between 48 and 72 h at 28.5°C, depending on the thickness of the chorion and the muscular activity of the embryo inside, both of which can vary within a group of embryos (5). Immediately after hatching, the larvae (measuring ~3 mm) attach to hard surfaces by means of small secretory cells in the epidermis of the head (54). Attachment at progressively higher levels enables them to reach the surface to which they need to gain access in order to inflate their swim bladders (100). This process occurs from about 72 h post-fertilisation, whereupon swimming, feeding and active avoidance behaviours commence (5).

4.2. Mating Behaviour

It is well known that spawning in domesticated zebrafish is influenced by photoperiod (101). Zebrafish show a distinct diurnal activity pattern, synchronised with the light/dark and feeding cycles. The first activity peak occurs immediately after illumination with two further peaks in the early afternoon and the last hour of light (61, 102). Spawning activity coincides with the first activity peak and usually commences within the first minute of exposure to light following darkness, continuing for about an hour (103). Field observations have shown that spawning in zebrafish under natural conditions is also largely limited to a short period at dawn (104). Notably, wild-caught zebrafish held in captivity are more likely than domesticated strains to spawn at times other than first light (R. Spence, pers. obs.). Extended day length may be a contributory factor in the seasonal onset of spawning in nature. It was noted by Breder and Rosen (101) that adding a dash of cold water to aquaria could encourage spawning in zebrafish. Thus, it may be that a drop in water temperature or an increase in water level may be additional cues used by zebrafish. In nature, zebrafish spawn during periods of heavy rain (R. Spence, pers. obs.).

Courtship behaviour in zebrafish consists of a male chasing the female rapidly, often nudging her flanks with his snout and attempting to lead her to a spawning site (see below), swimming around or in front of her in a tight circle, or figure of eight, with his fins raised. If she does not follow, he may alternate between circling the female and swimming back and forth between the female and the spawning site. Once over a spawning site he swims closely alongside the female, spreading his dorsal and caudal fins around her so that their genital pores are aligned, and may oscillate his body at high frequency and low amplitude. This behaviour triggers oviposition in the female and sperm is released simultaneously. This sequence of behaviours is repeated throughout the spawning period, females releasing between 5 and 20 eggs at a time. Male courtship behaviour is most active in the first 30 min and although it continues for about an hour, few females extrude eggs after the first 30 min (103). Wild zebrafish display similar courtship and territorial behaviours during spawning as have been described in domesticated strains (104). Under more natural conditions, courtship involves males actively pursuing females, who utilize the whole water column, alternately swimming towards the surface and then diving steeply down to the substrate to spawn. Small groups of 3–7 fish usually take part in these chases.

Courtship behaviour in the male is triggered by female pheromones. In a study by van den Hurk and Lambert (94) males, but not females, were attracted to ovarian extracts injected into the aquarium. Anosmic males failed to court females while control males only courted females that had ovulated. Further, anosmic males were extremely aggressive, suggesting that ovarian pheromones also inhibit aggression that is common in both sexes during foraging.

Zebrafish typify a basic mating pattern common to many cyprinid fishes; they are group spawners and egg scatterers (101). Females will spawn directly onto a bare substrate, but when provided with an artificial spawning site, such as a plastic box filled with marbles, will preferentially use it for oviposition (105). Some male zebrafish are territorial during mating (105). Both territorial and non-territorial males show the same courtship behaviour but whereas non-territorial males pursue females, territorial males confine their activities to within a few body lengths of a spawning site and chase other males away when they try to approach. A study by Spence and Smith (105) examined the effects of manipulating density and sex ratio on the behaviour of these territorial males. Aggression rates increased at higher densities. However, while courtship behaviour increased with density under a female-biased sex ratio, when the sex ratio was male-biased courtship rate decreased relative to that observed at low densities. A subsequent microsatellite parentage analysis showed that the reproductive success of territorial males was also density dependent (106). At low densities territorial males sired significantly more offspring than non-territorial males. However, at higher densities territorial males were no more successful than non-territorials. Thus male zebrafish display two distinct mating tactics, territorial defence and active pursuit of females, the adoption of which is flexible and may be density dependent. Another study (107) used a higher density level and found that territoriality broke down completely and aggression was reduced in consequence. Thus it is likely that aggression will be highest at intermediate densities, depending on the availability of defendable territories.

Density can also affect female reproductive success, mean *per capita* egg production decreasing at higher densities (105, 107). A parentage analysis indicated that this effect was due to females spawning smaller clutches, rather than some females being excluded from spawning (106). There are several possible explanations for reduced female egg production at high densities; increased male-male aggression may interfere with female oviposition attempts and/or competition may arise among females for access to spawning sites. Alternatively, reduced female egg production may arise through pheromonally mediated reproductive suppression. Females exposed to the pheromones of other females for several days prior to spawning have been shown to be significantly less likely to spawn compared to isolated females (97).

Further, dominant females produce more eggs than subordinates (97). In a study on female territoriality conducted in a large 2 × 2 m aquarium, Delaney et al. (108) showed that females avoid the presence and, therefore, also the direct exposure to pheromones of other females. Females have a significant preference to stay with one or several males over other females. Tested in a T-maze, an increasing concentration of chemical cues from female zebrafish elicited avoidance behaviour in other females (109). Thus, competition among both males and females may play a role in the zebrafish mating system.

4.3. Mate Choice

The existence and nature of female mating preferences can be difficult to demonstrate in species where male competition plays a significant part in the mating system; matings are likely to be determined by the dominant male excluding other males rather than females actively choosing mates. There is some evidence that female zebrafish prefer larger males (110), and body size tends to correlate with dominance in teleost fishes (111). When female egg production is used as a measure of preference, female zebrafish do appear to prefer some males over others (93). However, while these preferences do not correlate with male dominance, neither do females correspond in their choice of males (93). In view of the role played by pheromones in the reproductive behaviour of both sexes, it is possible that mating preferences may be based on olfactory cues. For instance, female zebrafish prefer the odour of unrelated males to unfamiliar brothers (112). In the zebrafish mating system, the two mechanisms of sexual selection, male-male competition and female preference, may operate in opposition. If females do not prefer dominant males, their preferences may undermine the ability of dominant males to monopolise matings. Further, competition among males for mating opportunities may be balanced by similar competition among females (97). Indeed, variance in reproductive success among females is equivalent to that among males, and consequently the opportunity for sexual selection is weak in zebrafish (106), borne out by the fact that they do not display striking sexual dimorphism.

4.4. Ovipostion Choice

Females are selective with respect to sites for oviposition. In choice tests conducted both with domesticated fish in the laboratory and with wild fish in a field-based mesocosm, females preferred a gravel substrate to silt (104). Territorial males were also observed to defend gravel-substrate spawning sites in preference to silt. This preference appears to relate to spawning site quality; egg survival is enhanced by incubation in a substrate that allows oxygenated water to circulate while protecting them from disturbance and cannibalism. In the laboratory, a preference for vegetation was also observed, although vegetation did not affect survival. Vegetation is thought to be important in the survival of larval zebrafish; they possess attachment organs that may assist them in reaching the surface to inflate their swim bladders (54). Sessa et al. (113) studied oviposition preference in relation to a depth gradient (0–4 cm) and found that females spawned preferentially in very shallow water. In the types of habitat where zebrafish are common, such as floodplain ponds, the substrate is often silty and zebrafish are thought to spawn in shallow vegetated areas that offer protection from predators (73, 104). Therefore, there may be competition for access to sites that afford better water circulation as well as protection for eggs and larvae. Choice of oviposition site is one of the few ways in which oviparous species with no parental care can maximize offspring survival. Thus, if females actively choose oviposition sites, males may increase their reproductive success by guarding such sites.

5. Social Behaviour

5.1. Shoaling Preferences

Zebrafish are a shoaling species; shoaling behaviour commences soon after hatching and increases with age (114) although shoaling preferences do not develop until fish reach the juvenile stage, c. 10 mm BL (115). Miller and Gerlai (116) showed that the average inter-individual distance between individual zebrafish in a shoal remained constant over multiple days. Group cohesion represents a balance between predator avoidance and competition for food. During feeding (when food was evenly dispersed) inter-individual distance increased; when presented with a model aerial predator, while in a bare tank with no hiding places, the shoal responded by scattering and quickly reassembling into a tight group (116).

Shoaling behaviour appears to be innate; fish reared in isolation quickly form shoals when placed together (117). McCann and Matthews (118) showed that zebrafish reared in isolation did not discriminate between shoals of conspecifics, pearl danios, *Danio albolineatus* (Blyth), or guppies, *Poecilia reticulata* (Peters), suggesting that species identification is learned. McCann and Carlson (119) tested this by cross-rearing zebrafish with the closely related unstriped pearl danio. Cross-reared individuals showed a reduced preference for associating with conspecifics. Engeszer et al. (120) showed that preferences for different intraspecific phenotypes are also learned. Wild-type zebrafish cross-reared with the stripeless pigment mutant *nacre* preferred the colour pattern of those with which they had been raised, irrespective of their own appearance. Based on a comparison of shoaling preferences among five different danio phenotypes, stripes appear to be a key shoaling cue (121). These studies suggest that species recognition in the zebrafish is mediated by a process of phenotype matching against a template based on early experience. Engeszer et al. (115) found that the visual preference of juvenile wild-type zebrafish for like phenotype remained even when their social environment was manipulated by placing individuals in groups of *nacre* shoalmates for 30 days. However, McCann and Carlson (119) found that the visual preference of cross-reared subjects was eroded after similar manipulation. These observations together suggest that template formation involves both genetic and learned components.

Zebrafish have also been shown to use olfactory cues in both species and kin recognition (112). In a series of odour flume choice tests, juvenile zebrafish preferred conspecifics to heterospecifics, unfamiliar kin to non-kin, and familiar to unfamiliar kin. Gerlach et al. (122) showed that kin recognition is based on olfactory imprinting, with a very specific 24-h developmental window requiring exposure to kin on day 6 post-fertilisation. There was no evidence of self-matching; larvae reared in isolation did not imprint on their own chemical cues. Exposure to non-kin at the critical stage did not result in imprinting which suggests some genetic involvement in the process. Thus, social preferences in zebrafish may be based on individual recognition as well as phenotype matching. Individual recognition may play a role in zebrafish since this species is known to establish dominance hierarchies (93, 97, 123). The mechanism underlying this olfactory recognition is not yet known.

Shoaling decisions in zebrafish are also influenced by shoal size and activity level. In a test of shoaling preferences, Pritchard et al. (124) showed that individuals generally preferred larger shoals. However, when shoal activity level was manipulated by changing the water temperature, fish preferred the more active shoal, regardless of size. Preferences also appear to differ between the sexes (125). Male zebrafish preferred to associate with female shoals compared to males but had no preference for shoal size. However, females preferred to associate with the larger shoal, regardless of whether it was composed of males or females. Zebrafish appear to be able to assess the nutritional state of conspecifics; food-deprived individuals preferred to shoal with well-fed conspecifics, and had increased foraging success than when shoaling with other food-deprived individuals (126).

Tests of shoaling preference based on visual cues have been conducted between wild-type zebrafish and various aquarium variants: leopard danios (127), longfin (128) and the transgenic Glofish^{TM}, which are genetically engineered to express red fluorescent proteins (129). No significant preference was detected in any of these tests. However, Engeszer et al. (130) compared shoaling preferences among 17 different pigment pattern mutants or closely related species and showed that, while wild and laboratory zebrafish exhibited similar preferences, there was a marked difference between the sexes. Male preferences were based on species and stripe patterning but female preferences did not correlate with a priori identifiable traits. While most tests of shoaling preference are based on dichotomous choice tests, Saverino and Gerlai (131) analysed video footage of shoals of test and stimulus fish swimming together, to determine inter-individual distances, and found that zebrafish shoaled more closely with conspecifics. They also presented fish with computer animated images of zebrafish, modifying their colour, location, pattern and body shape and found a preference for yellow and avoidance of elongated images.

5.2. Aggression and Dominance

Zebrafish of both sexes can establish dominance hierarchies. Aggressive interactions involve chasing and in some cases biting. Display behaviour involves pairs of fish orienting head to tail with their fins splayed and slowly circling one another while ascending (R. Spence, personal observation). This behaviour operates within and between the sexes; its function is not clear but it may be a means of individual recognition that reinforces dominance ranks. Once dominance relationships become established, aggression becomes less intense (53). When fish are housed in pairs, the dominant individual often appears darker and utilises the entire aquarium, while subordinates are pale and occupy a smaller area (53). Dominance relationships appear to be relatively stable over time, at least over the duration of 5-day experiments (105, 123). Moreover, males separated for 4 days have been shown to re-establish identical dominance ranks once reunited (G. Gerlach, unpublished data).

The sex of an individual does not appear to be an important factor in determining its dominance rank (123). The relationship between body size and dominance is unclear, partly because studies often control for size (93, 105, 123). However, in studies using fish of different sizes, Hamilton and Dill (132) found that size correlated positively with rank, while Basquill and Grant (133) found that it was not. Dominance has been demonstrated both during mating behaviour, where males establish territories around spawning sites (105) and foraging, where dominant individuals attempt to monopolise a food source (123, 132, 133). It is not known whether males that are territorial during spawning are also dominant during foraging.

In a study of zebrafish foraging behaviour, Gillis and Kramer (134) manipulated fish density and food patch profitability. Zebrafish formed shoals but aggressive interactions took place near feeding sites. The distribution of fish was affected by patch profitability, with more fish being concentrated around the most profitable food patch. However, the variability in the distribution between the three patches was greater when fish density was lower. At high densities, there were more fish in the least profitable patch and fewer in the most profitable patch than would be predicted by an ideal free distribution model (135). Aggressive interference did not fully explain the density-related reduction in foraging efficiency; aggressive interactions increased with patch profitability but decreased at high population densities. Thus, foraging distributions may also be influenced by non-aggressive interactions, while aggressive interactions are ameliorated at high densities.

Aggression and food monopolisation are also influenced by habitat structure. Basquill and Grant (133) compared levels of aggression in a vegetated *versus* a non-vegetated habitat. Aggression and food monopolisation by the dominant fish were lower in the vegetated habitat. This effect could be because the presence of vegetation makes the environment more difficult to defend. An alternative explanation is that a vegetated environment is perceived as safer; dominant fish may be more willing to forage in open habitats where predation risk is higher, while to subordinate fish the perceived benefit of shoaling in a risky habitat may outweigh the cost of reduced foraging efficiency. In order to test these two hypotheses, Hamilton and Dill (132) compared aggression and resource monopolisation among three habitats, open, vegetated and unvegetated with overhead cover. When allowed to choose, fish preferred to forage in the covered habitat and there was no effect of vegetation. There was no difference in aggression among habitats, but resource monopolisation was greater in the open "risky" habitat.

Rearing environment may also influence aggression and dominance. Marks et al. (136) found that fish raised in an hypoxic environment were less aggressive and spent more time in refugia than those reared in a normoxic environment. This result suggests that zebrafish offer a potential model for exploring phenotypic plasticity in behaviour, particularly developmental plasticity.

5.3. Exploratory Behaviour

Shoaling behaviour can increase the probability of an individual fish detecting and avoiding predators (137). A related behaviour is predator inspection, whereby individual fish leave a shoal briefly to approach a predator. These two traits are known to be at least partly genetically determined in zebrafish. Wright et al. (138) showed differences in "boldness" (defined as the propensity to approach a novel object, in the shape of a black cylinder suspended in an experimental aquarium) among laboratory raised wild (F_2) zebrafish from four different populations. An intra-population study indicated a genetic component to shoaling tendency (the time an individual fish spent associating with a stimulus shoal), although there was no equivalent inter-population difference. In a further study, Wright et al. (34) compared boldness and shoaling tendency between wild (F_2) and laboratory zebrafish (AB line). The AB fish showed reduced shoaling tendency and increased boldness compared to wild fish, presumably as a result of relaxed selection for anti-predator behaviours. Robison and Rowland (33) similarly compared the Nadia wild (F_5) strain with a transgenic line TMI, which contains a green fluorescent protein transgene, allowing them to be visually distinguished from other strains in a mixed aquarium. They found that Nadia were less surface orientated, were more likely to freeze on the bottom of the aquarium when presented with a novel object, and were less likely to inspect novel objects compared to TMI fish. Hybrids between the two strains showed intermediate responses and inter-strain differences were still apparent among strains reared in mixed tanks, suggesting that the behaviour was not learned.

It is also possible that the results of both these studies reflect pre-existing strain differences and are not related to domestication. A further study using Nadia, TMI and an additional domesticated strain (SH) revealed significant inter-strain differences across five behavioural measures, although the observed relationships within strains were relatively weak and occasionally inconsistent (139, 140). These observations, together with the inter-population differences among wild fish identified by Wright et al. (138) indicate the need for caution in interpreting behaviours as indicative of particular behavioural patterns such as domestication.

6. Cognitive Behaviour

Learning mediates many aspects of animal behaviour, including social interactions, foraging, navigation and predator avoidance. In zebrafish, the preference for associating with other fish is innate, while the preference for particular colour patterns is based on learned behaviour. Individuals raised in isolation do not display colour pattern preferences whereas cross-reared individuals prefer to associate with the colour pattern with which they were raised (120, 127). The preference effect of cross-rearing does not persist once fish are housed in groups of the same colour pattern, so the early learned preference can be modified by later experience (119). However, zebrafish reared with others of the same colour pattern retain the preference even when subsequently housed with an alternative colour pattern, indicating that there may be some genetic involvement in colour pattern preference (115). Learned preferences are mediated by olfactory as well as visual cues; zebrafish can differentiate between familiar and unfamiliar conspecifics on the basis of odour, and thus, appear capable of individual recognition (112, 122).

The response shown by zebrafish to alarm substance (*see* **Section 3.5**) is also innate but appears to function as a means of learned predator recognition (86). Alarm substance can initiate a conditioned response to an innocuous odour, such as morpholine, when the two are presented simultaneously (141). Hall and Suboski (142) further elicited a learned response to a visual cue by pairing alarm substance with a red light as well as with morpholine. Thus, conditioning can operate across different sensory modalities. Hall and Suboski (143) also demonstrated second order conditioning whereby fish conditioned with alarm substance to respond to either light or morpholine, then learned to react to the second neutral stimulus when presented in combination with the first conditioned stimulus in the absence of alarm substance. The mechanism for communicating learned predator recognition appears to be classical conditioning, pairing of an unconditioned stimulus (alarm substance) with a conditioned stimulus (light or morpholine) to produce a conditioned response (alarm reaction). Conditioned responses can develop after a single trial, and a response can be obtained even when there is a time delay of several minutes between presentation of the unconditioned and conditioned stimuli (144). Furthermore, conditioned responses can be passed on to naïve fish, a process known as social facilitation. Naïve fish exposed to morpholine when in the company of morpholine-sensitized fish subsequently display an alarm reaction to morpholine. The naïve fish retain this learned response when solitary or in the company of a new group of naïve fish (141).

An alternative approach to studying learning is to use an operant conditioning paradigm, whereby fish are trained to swim in a specific direction for a food reward paired with a visual cue. This approach has been used to study spatial memory, landmark use and orientation in other species (145), and the few studies available indicate that zebrafish are potentially a useful model for research in this area. In a study to investigate spatial learning and memory, Williams et al. (146) trained adult zebrafish to swim alternately to one or other side of a divided aquarium to receive a food reward. Once trained, the fish could remember the task after a 10-day period during which they were fed ad libitum in another aquarium. Zebrafish were also able to learn to swim into one of three compartments when the one containing the reward was cued by a white light (147). A three-choice design provides better evidence of learning than a two-choice design, as the level of a chance response is reduced to a third. Williams et al. (146) reported that fish learned the task in approximately 14 trials, although Bilotta et al. (147) reported wide individual variability in speed of learning. When food rewards were withheld, the training effect was quickly lost (146, 148). Given the strong shoaling instinct of zebrafish, an alternative reward shown to be effective in associative learning is the sight of other fish, or even of computer-generated images of fish (149, 150).

Little is known about the development of learning capacity. Williams et al. (146) found that age affected acquisition of conditioned responses in zebrafish. Juveniles of 6–8 weeks learned the task as well or better than adults, whereas those of 3–4 weeks were not able to do so. It was not clear whether this was a result of limited cognitive capacity or because the task presented to the fish was too physically demanding. A related question, which has not been investigated in zebrafish, is the extent to which habitat complexity during rearing influences cognitive development. Research with other fishes and comparisons among populations suggests that learning in fish may be related to the demands of their environment (151).

The majority of studies of learning involve testing individual fish (152). However, in a shoaling species like zebrafish, fish may perform better in groups; the stress of being isolated may inhibit learning ability in isolated individuals. Moreover, fish are known to be able to learn by watching others (153). However, Gleason et al. (154) found that while zebrafish learned an avoidance response to an electric shock fastest in groups of five or more, single fish learned faster than pairs. Thus the relationship between learning and group size may not be straightforward. Steele et al. (155) obtained similar results in exploratory feeding behaviour in response to alanine, a ubiquitous amino acid in the aquatic environment that functions as a chemical attractant and is the primary constituent of many prey odours. They found that the fastest response was elicited in groups of four fish, but single fish responded faster than groups of two, six or eight. Group size has not been studied in relation to spatial learning in zebrafish.

Miklósi and Andrew (156) used beads of different colours and patterns to study the effects of habituation to stimuli. Based on video footage of zebrafish biting responses, they concluded that habituation is mediated by cerebral lateralisation of function; responses are controlled by different cerebral hemispheres under different circumstances. In trials, fish initially approached the bead with the right eye but in subsequent trials, once the object was familiar, used their left eye. Miklósi and Andrew concluded that right hemisphere control (i.e. left eye) mediates escape/attack responses (automatic behaviour), whereas left hemisphere (right eye) control is used in assessing novel stimuli and involves the inhibition of Mauthner cell discharge.

Many studies of learning are based on the use of neutral stimuli. However, in many species, innate receiver biases have evolved that cause them to respond more strongly to certain stimuli, and thus affect learning outcomes. Biases can exist at any level along the signal reception and processing continuum, from stimulation of a primary sensory receptor to synthesis at higher levels of integration, including learning, memory and decision making (157). Both learned preferences and innate receiver biases operate in the context of foraging. Spence and Smith (158) raised groups of fish on diets consisting solely of one colour: red, blue, green or white. When fish were subsequently tested for their colour preferences in a foraging context, each group responded most strongly to red, irrespective of the colour of food with which they had been conditioned. However, there was also a significant effect of conditioning. The observed sensory bias towards red may have evolved as a function of the nature of the transmission environment that zebrafish inhabit, in combination with an adaptive preference for carotenoid compounds in their diet (158).

Different tasks have been shown to elicit different preferences. Colwill et al. (148) used a T-maze with different coloured arms (green *versus* purple or red *versus* blue) to assess visual discrimination learning in zebrafish. They found that while fish could be trained to swim down whichever coloured arm was associated with a food reward, they learned faster and retained the response longer when the colour associated with the reward was purple or blue than when it was green or red. Thus, not only were the stimuli not perceived as equal, but the colour preferences shown in this context differed from those in the foraging study by Spence and Smith (158). Similarly, two studies reached different conclusions about whether zebrafish prefer a dark or light environment. Serra et al. (159) found that zebrafish spent more time in a black chamber than a white one and concluded that they have an innate preference for dark environments. In contrast, Gerlai et al. (160) concluded that zebrafish did not prefer a dark environment; fish initially avoided a dark chamber and on habituation spent equal amounts of time in illuminated and dark chambers. Clearly, the existence of innate preferences needs to be understood when designing behavioural protocols for learning studies in zebrafish.

7. Genetic Basis of Behaviour

The relationship between genes and complex behaviours is not straightforward (161). Behavioural syndromes comprise numerous individual components, involving multiple, interacting genes whose expression is influenced by the environment. The starting point in such research is to identify behavioural syndromes that can be quantified, with simple, reliable protocols that allow high-throughput screening, either for mutagenesis or naturally occurring behavioural variation. Much of the pioneering behavioural genetics research has utilised *Drosophila* as a model;

genes have been identified that control complex syndromes, such as learning and memory, mating behaviour and circadian rhythms (161, 162). The advent of functional genomics has enabled research to be extended to other species (163, 164).

In comparison, there is a paucity of studies on complex behaviour in zebrafish, although it is recognised as having great potential as a model for understanding the genetic basis of human behavioural disorders (18, 19). One area of interest has been the effect of drugs of abuse on behaviour. Darland and Dowling (165) conducted a behavioural screen for cocaine addiction using the conditioned place preference paradigm (CPP), whereby the drug is paired with a neutral stimulus in one compartment of the aquarium and the amount of time the fish spends in each compartment is measured before and after administration of the drug. Three out of 18 families of mutagenised fish showed abnormal responses in the CPP and were subjected to further behavioural screens, testing spatial cognition in a T-maze, swimming behaviour, and sensitivity to light. Each family had different behavioural profiles, which were shown to be heritable, each supposed as representing a different single gene mutation that affected addiction (165). Lau et al. (166) used CPP to demonstrate a preference by wild-type zebrafish for both food and morphine as rewards. In contrast, the *too few* mutants, in which the basal forebrain DA and 5HT neurons are selectively reduced, lacked the morphine preference, while still displaying a preference for food. This result, whereby a single gene mutation can dissociate the preference for a natural reward and an addictive drug, indicates that the two preferences are controlled by different pathways.

Gerlai et al. (160) designed a series of simple, easily quantifiable tests to examine the effects of alcohol administered at different concentrations on locomotion, aggression, shoaling tendency, alarm response, light/dark preference and pigmentation. These tests could be used to identify individuals with abnormal responses to alcohol. Echevarria et al. (167) similarly used a battery of tests to examine the effects of NMDA and dopaminergic manipulation (using MK-801 and SKF 38393) on activity level and shoaling tendency. Several studies have also compared the effects of acute and chronic alcohol administration among different zebrafish strains. Inter-strain differences were detected in startle response, predator avoidance, aggression and shoal cohesion, suggesting that there is a genetic basis to both initial sensitivity and the development of tolerance to alcohol (168–170).

Zebrafish may also be a suitable model for studying the genetic basis of social behaviour. Larson et al. (53) showed that there are clear differences between dominant and subordinate fish in the expression of arginine vasotocin, a neurohormone known to mediate social behaviour such as aggression, courtship and parental behaviour in vertebrates, although the system varies among taxa. Dominance relationships are not fixed and must, therefore, involve differential expression of different genetic pathways.

Tropepe and Sive (171) suggested that a forward genetics screening approach might be employed to model the behavioural deficits involved in autism using zebrafish. As deficits in social behaviour are strongly characteristic of autism, behaviours such as courtship and shoaling may represent a suitable paradigm for sociability. In mice, tests of exploratory behaviour have been used as a paradigm for anxiety and fear, exploratory behaviour tending to be negatively correlated with anxiety (172). Using a similar approach, Wright et al. (34) utilised the pronounced differences between wild and laboratory strains of zebrafish in willingness to approach an unfamiliar object (boldness) and attempted to identify quantitative trait loci associated with these phenotypes (see Chapter 5 in this book).

Other complex behaviours that offer potential for genetic analysis are learning and memory. Protocols where fish are trained to swim in a particular direction for a food reward can be used to assess speed of learning and retention time between different strains of fish, fish reared under different conditions, or known behavioural mutants. These protocols have also been used to assess the effects of drugs of abuse on learning and memory (173, 174). Yu et al. (175) studied cognitive aging in zebrafish, comparing 1, 2 and 3-year-old fish. They found that the younger fish performed better in both temporal and spatial learning and that CPP could be established more quickly. In addition, cognitive aging was accelerated in mutant and gamma-irradiated fish. Genetic analysis of cerebral lateralisation of function may offer insights into the molecular basis of habituation. For instance, the mutant *frequent situs inversus* (*fsi*), which shows reversal of asymmetry in many cerebral and visceral organs, showed reversal of behavioural asymmetry in some tests but not others, suggesting that at least two different mechanisms are involved in lateralisation of function (176).

Zebrafish have also been used to investigate the effects of anthropogenic disturbance on fish behaviour. Larsen et al. (177) studied the effects of endocrine disrupting chemicals on zebrafish sexual development and courtship behaviour. Exposure to environmentally realistic concentrations of 17α-ethinyloestradiol (EE2) from egg until sexual maturity resulted in a female-biased sex ratio, while males displayed female secondary sexual characteristics such as the development of urogenital papillae, rounder body shape and smaller, less distinctly patterned anal fins. Male courtship behaviour proved more resistant to the effects of EE2 and only a few biological males at the highest concentration treatment were unable to induce spawning. Another study investigated the effects of anthropogenic disturbance on chemosensory ability (178). Elevated levels of humic acid (HA) impaired the ability of adult zebrafish to use olfactory cues to distinguish between conspecifics and heterospecifics. The short generation time of zebrafish allows the effects of lifetime exposure to chemical disruption to be studied in a relatively short time and the results can then be extrapolated to longer-lived fish species.

8. Research Priorities

8.1. Field-Based Studies

While the three surveys reported here have provided basic information about zebrafish ecology (64, 66, 70), there is a need for more field-based studies. The current distribution status of the zebrafish is unknown, as recent studies have concentrated solely on the Ganges and Brahmaputra river systems. In addition, sampling from a wide range of populations would enable the cataloguing of natural variation in morphological, physiological and behavioural traits. This should include comparing the behaviour of zebrafish from a number of populations with different environmental parameters and predator regimes, as has been done with guppies (179). For instance, inter-population differences in anti-predator behaviours (such as those identified by Wright et al. (138)) may relate to actual differences in natural predator regime.

Field observations of zebrafish behaviour would also prove invaluable and would supplement more detailed laboratory observations, providing definitive data on intra- and inter-specific interactions. This objective requires the identification of field sites suitable in terms of accessibility and water clarity.

8.2. Behavioural Studies

The number of behavioural studies of zebrafish looks set to increase, and many researchers whose primary expertise is in genetics or developmental biology are utilising behavioural protocols such as CPP as a paradigm for testing the reinforcing properties of drugs of abuse. One of the problems with designing and conducting behavioural experiments is demonstrating that the results are a valid measure of the behaviour under consideration. Thus there is a need for adequate controls, in order to ensure that the results are not due to unrelated artifacts (180); slight differences in experimental design and set-up can produce different results. For instance, preference tests need to take innate biases into account. Precision of measurement may also determine whether a preference is detected. Other sources of error include outside disturbance, either visual or auditory, and general handling of subjects, including acclimatisation. Behaviour may vary according to time of day at which observations are recorded, especially in relation to mating behaviour and feeding regime.

There is growing evidence for behavioural differences among zebrafish populations, even among domesticated strains. A number of studies have found strain- and dose-dependent differences in sensitivity to ethanol exposure (168, 169, 170, 181). Moretz et al. (139, 140) found differences among three strains, one wild-derived (Nadia) and two domesticated (TM1 and SH) in shoaling, activity level, predator approaches, latency to feed after disturbance and biting at a mirror stimulus. Thus observed differences between wild and domesticated strains cannot all be ascribed to the effects of domestication. Relatively few studies have compared behavioural differences between wild and domesticated strains of zebrafish. Given that genetic variability is higher among wild zebrafish but reduces over a few generations, more emphasis should be placed on studies using wild fish and specifying number of generations removed from the wild (182).

In addition to the need for adequate controls, behavioural results are also dependent on a degree of experimenter interpretation, and this is perhaps the most difficult aspect to validate. Mating behaviour is the most straightforward to validate as observed behaviours can be correlated with egg production and parentage determined if necessary. Other behaviours are more difficult to validate. For instance, is biting at a mirror indicative of aggression (160) or purely an attempt to interact with a conspecific? (139).

8.3. Husbandry and Welfare

All areas of zebrafish research would greatly benefit from improvements and standardisation of husbandry practices (183). In addition to achieving greater production and efficiency in research, the establishment of biologically justifiable practices for zebrafish culture would also address important concerns regarding zebrafish welfare. Research based on a knowledge of zebrafish natural history may inform practices in a number of areas: water chemistry, nutrition, breeding, larval rearing, tank design and optimal fish densities.

One of the most important aspects of zebrafish husbandry is the induction of spawning in captivity. Research has demonstrated the role of density, spawning substrate and water depth in spawning (104, 105, 113). Commonly utilized spawning methods and equipment (184, 185) may not take account of these factors, and thereby may result in reduced breeding efficiency, and/or the production of embryos of suboptimal or inconsistent quality. While it may not always be possible to incorporate behavioural and natural history data into breeding protocols (as, for instance, in the case of genetic studies which require sib mating), even simple efforts to replicate natural situations and facilitate behavioral preferences, for example by the addition of spawning substrate and plastic plants (104) and the presentation of shallow areas in which to spawn (113), may improve productivity.

The social environment of captive zebrafish is another factor to be considered in relation to their husbandry and welfare. In the majority of zebrafish research facilities fish are kept in bare tanks and densities may be high, generally determined by growth rates. As a shoaling species, zebrafish benefit from being kept in groups, although they do exhibit antagonistic behaviour and form dominance hierarchies (53, 94). While aggression generally seems to be inversely correlated with density (134, 186), zebrafish also show an elevation in circulating levels of the stress hormone cortisol when they are subjected to crowded conditions (187), suggesting that intermediate densities may be the most favourable. The provision of refugia, such as artificial plants, in holding tanks may further offset the potentially negative effects of aggression. In nature, zebrafish are often associated with aquatic vegetation (66), a preference that is also seen in laboratory populations (108) while aggression and monopolisation of food resources by dominant individuals is decreased in structurally complex environments (133).

It is a commonly held perception that environmental enrichment, such as adding plants to aquaria, will have beneficial welfare outcomes, yet little research has been conducted on the effects of enrichment in fish. In an experiment to test the effects of rearing environment on cognitive development (Spence, Magurran and Smith, unpublished data), two strains of zebrafish (WIK and a second generation wild strain) were reared in either a structurally simple or complex environment and compared in their ability to locate a food reward in a five-chambered maze. There was a significant difference in spatial learning between strains but not between rearing environments. Notably, while both strains learned the task in the same number of trials, wild fish were initially slower in locating the reward, which may reflect differences in boldness between strains rather than spatial cognition *per se*. In addition, fish of both strains reared in a spatially complex environment were smaller than those reared in a simple environment, although performance in the learning task was not related to size. These results do not mean that the addition of plants will not have other welfare benefits, such as reduction of aggression or improvements in breeding efficiency. However, it does demonstrate the need for empirical testing of the various outcomes of different husbandry practices.

9. Summary

Zebrafish appear to be primarily a floodplain species, inhabiting shallow ponds and ditches or the slower reaches of streams. They are an abundant species and are among the smallest fish species in the assemblages in which they occur. Their diet, based on gut content analysis, consists primarily of insects (of both terrestrial and aquatic origin) and zooplankton, as well as inorganic material. These results indicate that they feed throughout the water column, consistent with observations of their vertical distribution and the finding that they tend to be confined to the shallow margins of water bodies.

The zebrafish is known for its rapid development in the laboratory. Length-frequency analysis indicates that under natural conditions the zebrafish is an annual species and recruitment is linked to the monsoon, which is also the period of the year with the highest temperatures. The most rapid growth takes place in the first 3 months, and slows thereafter, virtually ceasing by about 18 months. Breeding may be dependent on food availability rather than season, as gravid females have been found in Bangladesh in winter and wild-caught zebrafish breed all year round in the laboratory.

Zebrafish reproductive behaviour has been studied almost exclusively in the laboratory on domesticated strains, although an experiment conducted with wild-caught fish under semi-natural conditions confirms that the mating behaviours described are broadly applicable in nature. Zebrafish have previously been characterised as group spawners and egg scatterers, although there is evidence that the mating system is influenced by both intrasexual competition and female mate preferences. Further, competition for high quality sites for oviposition may be a key feature of mating behaviour in nature. Given the role of pheromones in zebrafish reproduction and evidence from other published behavioural studies (97, 112), these may play a role in mate choice; in particular, the zebrafish may be a suitable model for studying the role of MHC in mate choice. However, the zebrafish has little to offer as a model for sexual selection compared to other fish behavioural models such as guppies, sticklebacks or bitterling. The opportunity for sexual selection appears to be weak in zebrafish, as might be predicted from their lack of marked sexual dimorphism.

The greatest advantage of the zebrafish as a model system comes from its well-characterised genetics, genetic and developmental techniques and tools, and the availability of well-characterised mutants. Zebrafish are also a tractable species for behavioural experiments, readily acclimatising to new environments, being constantly active and little disturbed by the presence of observers. In order for the zebrafish to be more widely adopted as a model by the behavioural ecology community, there is a need for more behavioural and field-based studies in order to catalogue natural variation in morphological, physiological and behavioural traits. The zebrafish appears ideally suited to studies of social and cognitive behaviour, and it is surprising that it has been so little utilised for this purpose. There is increasing interest in employing social and cognitive tests with zebrafish to study the genetic basis of behaviour, and a growing need for more comprehensive and better controlled studies in this area.

Web Citations

The Wellcome Trust Sanger Institute: http://www.sanger.ac.uk

Zebrafish Information Network (ZFIN), The Zebrafish International Resource Center, University of Oregon: http://www.zfin.org/

References

1. Amsterdam, A. & Hopkins, N. (2006) Mutagenesis strategies in zebrafish for identifying genes involved in development and disease. *Trends Genet* 22, 473–478.
2. Grunwald, D. J. & Eisen, J. S. (2002) Headwaters of the zebrafish – emergence of a new model vertebrate. *Nat Rev Genet* 3, 717–724.
3. Rubinstein, A. L. (2003) Zebrafish: from disease modelling to drug discovery. *Curr Opin Drug Discov Devel* 6, 218–223.
4. Vascotto, S. G., Beckham, Y. & Kelly, G. M. (1997) The zebrafish's swim to fame as an experimental model in biology. *Biochem Cell Biol* 75, 479–485.
5. Kimmel, C. B., Ballard, W. W., Kimmel, S. R., Ullmann, B. & Schilling, T. F. (1995) Stages of embryonic development of the zebrafish. *Dev Dyn* 203, 253–310.
6. Creaser, C. W. (1934) The technique of handling the zebrafish (*Brachydanio rerio*) for the production of eggs which are favourable for embryological research and are available at any specified time throughout the year. *Copeia* 1934, 159–161.
7. Streisinger, G., Walker, C., Dower, N., Knauber, D. & Singer, F. (1981) Production of clones of homozygous diploid zebra fish (*Brachydanio rerio*). *Nature* 291, 293–296.
8. Kimmel, C. B. (1989) Genetics and early development of zebrafish. *Trends Genet* 5, 283–288.
9. Kimmel, C. B. (1993) Patterning the brain of the zebrafish embryo. *Annu Rev Neurosci* 16, 707–732.
10. Kimmel, C. B., Warga, R. M. & Schilling, T. F. (1990) Origin and organization of the zebrafish fate map. *Development* 108, 581–594.
11. Granato, M. & Nüsslein-Volhard, C. (1996) Fishing for genes controlling development. *Curr Opin Genet Dev* 6, 461–468.
12. Driever, W., Solnica-Krezel, L., Schier, A. F., Neuhauss, S. C. F., Malicki, J., Stemple, D. L., Stainier, D. Y. R., Zwartkruis, F., Abdelilah, S., Rangini, Z., Belak, J. & Boggs, C. (1996) A genetic screen for mutations affecting embryogenesis in zebrafish. *Development* 123, 37–46.
13. Haffter, P., Granato, M., Brand, M., Mullins, M. C., Hammerschmidt, M., Kane, D. A., Odenthal, J., Van Eeden, F. J. M., Jiang, Y. J., Heisenberg, C. P., Kelsh, R. N., Furutani-Seiki, M., Vogelsang, E., Beuchle, D., Schach, U., Fabian, C. & Nüsslein-Volhard, C. (1996a) The identification of genes with unique and essential function in the development of the zebrafish, *Danio rerio*. *Development* 123, 1–36.
14. Amsterdam, A., Burgess, S., Golling, G., Chen, W., Sun, Z., Townsend, K., Farrington, S., Haldi, M. & Hopkins, N. (1999) A large scale insertional mutagenesis screen in zebrafish. *Genes Dev* 13, 2413–2724.
15. Chen, E. & Ekker, S. C. (2004) Zebrafish as a genomics research model. *Curr Pharm Biotechnol* 5, 409–413.
16. Eggert, U. S. & Mitchison, T. J. (2006) Small molecule screening by imaging. *Curr Opin Chem Biol* 10, 232–237.
17. Golling, G., Amsterdam, A., Sun, Z., Antonelli, M., Maldonado, E., Chen, W., Burgess, S., Haldi, M., Artzt, K., Farrington, S., Lin, S., Nissen, R. & Hopkins, N. (2002) Insertional mutagenesis in zebrafish rapidly identifies genes essential for early vertebrate development. *Nat Genet* 31, 135–140.

18. Guo, S. (2004) Linking genes to brain, behavior and neurological diseases: what can we learn from zebrafish? *Genes Brain Behav* 3, 63–74.
19. Patton, E. E. & Zon, L. I. (2001) The art and design of genetic screens: zebrafish. *Nat Rev Genet* 2, 956–966.
20. Postlethwait, J. H. & Talbot, W. S. (1997) Zebrafish genomics: from mutants to genes. *Trends Genet* 13, 183–190.
21. Sood, R., English, M. A., Jones, M., Mullikin, J., Wang, D. M., Anderson, M., Wu, D., Chandrasekharappa, S. C., Yu, J., Zhang, J. & Paul Liu, P. (2006) Methods for reverse genetic screening in zebrafish by resequencing and TILLING. *Methods* 39, 220–227.
22. Udvadia, A. J. & Linney, E. (2003) Windows into development: historic, current, and future perspectives on transgenic zebrafish. *Dev Biol* 256, 1–17.
23. Dooley, K. & Zon, L. I. (2000) Zebrafish: a model system for the study of human disease. *Curr Opin Genet Dev* 10, 252–256.
24. Lieschke, G. J. & Currie, P. D. (2007) Animal models of human disease: zebrafish swim into view. *Nat Rev Genet* 8, 352–367.
25. Shin, J. T. & Fishman, M. C. (2002) From zebrafish to human: modular medical models. *Annu Rev Genom Hum Genet* 3, 311–340.
26. Berghmans, S., Jette, C., Langenau, D., Hsu, K., Stewart, R., Look, T. & Kanki, J. P. (2005) Making waves in cancer research: new models in the zebrafish. *Biotechniques* 39, 227–237.
27. Guyon, J. R., Steffen, L. S., Howell, M. H., Pusack, T. J., Lawrence, C. & Kunkel, L. M. (2006) Modeling human muscle disease in zebrafish. *Biochim Biophys Acta* 1772, 205–215.
28. Rubinstein, A. L. (2006) Zebrafish assays for drug toxicity screening. *Expert Opin Drug Metab Toxicol* 2, 231–240.
29. Barbazuk, W. B., Korf, I., Kadavi, C., Heyen, J., Tate, S., Wun, E., Bedell, J. A., McPherson, J. A. & Johnson, S. L. (2000) The syntenic relationship of the zebrafish and human genomes. *Genome Res* 10, 1351–1358.
30. Postlethwait, J. H., Yan, Y.-L., Gates, M. A., Horne, S., Amores, A., Brownlie, A., Donovan, A., Egan, E. S., Force, A., Gong, Z., Goute, C., Fritz, A., Kelsh, R., Knapik, E., Liao, E., Paw, B., Ransom, D., Singer, A., Thomson, M., Abduljabbar, T. S., Yelick, P., Beier, D., Joly, J.-S., Larhammar, D., Rosa, F., Westerfield, M., Zon, L. I., Johnson, S. L. & Talbot, W. S. (1998) Vertebrate genome evolution and the zebrafish gene map. *Nat Genet* 18, 345–349.
31. Gerlai, R. (2003) Zebrafish: an uncharted behavior genetic model. *Behav Genet* 33, 461–468.
32. Miklósi, A. & Andrew, R. J. (2006) The zebrafish as a model for behavioural studies. *Zebrafish* 3, 227–234.
33. Robison, B. D. & Rowland, W. (2005) A potential model system for studying the genetics of domestication: behavioral variation among wild and domesticated strains of zebra danio (*Danio rerio*). *Can J Fish Aquat Sci* 62, 2046–2054.
34. Wright, D., Nakamichi, R., Krause, J. & Butlin, R. K. (2006a) QTL analysis of behavioural and morphological differentiation between wild and laboratory zebrafish (*Danio rerio*). *Behav Genet* 36, 271–284.
35. Nelson, J. S. (2006) *Fishes of the World*. 4th edn. New York, NY, Wiley.
36. Fang, F. (2001) Phylogeny and species diversity of the South and Southeast Asian cyprinid genus Danio Hamilton (Teleostei, Cyprinidae). PhD Thesis, Stockholm University, Stockholm, Sweden.
37. Barman, R. P. (1991) A taxonomic revision of the Indo-Burmese species of *Danio rerio*. *Rec Zool Surv India Occas Pap* 137, 1–91.
38. Talwar, P. K. & Jhingran, A. G. (1991) *Inland Fishes of India and Adjacent Countries*. Calcutta, Oxford & I. B. H. Publishing.
39. Howes, G. J. (1991) Systematics and biogeography: an overview. In Winfield, I. J. & Nelson, J. S. (Eds.) *Cyprinid Fishes Systematics, Biology and Exploitation*. London, Chapman and Hall, pp. 1–33.
40. Weber, M. & de Beaufort, L. F. (1916) *The Fishes of the Indo-Australian Archipelago*. London, E. J. Brill.
41. Meyer, A., Biermann, C. H. & Orti, G. (1993) The phylogenetic position of the zebrafish (*Danio rerio*), a model system in developmental biology: an invitation to the comparative method. *Proc R Soc Lond Ser B* 252, 231–236.
42. Meyer, A., Ritchie, A. P. & Witte, K.-E. (1995) Predicting developmental processes from evolutionary patterns: a molecular phylogeny of the zebrafish (*Danio rerio*) and its relatives. *Philos Trans R Soc Lond Ser B* 349, 103–111.
43. McClure, M. (1999) Development and evolution of melanophore patterns in fishes of the genus Danio (Teleostei: Cyprinidae). *J Morphol* 241, 83–105.
44. Parichy, D. M. & Johnson, S. L. (2001) Zebrafish hybrids suggest genetic

mechanisms for pigment pattern diversification in Danio. *Dev Genes Evol* 211, 319–328.

45. Quigley, I. K., Roberts, R., Manuel, J. L., Nuckels, R. J., Herrington, E., MacDonald, E. L. & Parichy, D. M. (2005) Evolutionary diversification of pigment pattern in Danio fishes: differential fms dependence and stripe loss in *D. albolineatus. Development* 132, 89–104.
46. Quigley, I. K., Turner, J. M., Nuckels, R. J., Manuel, J. L., Budi, E. H., MacDonald, E. L. & Parichy, D. M. (2004) Pigment pattern evolution by differential deployment of neural crest and post-embryonic melanophore lineages in *Danio* fishes. *Development* 131, 6053–6069.
47. Sanger, T. J. & McCune, A. R. (2002) Comparative osteology of the Danio (Cyprinidae: Ostariophysi) axial skeleton with comments on Danio relationships based on molecules and morphology. *Zool J Linnean Soc* 135, 529–546.
48. Fang, F. (2003) Phylogenetic analysis of the Asian cyprinid genus Danio (Teleostei, Cyprinidae). *Copeia* 4, 714–728.
49. Mayden, R. L., Tang, K. L., Conway, K. W., Freyhof, J., Chamberlain, S., Haskins, M., Schneider, L., Sudkamp, M., Wood, R. M., Agnew, M., Bufalino, A., Sulaiman, Z., Miya, M., Saitoh, K. & He, S. (2007) Phylogenetic relationships of *Danio* within the order cypriniformes: a framework for comparative and evolutionary studies of a model species. *J Exp Zool (Mol Dev Evol)* 308B, 642–654.
50. Parichy, D. M. (2006a) Evolution of danio pigment pattern development. *Heredity* 97, 200–210.
51. Parichy, D. M. (2006b) Homology and the evolution of novelty during Danio adult pigment pattern development. *J Exp Zool* 306B, 1–13.
52. Hatamoto, K. & Shingyoji, C. (2008) Cyclical training enhances the melanophore responses of zebrafish to background colours. *Pigment Cell Melanoma Res* 21, 397–406.
53. Larson, E. T., O'Malley, D. M. & Melloni, R. H., Jr. (2006) Aggression and vasotocin are associated with dominant-subordinate relationships in zebrafish. *Behav Brain Res* 167, 94–102.
54. Laale, H. W. (1977) The biology and use of zebrafish, *Brachydanio rerio* in fisheries research. A literature review. *J Fish Biol* 10, 121–173.
55. Schilling, T. (2002) The morphology of larval and adult zebrafish. In Nüsslein-Volhard, C. & Dahm, R. (Eds.) *Zebrafish.* Oxford, Oxford University Press, pp. 59–94.
56. Meinken, H. (1963) Brachydanio frankei spec. nov., der Leopard danio. *Aquarium Terrararien* 10, 39–43.
57. Fang, F. (1998) Danio kyathit, a new species of cyprinid fish from Myitkyina, northern Myanmar. *Ichthyol Explor Freshw* 8, 273–280.
58. Petrovicky, I. (1966) Hybridization between *Brachydanio rerio* (Hamilton-Buchanan) and *Brachydanio frankei* (Meinken). *Ichthyologica* 37, 53–62.
59. Haffter, P., Odenthal, J., Mullins, M. C., Lin, S., Farrell, M. J., Vogelsang, E., Haas, F., Brand, M., van Eeden, F. J. M., Furutani-Seiki, M., Granato, M., Hammerschmidt, M., Heisenberg, C. P., Jiang, Y. J., Kane, D. A., Kelsh, R. N., Hopkins, N. & Nüsslein-Volhard, C. (1996b) Mutations affecting pigmentation and shape of the adult zebrafish. *Dev Genes Evol* 206, 260–276.
60. Watanabe, M., Iwashita, M., Ishii, M., Kurachi, Y., Kawakami, A., Kondo, S. & Okada, N. (2006) Spot pattern of leopard *Danio* is caused by mutation in the zebrafish connexin41.8 gene. *EMBO Rep* 7, 893–897.
61. Plaut, I. (2000) Effects of fin size on swimming performance, swimming behaviour and routine activity of zebrafish *Danio rerio. J Exp Biol* 203, 813–820.
62. Kelsh, R. N., Brand, M., Jiang, Y. J., Heisenberg, C. P., Lin, S., Haffter, P., Odenthal, J., Mullins, M. C., van Eeden, F. J. M., Furutani-Seiki, M., Granato, M., Hammerschmidt, M., Kane, D. A., Warga, R. M., Beuchle, D., Vogelsang, L. & Nusslein-Volhard, C. (1996) Zebrafish pigmentation mutations and the processes of neural crest development. *Development* 123, 369–389.
63. Kawakami, K., Amsterdam, A., Nobuyoshi, S., Becker, T., Mugg, J., Shima, A. & Hopkins, N. (2000) Proviral insertions in the zebrafish hagoromogene, encoding an F-box/WD40-repeat protein, cause stripe pattern anomalies. *Curr Biol* 10, 463–466.
64. Lamason, R. L., Mohideen, M. A., Mest, J. R., Wong, A. C., Norton, H. L., Aros, M. C., Jurynec, M. J., Mao, X., Humphreville, V. R., Humbert, J. E., Sinha, S., Moore, J. L., Jagadeeswaran, P., Zhao, W., Ning, G., Makalowska, I., McKeigue, P. M., O'donnell, D., Kittles, R., Parra, E. J., Mangini, N. J., Grunwald, D. J., Shriver, M. D., Canfield, V. A. & Cheng, K. C. (2005) SLC24A5, a putative cation exchanger, affects pigmentation in zebrafish and humans. *Science* 310, 1782–1786.
65. Lister, J. A., Robertson, C. P., Lepage, T., Johnson, S. L. & Raible, D. W. (1999) nacre encodes a zebrafish microphthalmia-related protein that regulates neural

crest-derived pigment cell fate. *Development* 126, 3757–3767.
66. Spence, R., Fatema, M. K., Reichard, M., Huq, K. A., Wahab, M. A., Ahmed, Z. F. & Smith, C. (2006a) The distribution and habitat preferences of the zebrafish in Bangladesh. *J Fish Biol* 69, 1435–1448.
67. Pritchard, V. L. (2001) Behaviour and morphology of the zebrafish, *Danio rerio*. PhD thesis, University of Leeds.
68. McClure, M. M., McIntyre, P. B. & McCune, A. R. (2006) Notes on the natural diet and habitat of eight danionin fishes, including the zebrafish, *Danio rerio*. *J Fish Biol* 69, 553–570.
69. Dutta, S. P. S. (1993) Food and feeding habits of *Danio rerio* (Ham. Buch.) inhabiting Gadigarh stream, Jammu. *J Freshw Biol* 5, 165–168.
70. Sterba, G. (1962) *Freshwater Fishes of the World*. London, Vista Books, Longacre Press.
71. Jayaram, K. C. (1999) *The Freshwater Fishes of the Indian Region*. Delhi, Narendra Publishing House.
72. Daniels, R. J. R. (2002) *Freshwater Fishes of Peninsula India*. Hyderabad, Universities Press.
73. Engeszer, R. E., Patterson, L. B., Rao, A. A. & Parichy, D. M. (2007a) Zebrafish in the wild: a review of natural history and new notes from the field. *Zebrafish* 4, 21–40.
74. Spence, R., Fatema, M. K., Ellis, S., Ahmed, Z. F. & Smith, C. (2007b) The diet, growth and recruitment of wild zebrafish (*Danio rerio*) in Bangladesh. *J Fish Biol* 71, 304–309.
75. Eaton, R. C. & Farley, R. D. (1974a) Growth and reduction of depensation of the zebrafish *Brachydanio rerio*, reared in the laboratory. *Copeia* 1974, 204–209.
76. Wright, D., Butlin, R. K. & Carlborg, Ö. (2006b) Epistatic regulation of behavioural and morphological traits in the zebrafish (*Danio rerio*). *Behav Genet* 36, 914–922.
77. Eaton, R. C. & Farley, R. D. (1974b) Spawning cycle and egg production of zebrafish, *Brachydanio rerio*, in the laboratory. *Copeia* 1974, 195–204.
78. Talling, J. F. & Lemoalle, J. (1998) *Ecological Dynamics of Tropical Inland Waters*. Cambridge, Cambridge University Press.
79. Gerhard, G. S., Kauffman, E. J., Wang, X., Stewart, R., Moore, J. L., Kasales, C. J., Demidenko, E. & Cheng, K. C. (2002) Life spans and senescent phenotypes of zebrafish (*Danio rerio*). *Exp Gerontol* 37, 1055–1068.
80. Kishi, S., Uchiyama, J., Baughman, A. M., Goto, T., Lin, M. C. & Tsai, S. B. (2003) The zebrafish as a vertebrate model of functional aging and very gradual senescence. *Exp Gerontol* 38, 777–786.
81. Dill, L. M. (1974a) The escape response of the zebra danio (*Brachydanio rerio*) I. The stimulus for escape. *Anim Behav* 22, 711–722.
82. Dill, L. M. (1974b) The escape response of the zebra danio (*Brachydanio rerio*) II. The effect of experience.. *Anim Behav* 22, 723–730.
83. Bass, S. L. S. & Gerlai, R. (2008) Zebrafish (*Danio rerio*) responds differentially to stimulus fish: the effects of sympatric and allopatric predators and harmless fish. *Behav Brain Res* 186, 107–117.
84. Pfeiffer, W. (1974) Pheromones in fish and amphibia. In Birch, M. C. (Ed.) *Pheromones*. Amsterdam, North Holland Publishing Company, pp. 269–296.
85. Wisenden, B. D., Vollbrecht, K. A. & Brown, J. L. (2004) Is there a fish alarm cue? Affirming evidence from a wild study. *Anim Behav* 67, 59–67.
86. Speedie, N. & Gerlai, R. (2008) Alarm substance induced behavioral responses in zebrafish (*Danio rerio*). *Behav Brain Res* 188, 168–177.
87. Rehnberg, B. G. & Smith, R. J. F. (1988) The influence of alarm substance and shoal size on the behaviour of zebra danios, *Brachydanio rerio* (Cyprinidae). *J Fish Biol* 33, 155–163.
88. Spitzbergen, J. M. & Kent, M. L. (2003) The state of the art of the zebrafish model for toxicology and toxicologic pathology research – advantages and current limitations. *Toxicol Pathol* 31S, 62–87.
89. Kent, M. L., Bishop-Stewart, J. K., Matthews, J. L. & Spitzbergen, J. M. (2002) Pseudocapillaria tomentosa, a nematode pathogen and associated neoplasms of zebrafish (*Danio rerio*) kept in research colonies. *Comp Med* 52, 354–358.
90. Devlin, R. H. & Nagahama, Y. (2002) Sex determination and sex differentiation in fish: an overview of genetic, physiological, and environmental influences. *Aquaculture* 208, 191–364.
91. Maack, G. & Segner, H. (2003) Morphological development of the gonads in zebrafish. *J Fish Biol* 62, 895–906.
92. Lawrence, C., Ebersole, J. P. & Kesseli, R. V. (2007) Rapid growth and out-crossing promote female development in zebrafish (*Danio rerio*). *Environ Biol Fishes*, DOI: 10.1007/s10641-007-9195-8.
93. Spence, R. & Smith, C. (2006) Mating preference of female zebrafish, *Danio rerio*, in

relation to male dominance. *Behav Ecol* 17, 779–783.
94. van den Hurk, R. & Lambert, J. G. D. (1983) Ovarian steroid glucuronides function as sex pheromones for male zebrafish, *Brachydanio rerio*. *Can J Zool* 61, 2381–2387.
95. van den Hurk, R., Schoonen, W. G. E. J., van Zoelen, G. A. & Lambert, J. G. D. (1987) The biosynthesis of steroid glucuronides in the testis of the zebrafish, *Brachydanio rerio*, and their pheromonal function as ovulation inducers. *Gen Comp Endocrinol* 68, 179–188.
96. Hisaoka, K. K. & Firlit, C. F. (1962) Ovarian cycle and egg production in the zebrafish, *Brachydanio rerio* (Hamilton-Buchanan). *Copeia* 1962, 788–792.
97. Gerlach, G. (2006) Pheromonal regulation of reproductive success in female zebrafish: female suppression and male enhancement. *Anim Behav* 72, 1119–1124.
98. Bloom, H. D. & Perlmutter, A. (1977) A sexual aggregating pheromone system in the zebrafish, *Brachydanio rerio* (Hamilton-Buchanan). *J Exp Biol* 199, 215–226.
99. Lee, K. W., Webb, S. E. & Miller, A. L. (1999) A wave of free cytosolic calcium traverses zebrafish eggs on activation. *Dev Biol* 214, 168–180.
100. Goolish, E. M. & Okutake, K. (1999) Lack of gas bladder inflation by the larvae of zebrafish in the absence of an air-water interface. *J Fish Biol* 55, 1054–1063.
101. Breder, C. M. & Rosen, D. E. (1966) *Modes of Reproduction in Fishes*. New York, NY, Natural History Press.
102. Baganz, D., Siegmund, R., Staaks, G., Pflugmacher, S. & Steinberg, C. E. W. (2005) Temporal pattern in swimming activity of two fish species (*Danio rerio* and Leucaspius delineatus) under chemical stress conditions. *Biol Rhythm Res* 36, 263 - 276.
103. Darrow, K. O. & Harris, W. A. (2004) Characterisation and development of courtship in zebrafish, *Danio rerio*. *Zebrafish* 1, 40–45.
104. Spence, R., Ashton, R. L. & Smith, C. (2007a) Adaptive oviposition decisions are mediated by spawning site quality in the zebrafish, *Danio rerio*. *Behaviour* 144, 953–966.
105. Spence, R. & Smith, C. (2005) Male territoriality mediates density and sex ratio effects on oviposition in the zebrafish (*Danio rerio*). *Anim Behav* 69, 1317–1323.
106. Spence, R., Jordan, W. C. & Smith, C. (2006b) Genetic analysis of male reproductive success in relation to density in the zebrafish, *Danio rerio*. *Front Zool* 3, 5.
107. Paull, G. A., van Look, K. J. W., Santos, E. M., Filby, A. L., Gray, M., Nasha, J. P. & Tyler, C. R. (2008) Variability in measures of reproductive success in laboratory-kept colonies of zebrafish and implications for studies addressing population-level effects of environmental chemicals. *Aquat Toxicol* 87, 115–126.
108. Delaney, M., Follet, C., Ryan, N., Hanney, N., Lusk-Yablick, J. & Gerlach, G. (2002) Social interaction and distribution of female zebrafish (*Danio rerio*) in a large aquarium. *Biol Bull* 203, 240–241.
109. Golubev, A. V. (1984) Role of chemical stimuli in group and spawning behavior of zebrafish, *Brachydanio rerio*. *Voprosy Ikhtiologii* 6, 1020–1027.
110. Pyron, M. (2003) Female preferences and male-male interactions in zebrafish (*Danio rerio*). *Can J Zool* 81, 122–125.
111. Wootton, R. J. (1998) *The Ecology of Teleost Fishes*. Dordrecht, Kluwer.
112. Gerlach, G. & Lysiak, N. (2006) Kin recognition and inbreeding avoidance in zebrafish, *Danio rerio*, is based on phenotype matching. *Anim Behav* 71, 1371–1377.
113. Sessa, A. K., White, R., Houvras, Y., Burke, C., Pugach, E., Baker, B., Gilbert, R., Look, A. T. & Zon, L. I. (2008) The effect of a depth gradient on the mating behaviour, oviposition site preference and embryo production in the zebrafish, *Danio rerio*. *Zebrafish* 5, 335–360.
114. Gerlach, G., Hodgins-Davis, A., MacDonald, B. & Hannah, R. C. (2007) Benefits of kin association: related and familiar zebrafish larvae (*Danio rerio*) show improved growth. *Behav Ecol Sociobiol* 6, 1765–1770.
115. Engeszer, R. E., da Barbiano, L. A., Ryan, M. J. & Parichy, D. M. (2007b) Timing and plasticity of shoaling behaviour in the zebrafish, *Danio rerio*. *Anim Behav* 74, 1269–1275.
116. Miller, N. & Gerlai, R. (2007) Quantification of shoaling behaviour in zebrafish (*Danio rerio*). *Behav Brain Res* 184, 157–166.
117. Kerr, J. P. (1963) Grouping behaviour of the zebrafish as influenced by social isolation. *Am Zool* 2, 532–533.
118. McCann, L. I. & Matthews, J. J. (1974) The effects of lifelong isolation on species identification in zebra fish (*Brachydanio rerio*). *Dev Psychobiol* 7, 159–163.
119. McCann, L. I. & Carlson, C. C. (1982) Effect of cross-rearing on species identification in zebra fish and pearl danios. *Dev Psychobiol* 15, 71–74.

120. Engeszer, R. E., Ryan, M. J. & Parichy, D. M. (2004) Learned social preference in zebrafish. *Curr Biol* 14, 881–884.
121. Rosenthal, G. G. & Ryan, M. J. (2005) Assortative preferences for stripes in danios. *Anim Behav* 70, 1063–1066.
122. Gerlach, G., Hodgins-Davis, A., Avolio, C. & Schunter, C. (2008) Kin recognition in zebrafish: a 24-hour window for olfactory imprinting. *Proc R Soc Lond B* 275, 2165–2170.
123. Grant, J. W. A. & Kramer, D. L. (1992) Temporal clumping of food arrival reduces its monopolization and defense by zebrafish, *Brachydanio rerio*. *Anim Behav* 44, 101–110.
124. Pritchard, V. L., Lawrence, J., Butlin, R. K. & Krause, J. (2001) Shoal choice in zebrafish, *Danio rerio*: the influence of shoal size and activity. *Anim Behav* 62, 1085–1088.
125. Rhul, N. & McRobert, S. P. (2005) The effect of sex and shoal size on shoaling behaviour in *Danio rerio*. *J Fish Biol* 67, 1318–1326.
126. Krause, J., Hartmann, N. & Pritchard, V. L. (1999) The influence of nutritional state on shoal choice in zebrafish, *Danio rerio*. *Anim Behav* 57, 771–775.
127. Spence, R. & Smith, C. (2007) The role of early learning in determining social preferences based on visual cues in the zebrafish, *Danio rerio*. *Ethology* 113, 62–67.
128. Kitevski, B. & Pyron, M. (2003) Female zebrafish (*Danio rerio*) do not prefer mutant longfin males. *J Freshw Ecol* 18, 501–502.
129. Snekser, J. L., McRobert, S. P., Murphy, C. E. & Clotfelter, E. D. (2006) Aggregation behavior in wildtype and transgenic zebrafish. *Ethology* 112, 181–187.
130. Engeszer, R. E., Wang, G., Ryan, M. J. & Parichy, D. M. (2008) Sex-specific perceptual spaces for a vertebrate basal social aggregative behaviour. *Proc Natl Acad Sci USA* 105, 929–933.
131. Saverino, C. & Gerlai, R. (2008) The social zebrafish: behavioral responses to conspecific, heterospecific, and computer animated fish. *Behav Brain Res* 191, 77–87.
132. Hamilton, I. M. & Dill, L. M. (2002) Monopolization of food by zebrafish (*Danio rerio*) increases in risky habitats. *Can J Zool* 80, 2164–2169.
133. Basquill, S. P. & Grant, J. W. A. (1998) An increase in habitat complexity reduces aggression and monopolization of food by zebra fish (*Danio rerio*). *Can J Zool* 76, 770–772.
134. Gillis, D. M. & Kramer, D. L. (1987) Ideal interference distributions: population density and patch use by zebrafish. *Anim Behav* 35, 1875–1882.
135. Fretwell, S. D. & Lucas, H. L. (1970) On territorial behaviour and other factors influencing habitat distribution in birds. I Theoretical development. *Acta Biotheor* 19, 16–36.
136. Marks, C., West, T. N., Bagatto, B. & Moore, F. B. G. (2005) Developmental environment alters conditional aggression in zebrafish. *Copeia* 2005, 901–908.
137. Pitcher, T. J. & Parrish, J. K. (1993) Functions of shoaling behaviour in teleosts. In Pitcher, T. J. (Ed.) *Behaviour of Teleost Fishes*. London, Chapman and Hall, pp. 363–439.
138. Wright, D., Rimmer, L. B., Pritchard, V. L., Krause, J. & Butlin, R. K. (2003) Inter and intra-population variation in shoaling and boldness in the zebrafish (*Danio rerio*). *Naturwissenschaften* 90, 374–377.
139. Moretz, J. A., Martins, E. P. & Robison, B. (2006) The effects of early and adult social environment on zebrafish (*Danio rerio*) behavior. *Environ Biol Fishes* 80, 91–101.
140. Moretz, J. A., Martins, E. P. & Robison, B. (2007) Behavioral syndromes and the evolution of correlated behaviour in zebrafish. *Behav Ecol* 18, 556–562.
141. Suboski, M. D., Bain, S., Carty, A. E., McQuoid, L. M., Seelen, M. I. & Seifert, M. (1990) Alarm reaction in acquisition and social transmission of simulated-predator recognition by zebra danio fish (*Brachydanio rerio*). *J Comp Psychol* 104, 101–112.
142. Hall, D. & Suboski, M. D. (1995a) Visual and olfactory stimuli in learned release of alarm reactions by zebra danio fish (*Brachydanio rerio*). *Neurobiol Learn Mem* 63, 229–240.
143. Hall, D. & Suboski, M. D. (1995b) Sensory preconditioning and second order conditioning of alarm reactions in zebra danio fish. *J Comp Psychol* 109, 76–84.
144. Korpi, N. L. & Wisenden, B. D. (2001) Learned recognition of novel predator odour by zebra danios, *Danio rerio*, following time-shifted presentation of alarm cue and predator odour. *Environ Biol Fishes* 61, 205–211.
145. Braithwaite, V. A. (1998) Spatial memory, landmark use and orientation in fish. In Healy, S. (Ed.) *Spatial Representation in Animals*. Oxford, Oxford University Press, pp. 86–102.
146. Williams, F. E., White, D. & Messer, W. S., Jr. (2002) A simple spatial alternation task for assessing memory function in zebrafish. *Behav Processes* 58, 125–132.
147. Bilotta, J., Risner, M. L., Davis, E. C. & Haggbloom, S. J. (2005) Assessing appetitive

choice discrimination learning in zebrafish. *Zebrafish* 2, 251–268.
148. Colwill, R. M., Raymond, M. P., Ferreira, L. & Escudero, H. (2005) Visual discrimination learning in zebrafish (*Danio rerio*). *Behav Processes* 70, 19–31.
149. Al-Imari, L. & Gerlai, R. (2008) Sight of conspecifics as reward in associative learning in zebrafish (*Danio rerio*). *Behav Brain Res* 189, 216–219.
150. Pather, S. & Gerlai, R. (2009) Shuttle box learning in zebrafish (*Danio rerio*). *Behav Brain Res* 196, 323–327.
151. Odling-Smee, L. & Braithwaite, V. A. (2003) The role of learning in fish orientation. *Fish Fish* 4, 234–246.
152. Braithwaite, V. A. (2005) Cognitive ability in fish. *Fish Physiol* 24, 1–37.
153. Brown, C. & Laland, K. (2003) Social learning in fishes: a review. *Fish Fish* 4, 280–288.
154. Gleason, P. E., Weber, P. G. & Weber, S. P. (1977) Effect of group size on avoidance learning in zebrafish, *Brachydanio rerio*, (Pisces: Cyprinidae). *Anim Learn Behav* 5, 213–216.
155. Steele, C. W., Scarfe, A. D. & Owens, D. W. (1991) The effects of group size on the responsiveness of zebrafish *(Brachydanio rerio)* to alanine, a chemical attractant. *J Fish Biol* 38, 553–564.
156. Miklósi, A. & Andrew, R. J. (1999) Right eye use associated with decision to bite in zebrafish. *Behav Brain Res* 105, 199–205.
157. Basolo, A. L. (2000) Problems in studying receiver biases and their effects on signal evolution. In Epsmark, Y., Amundsen, T. & Rosenqvist, G. (Eds.) *Animal Signals: Signalling and Signal Design in Animal Communication.* Trondheim, Tapir Academic Press, pp. 177–193.
158. Spence, R. & Smith, C. (2008) Innate and learned colour preference in the zebrafish, *Danio rerio. Ethology* 114, 582–588.
159. Serra, E. L., Medalha, C. C. & Mattioli, R. (1999) Natural preference of zebrafish (*Danio rerio*) for a dark environment. *Braz J Med Biol Res* 32, 1551–1553.
160. Gerlai, R., Lahav, M., Guo, S. & Rosenthal, A. (2000) Drinks like a fish: zebrafish (*Danio rerio*) as a behavior genetic model to study alcohol effects. *Pharmacol Biochem Behav* 67, 773–782.
161. Sokolowski, M. B. (2001) Drosophila: genetics meets behaviour. *Nat Rev Genet* 2, 879–890.
162. Anholt, R. R. H. & Mackay, T. F. C. (2004) Quantitative genetic analysis of complex behaviours in Drosophila. *Nat Reviews Genet* 5, 838–849.
163. Fitzpatrick, M. J., Ben-Shahar, Y., Smid, H. M., Vet, L. E. M., Robinson, G. E. & Sokolowski, M. B. (2000) Candidate genes for behavioural ecology. *Trends Ecol Evol* 20, 96–104.
164. Vasemägi, A. & Primmer, C. R. (2005) Challenges for identifying functionally important genetic variation: the promise of combining complementary research strategies. *Mol Ecol* 14, 3623–3642.
165. Darland, T. & Dowling, J. E. (2001) Behavioral screening for cocaine sensitivity in mutagenized zebrafish. *Proc Natl Acad Sci USA* 98, 11691–11696.
166. Lau, B., Bretaud, S., Huang, Y., Lin, E. & Guo, S. (2006) Dissociation of food and opiate preference by a genetic mutation in zebrafish. *Genes Brain Behav* 5, 497–505.
167. Echevarria, D. J., Hammack, C. M., Pratt, D. W. & Hosemann, J. D. (2008) A novel behavioral test battery to assess global drug effects using the zebrafish. *Int J Comp Psychol* 21, 19–34.
168. Dlugos, C. A. & Rabin, R. A. (2003) Ethanol effects on three strains of zebrafish: model system for genetic investigations. *Pharmacol Biochem Behav* 74, 471–480.
169. Gerlai, R., Ahmad, F. & Prajapati, S. (2008) Differences in acute alcohol-induced behavioral responses among zebrafish populations. *Alcohol Clin Exp Res* 32, 1763–1773.
170. Gerlai, R., Chatterjee, D., Pereira, T., Sawashima, T. & Krishnannair, R. (2009) Acute and chronic alcohol dose: population differences in behavior and neurochemistry of zebrafish. *Genes Brain Behav*, DOI: 10.1111/j.1601-183X.2009.00488.x.
171. Tropepe, V. & Sive, H. L. (2003) Can zebrafish be used as a model to study the neurodevelopmental causes of autism? *Genes Brain Behav* 2, 268–281.
172. Flint, J., Corley, J. C., DeFries, J. C., Fulker, D. W., Gray, J. A., Miller, S. & Collins, A. C. (1995) A simple genetic basis for a complex psychological trait in laboratory mice. *Science* 269, 1432–1435.
173. Levin, E. D. & Chen, E. (2004) Nicotinic involvement in memory function in zebrafish. *Neurotoxicol Teratol* 26, 731–735.
174. Swain, H. A., Sigstand, C. & Scalzo, F. M. (2004) Effects of dizocilpine (MK-801) on circling behaviour, swimming activity, and place preference in zebrafish (*Danio rerio*). *Neurotoxicol Teratol* 26, 725–729.
175. Yu, L., Tucci, V., Kishi, S. & Zhdanova, I. V. (2006) Cognitive aging in zebrafish. *PLoS* 1, e14, doi: 10.1371/journal.pone.0000014.
176. Barth, K. A., Miklósi, A., Watkins, J., Bianco, I. H., Wilson, S. W. & Andrew, R. J. (2005)

fsi zebrafish show concordant reversal of laterality of viscera, neuroanatomy, and a subset of behavioral responses. *Curr Biol* 15, 844–850.

177. Larsen, M. G., Hansen, K. B., Henriksen, P. G. & Baatrup, E. (2008) Male zebrafish (*Danio rerio*) courtship behaviour resists the feminising effects of 17-ethinyloestradiol—morphological sexual characteristics do not. *Aquat Toxicol* 87, 234–244.
178. Fabian, N. J., Albright, L. B., Gerlach, G., Fisher, H. S. & Rosenthal, G. (2007) Humic acid interferes with species recognition in zebrafish (*Danio rerio*). *J Chem Ecol* 33, 2090–2096.
179. Magurran, A. E. (2005) *Evolutionary Ecology: The Trinidadian Guppy.* Oxford Series in Ecology & Evolution, Oxford, Oxford University Press.
180. Ninkovic, J. & Bally-Cuif, L. (2006) The zebrafish as a model system for assessing the reinforcing properties of drugs of abuse. *Methods* 39, 262–274.
181. Loucks, E. & Carvan, M. J. (2004) Strain-dependent effects of developmental ethanol exposure in zebrafish. *Neurotoxicol Teratol* 26, 745–755.
182. Coe, T. S., Hamilton, P. B., Griffiths, A. M., Hodgson, D. J., Wahab, M. A. & Tyler, C. R. (2009) Genetic variation in strains of zebrafish (*Danio rerio*) and the implications for ecotoxicology studies. *Ecotoxicology* 18, 144–150.
183. Lawrence, C. (2007) The husbandry of zebrafish (*Danio rerio*): a review. *Aquaculture* 269, 1–20.
184. Mullins, M. C., Hammerschmidt, M., Haffter, P. & Nüsslein-Volhard, C. (1994) Large-scale mutagenesis in the zebrafish: in search of genes controlling development in a vertebrate. *Curr Biol* 4, 189–202.
185. Westerfield, M. (1995) *The Zebrafish Book; A Guide for the Laboratory Use of Zebrafish (Brachydanio rerio).* Eugene, OR, University of Oregon Press.
186. Chapman, M. R. & Kramer, D. L. (1996) Guarded resources: the effect of intruder number on the tactics and success of defenders and intruders. *Anim Behav* 52, 83–94.
187. Ramsay, J. M., Feist, G. W., Varga, Z. M., Westerfield, M., Kent, M. L. & Schreck, C. B. (2006) Whole-body cortisol is an indicator of crowding stress in adult zebrafish, *Danio rerio. Aquaculture* 258, 565–574.

Chapter 2

Olfactory Behavior: Making Scents of a Changing World

Kathleen E. Whitlock

Abstract

The olfactory sensory system is a part of the nervous system that has something for everyone; with as many as 1,000 genes coding for olfactory receptors it sports the largest gene family in the vertebrate genome; the olfactory sensory neurons regenerate throughout life; the sensory neurons send axons directly into the nervous system with the first synaptic contact occurring within the olfactory bulb; and it is the functional unit for essential behaviors such as courtship, predator avoidance and localization of food sources. Olfactory behaviors are unique in that the sensory coding of the system is not understood in as much detail as other sensory systems such as the visual and auditory systems, and the central projections are processed differently within the central nervous system. Here I review aspects of olfactory behaviors in fish, with an emphasis on zebrafish, and ponder the future of olfactory behavior research in the coming decade.

Key words: Olfactory system, olfactory imprinting, olfactory receptors, olfactory neurons, olfactory-directed movement, self-recognition, Immediate Early Genes, hormones, fluid movement, sensory integration, imprinting.

1. Features of the Olfactory Sensory System

1.1. Olfactory Receptors

The olfactory sensory system presents a unique window through which the world is viewed, and this view of the world is somewhat off limits to humans. The ability to sense the information-rich world of odorants is dependent upon the mechanisms for detection, and vertebrates express anywhere from 100 different olfactory receptors (fish; (1, 2)) to around 1,000 olfactory receptors (mice; (3)). Yet when comparing the mouse and human olfactory receptor (OR) super families we humans have a pauperized world with regard to the detection of odors because the gene to

A.V. Kalueff, J.M. Cachat (eds.), *Zebrafish Models in Neurobehavioral Research*, Neuromethods 52,
DOI 10.1007/978-1-60761-922-2_2, © Springer Science+Business Media, LLC 2011

pseudogene ratio in humans is the lowest relative to other mammals. Indeed, approximately 50% of the OR genes in humans are pseudogenes: 390 putatively functional genes and 465 pseudogenes (4) compared to mice with 913 intact OR genes and 296 OR pseudogenes (5). This loss of functional OR genes in primates has been attributed to the decreased need for olfaction to survive, as well as the acquisition of full trichromatic color vision (6).

The mammalian olfactory system contains not only the large olfactory receptor family, but also trace amino-associated receptors (TAARs: <20 genes; (7)), vomeronasal receptors type one (V1R: ~150 genes; (8)), and vomeronasal receptors type two (V2R: ~60 genes; (9)). Fish do not have a separate vomeronasal system as is observed in terrestrial vertebrates, but they do express these four classes of olfactory receptors in the main olfactory epithelia. In zebrafish the olfactory receptor genes have been analyzed (1) for number and species specificity: ORs (~102 genes), TAARs (~109 genes), V1R (6 genes), and V2R (46 genes) (2).

Little is known about the ligand specificity of these receptors. Analysis of a V2R-like receptor in goldfish (5.24) has shown that neurons expressing this receptor respond to all 20 amino acids, though it binds long-chain amino acids lysine and arginine with greater affinity (10, 11). More recent analysis has identified, through computational screening, additional agonists of receptor 5.24 (12). The promiscuity of this receptor is characteristic of the nature of odorant binding in that the receptors are broadly tuned allowing the olfactory sensory system to detect a variety of odors that exceeds the number of actual receptors expressed.

1.2. Olfactory Sensory Neurons

The olfactory receptors are localized in the dendrites of the olfactory sensory neurons (OSNs), and in fishes OSNs are of three distinct types: ciliated OSNs, microvillous OSNs, and crypt cells. The ciliated OSNs of fishes express OR (13), the microvillous express primarily VNR (13, 14), and crypt cells at least VNR (15, 16). These cell types are stratified within the olfactory epithelium (OE) where the crypt cells – having few cilia and microvilli – sit most apically, the microvillous sensory neurons – having short dendrites and microvilli – lie in the intermediate level, and the ciliated cells – having long dendrites and few cilia –sit most basally in the OE. After leaving the olfactory bulb the post-synaptic projections segregate into specific tracts (**Fig. 2.1**) (15). Based on correlation with tract projections within the central nervous system (CNS), it has been proposed that these three types of olfactory sensory neurons in the fish epithelia respond to different classes of odorants: ciliated OSNs respond to bile salts and alarm substances, the microvillous sensory neurons to food odors, and the crypt cells to sex pheromones. Analyses of electro-olfactogram (EOG) responses support the model where microvillous sensory neurons respond to single or mixtures of amino acids (17), cil-

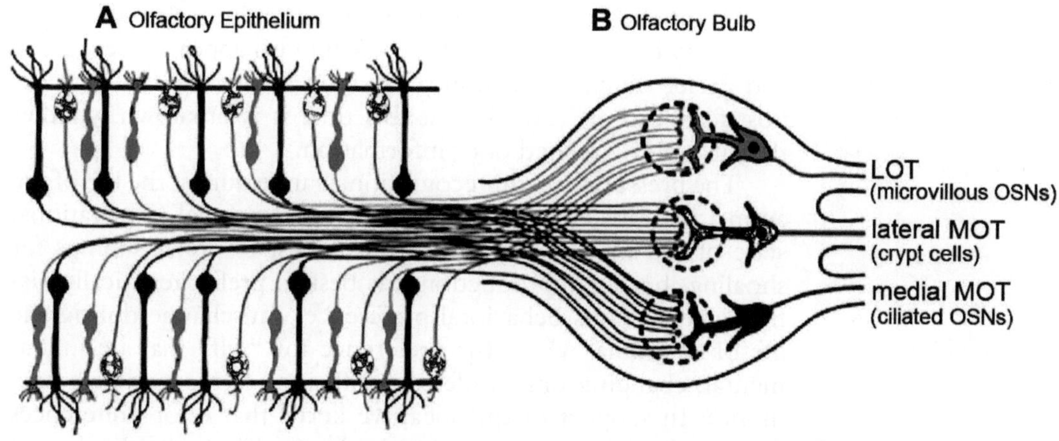

Fig. 2.1. Olfactory sensory neuron cell types in the fish olfactory system. The olfactory epithelia **a** has three types of olfactory sensory neurons: the microvillous (*gray*), ciliated (*black*), and the crypt cells (*mottled*). The sensory information from these different classes of sensory neurons is segregated such that it leaves the olfactory bulb **b** in the lateral olfactory tract (LOT, microvillous input), lateral-medial olfactory tract (lateral MOT, crypt cell input), and the medial MOT (ciliated input). Modified from Hamdani and Døving (15).

iated sensory neurons respond to amino acids and urine odors, and crypt cells respond to at least amino acids (18). With these three classes of olfactory sensory neurons and a variety of olfactory receptors, fish can detect a wide range of odorants including food odors as well as olfactory cues that coordinate reproduction, and can convey information about kin, self-recognition, suitability of spawning sites, and danger.

2. Olfaction and Self-Recognition

The ability to discriminate between related and non-related conspecifics is essential for shoaling, mate choice, and alarm response in fishes. This ability, called kin recognition, has been demonstrated in a wide range of vertebrates including fishes, birds, and primates. Fundamental cues in the discrimination of individuals are the proteins of the major histocompatibility complex (MHC), which are encoded by a group of genes essential for immune response and show the greatest sequence variation among individuals (19, 20). Female mice choose mates based on assessment of genetic relatedness using olfactory cues including those of MHC proteins (21). Ability to detect genetic relatedness is also evident in sticklebacks which also use similar olfactory cues (22, 23). Although important, MHC odors are not the sole determinant of olfactory driven behaviors. Both sticklebacks and juvenile char appear to use not only MHC-based odor cues but also additional odor information to determine relatedness and familiarity of conspecifics (24, 25). Zebrafish appear to recognize and prefer siblings over non-related kin as juveniles (6–8 weeks post-fertilization) (26), but at this time it is unknown whether they use MHC-related odor information.

The pressures for kin recognition change during the life of the animal and are dependent upon the maturational/motivational state of the animal. For example, it is best to prefer siblings for shoaling, but for reproduction it is best to prefer genetically distinct mates. Thus behavioral preferences can change during the life of an animal where the preference for "self" may be fundamental, changing only under the influence of reproductive hormones. In support of this idea, we know that odor preferences are not fixed but plastic: fish (char) raised in isolation do not show kin preference (24) and mice mate choice based on self-referent matching can be manipulated by imprinting mouse pups on a mother of a different MHC genotype (20, 21). The ability to recognize conspecifics also plays a role in fright response as predators that eat your conspecifics emit odor cues important for learning the fright response (see below). Therefore, the ability to distinguish self, conspecifics, and predators is mediated through odor cues and can be modulated by the context in which the odor is presented.

3. The Nose Does Not Work Alone

Information entering through the olfactory sensory system does not create a reflex-like response; rather it is constantly integrated with other information entering the nervous system from the outside world. For example, in male moths, the olfactory behavioral response to the pheromone released by the female "blinds" the male moth to the sound of predators (bats) by raising the threshold of response to bat cries, thus leaving the male moth open to predation when tracking the pheromone odor plume (27). Sharks use information from their lateral line to extract directional information from odor plumes which are dynamic and complex structures (see below). Animals with lesioned lateral lines display increased search time in the odor plume (28). Fish, unlike humans, have a visual spectral response that extends into the UV (300–400 nm), thus they (again) have a privileged view of the sensory world to which we are blind (Human range ~400–700 nm). The information obtained from the UV spectra is used in social contexts such as shoaling behavior in sticklebacks (29), a behavior that also uses olfactory information. Thus as a note of caution, it is important to take into account the characteristics of other sensory modalities (light, touch, vibration) when designing our experiments to ensure that non-olfactory sensory information remains constant across experiments.

The interaction of visual response with olfactory information may be mediated through a cranial nerve that, in fishes, sends efferent projections to the retina of the eye, thus suggesting an important link between the olfactory sensory system and the visual system (**Fig. 2.2**). This cranial nerve, the terminal nerve (TN), is the most enigmatic of the cranial nerves in vertebrates in that its function is poorly understood. The appearance of the terminal nerve is conserved across the jawed vertebrates where it is associated with the olfactory nerve (30). The cell bodies of the terminal nerve are located in clusters (ganglia) associated with the olfactory nerve and bulb, and whose position varies according to the species. In fishes and some amphibians, the terminal nerve contains a unique population of neurons whose axons terminate in the interplexiform layer of the retina (31). This subset

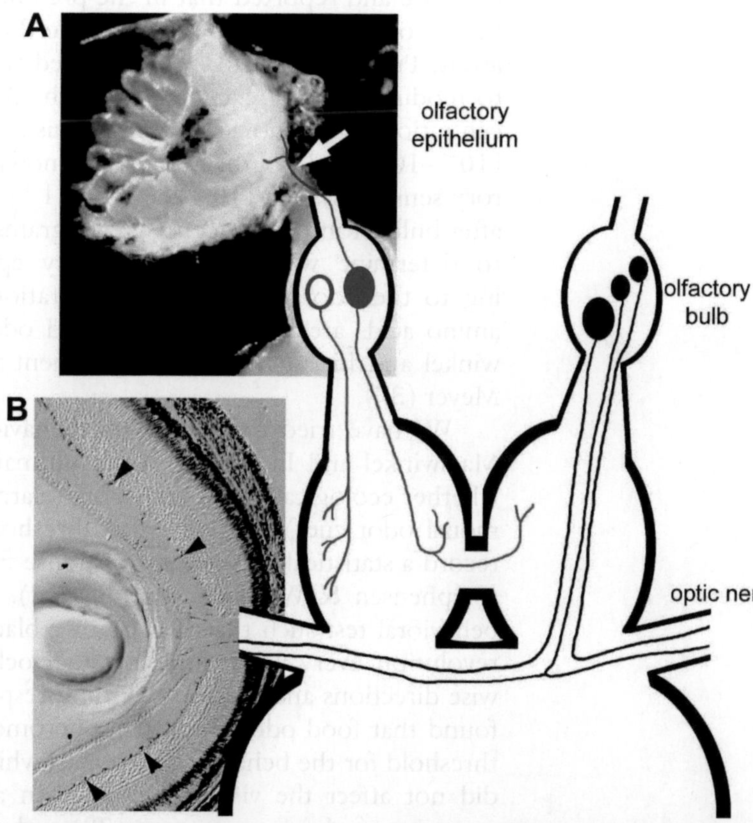

Fig. 2.2. The neurons of the terminal nerve have extensive axonal projections in the central nervous system. A subset of the neurons within the terminal has been reported to have dendrites within the olfactory epithelium (**a**, *arrow*). The terminal nerve has connection with the retina of the eye (**b**) where the axons terminate in the interplexiform layer. Adapted from Whitlock (30).

of the neurons contains various neuroactive peptides such as neuropeptide Y (NPY) and FMRFamide, as well as the neuroendocrine decapeptide gonadotropin releasing hormone (GnRH). It has been shown that GnRH release stimulates dopamine release from the interplexiform cells and FMRFamide can antagonize the actions of GnRH (32). The dendrites of specific neurons within the TN containing GnRH have been reported to terminate in the OE of the dwarf gourami (33). Thus the terminal nerve is proposed to modulate both olfactory acuity (34) as well as visual acuity (35) in response to sensory cues. To test this idea, Maaswinkel and Li (36), measured changes in visual sensitivity in response to odorants using adult zebrafish. In these experiments the test is a mediated escape response when encountering a threatening stimulus. A rotating drum with a black segment marked on white paper served as the threatening stimulus. Normally the fish swims around the circular container, but when the fish encounters this black segment it reverses direction. The researchers measured the threshold of light intensity at which the fish lost the response and reported that in the presence of amino acids, used here as odorants, the fish responded to the stimulus at lower light levels. Previously it had been reported that food odors are able to modulate retinal excitability in fish (37). Curiously, the concentrations of amino acids used in this study were much greater [10^{-5}–10^{-3} mol l^{-1}] than the range known to stimulate the olfactory sensory system [10^{-7}–10^{-9} mol l^{-1}]. The response was lost after bulbectomy, but electro-olfactograms were never performed to determine whether the olfactory epithelia were responding to the odorants in the concentration range used. Because amino acids are components of food odor, the study of Masswinkel and Li is generally in agreement with that of Weiss and Meyer (37).

We have tried to employ the behavioral paradigm used by Masswinkel and Li (36) with the ultimate goal of determining whether ecologically relevant odors (alarm pheromone and hormonal odor cues) affected visual threshold, but were unable to record a statistically significant response from our adult zebrafish (Stephensen & Whitlock, unpublished). Thus we modified the behavioral test such that the rotating black segment moved one revolution every 15 s randomly in clockwise or counterclockwise directions and the fish's turning response was recorded. We found that food odor and alarm pheromone decreased the light threshold for the behavioral response, while odors of conspecifics did not affect the visual threshold. In addition we tested the response on the *laure* mutant (38) and showed that food odor does not affect the visual threshold in this mutant, thus suggesting the effect is due to odor detection at the level of the olfactory sensory system (Stephenson & Whitlock, unpublished). In summary, there is clearly an interaction between the olfactory sensory system and retinal sensitivity, and the terminal nerve is the most likely candidate, though the exact circuitry and mechanisms are unknown at this time.

4. Olfactory Imprinting

Memory is the ability to recover information of past events or knowledge, and the assessment of whether a memory is formed and retained is generally assayed through behavioral tests. Behavioral imprinting is a type of memory that involves exposure to a stimulus during early development and a memory of the stimulus is retained long term in the absence of priming. A famous example of behavioral imprinting is that of visual imprinting shown by Konrad Lorenz using the Greylag geese (for which he was awarded a Nobel Prize). Fish show an equally dramatic example of behavioral imprinting using their olfactory sensory system. In salmon, the juveniles are born in fresh water, migrate to the ocean where they grow to adulthood, and then return to their native stream to reproduce and die (in the case of Pacific salmon). The salmon "sniff" their way home to their native stream using olfactory cues (which are not fully understood at this time) in the fresh water river system (39). Studies using Pacific salmon have shown that dissociated OSNs of fish imprinted on the artificial odor phenylethyl alcohol (PEA) show a statistically significant increase in physiological response to this odorant (40). We have developed zebrafish as a tractable model system to study olfactory imprinting. Using PEA, an odorant that has no olfactory behavioral response on it own yet triggers a physiological response, we tested whether zebrafish could imprint on this odor. We exposed zebrafish to this odorant daily for the first 3 weeks of life. These fish were allowed to grow to adulthood and then tested in a Y-maze to determine whether they had a preference for the PEA. We were able to show not only that the adults exposed as juveniles had a preference relative to their control sibling, but also that these fish show altered gene expression in the OE. Specifically the transcription factor *otx2* was up-regulated in the imprinted fish relative to their controls (41), suggesting that olfactory imprinting affects gene expression in the peripheral sensory system.

In a recent study using mice, it has been shown that juvenile mice exposed to octanol show changes in protein expression in both neuronal and non-neuronal cells in the OE (42). Thus there is growing evidence that stimulation of the olfactory sensory system early in life results in changes not only in the central nervous system but also in the peripheral olfactory sensory system.

In our study demonstrating that the odors can modulate gene expression in the developing zebrafish OE (41), an often asked question is what are the mechanisms that allow the environment to affect changes in gene expression. A natural candidate is the class of Immediate Early Genes (IEGs) whose expression is triggered by neuronal activity. We have cloned and analyzed the expression of several of these genes (*c-fos*, *c-jun*) during early development (43) and found changes in gene expression triggered by odorants (see below, hormones). More recent data (Maturana & Whitlock, unpublished) indicates that the anterior region of the OE (**Fig. 2.3**) is transcriptionally active in response to early olfactory sensory stimulation.

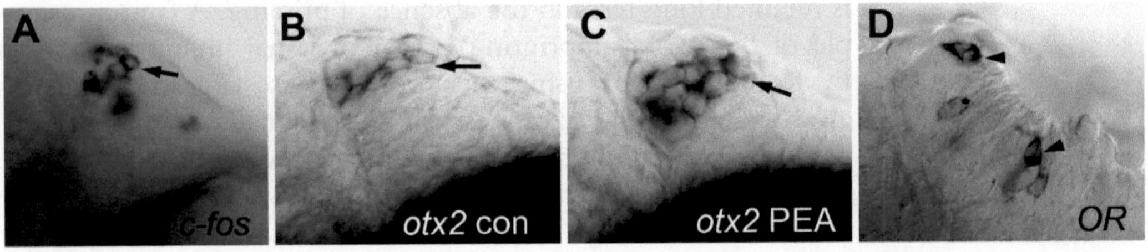

Fig. 2.3. The olfactory epithelium is suitable for analysis of gene expression in response to environmental odorants. The Immediate Early Gene (IEG) *c-fos* has a limited pattern of expression in the developing olfactory epithelium at 56 hpf (**a**). The region of expression corresponds to the expression domain of the transcription factor *otx2* (**b**), whose expression domain can be modified by odor exposure (**c**); (41). The olfactory receptor genes are expressed early during development (**d**) and their expression may be modifiable through odorant exposure.

A frequently asked question is why zebrafish would imprint on odors experienced as juveniles. In the case of the salmon, imprinting ensures that the fish return to an environment having a substrate suitable for spawning and growth of the offspring. Site fidelity is advantageous to a variety of fish species including oceanic reef fishes, such as cardinalfish and clownfish, where it has been shown that the pelagic larvae prefer the scents of their natal reef sites (44), and odors of vegetation (45) as well as conspecifics (46) are important olfactory cues. There is no reason to expect zebrafish to be different, though the stakes may not be as high as in a salmon or a juvenile reef fish, for they lack the dramatic migration and change in physiology evidenced in salmon. In zebrafish, we know from work in the field that they prefer shallow, still waters with silt bottoms and vegetation (47, 48), the latter most likely providing cover for the juveniles. These waters are generally seasonal and young zebrafish move back to the main streams as the seasonally high water recedes (47). Thus the fish retaining a memory of an odor bouquet of water containing vegetation and silty bottoms are more likely to quickly find suitable spawning sites as well as others of its species to reproduce with.

5. Hormones and Olfaction

Essential for the propagation of the species is the coordination of reproductive behaviors leading to successful fertilization of offspring. Zebrafish clearly have stereotyped reproduction (courting) behaviors and these have been described as falling into three distinct phases: initiatory, receptive/appetitive, and spawning (49). Fish use a variety of sensory cues to coordinate reproduction and important among these are olfactory cues conveying the reproductive state of the male and female fish. Using goldfish as a model system, it has been demonstrated that hormones are used as pheromones to coordinate reproduction. The production of gonadal hormones necessary for ovulation, triggered by a surge of luteinizing hormone (LH), results in the female fish releasing hormonal metabolites that are potent sex pheromones: prostaglandin $F_{2\alpha}$ $PGF_{2\alpha}$ and a progesterone (4-pregnen-17,20β-diol-3-one, 20-sulphate; 17, 20, βP). These hormones, released in the urine of ovulatory goldfish elicit behavioral and physiological responses in adult male goldfish (50, 51); 17, 20, βP and its metabolite 17, 20, βP-S are released 12 h before spawning as a "pre-ovulatory primer" which trigger sperm production and spawning behaviors, dependent upon the ratios of 17, 20, βP and its metabolite (50, 52). Post-ovulatory females also release F prostaglandins which act as a cue for spawning behaviors.

In our ongoing analysis of the interactions between the olfactory environment and the developing olfactory sensory system, we have used probes that recognize the IEG *c-fos*, *c-jun* as a molecular genetic readout for neuronal activity in the developing embryo. We cloned these genes and analyzed their expression pattern through in situ hybridization and have shown that, like other vertebrates, these genes are expressed in the olfactory bulb (53) and OE (54). Based on studies in a variety of fishes we initiated an analysis looking for changes in gene expression (by in situ hybridization) in response to olfactory stimuli. A cautionary note is that the expression of *c-fos* is sensitive to whether or not the embryos have been dechorionated prior to fixation (MacKenzie & Whitlock, unpublished) suggesting that the expression observed in fish maintained in the chorion is different from that of dechorionated fish incubated in straight embryo media. Based on the work of Sorensen et al. (51), we tested the response of *c-fos* to prostaglandin $F_{2\alpha}$ $PGF_{2\alpha}$, taurocholic acid (a bile acid) and a progesterone (4-pregnen-17,20β-diol-3-one,20-sulphate; 17, 20P), and showed that the number of *c-fos* expressing cells changed in the presence of taurocholic acid and $F_{2\alpha}$ $PGF_{2\alpha}$. Thus hormones play a role as olfactory cues in olfactory behaviors essential for reproduction, and this olfactory stimulus appears to trigger a genomic response in IEGs in the OE.

6. Direct Action of Hormones on the Olfactory Epithelium

Clearly, as evidenced by work in goldfish, hormones are employed by fish as pheromones. In the case of GnRH there is evidence that hormones may act directly on the olfactory epithelium. GnRH is both a neuromodulatory peptide and an essential reproductive hormone in vertebrates. There is evidence that the reproductive state of animals affects the sensitivity of the olfactory sensory system (55) and that GnRH systems are affected by olfactory behaviors (56).

Do circulating hormones affect the sensitivity of the OE directly, thus affecting the behavior of the animal? It has been shown previously that GnRH has the ability to modulate odorant response in the peripheral nervous system (57). To build evidence for the idea that hormones modulate neuronal activity in the peripheral nervous system, one would have to localize hormone receptors within the OE of fish (and other animals). We have shown using the ISPR3 antibody recognizing the CLEGKVSHSL motif in extracellular loop 3 GnRH-Receptor (**Fig. 2.4**) that there is expression in the OE in adult zebrafish. Zebrafish have a related motif in GnRH-R4, thus GnRH-R4 is most similar to the motif that was used to raise the antibody (59). These data suggest that the neuroendocrine decapeptide GnRH has, as one of its targets, the peripheral OE, and that the OE is plastic in respect to gene expression modulation. More recently the androgen receptor has been cloned in zebrafish and, through in situ hybridization, shown to be strongly expressed in the developing OE (60). These data suggest that circulating hormones may have effects on the development and sensitivity of the olfactory sensory system. Thus, one could imagine a model where circulating androgens (testosterone) affect the type of olfactory receptors expressed in the OE during development and that GnRH may play an "activational" role in the adult animal by triggering sensitivity to odorants when the animal is in a reproductive state.

7. Movement of Fish

The analysis of olfactory behavior in aquatic animals requires an understanding of the dynamics of fluid movement in order to use the correct analysis. When designing an experiment to test

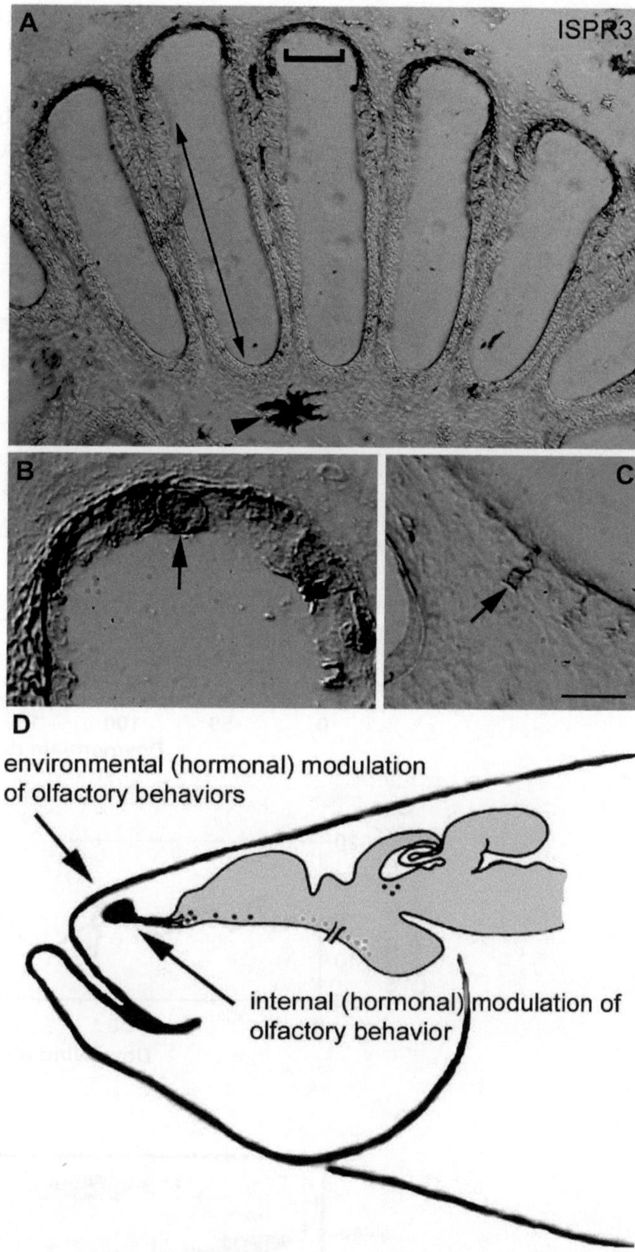

Fig. 2.4. Hormonal modulation of olfactory behaviors may result from external stimuli or internal circulating hormones. The antibody ISPR3 against GnRH receptor recognizes cells in the zebrafish olfactory epithelium including large (**a**), superficial cells (**b**) and smaller, neuron-like cells (**c**). These cells may respond to GnRH encountered in the external environment or circulating GnRH (**d**) that fluctuates with the reproductive cycle. Modified from Whitlock et al. (58).

an olfactory driven behavior, the response of a fish or group of fish can be measured using a simple metric, such as time spent in a predefined region of the holding tank, or they can be tested for orientation within a plume of odorant. Testing animals in an odor plume is more complicated as the odor source is moving. Odor plumes characteristically have varying gradients dependent upon the axis, where edges lateral to the direction of flow have distinct boundaries and the longitudinal axis (direction of flow)

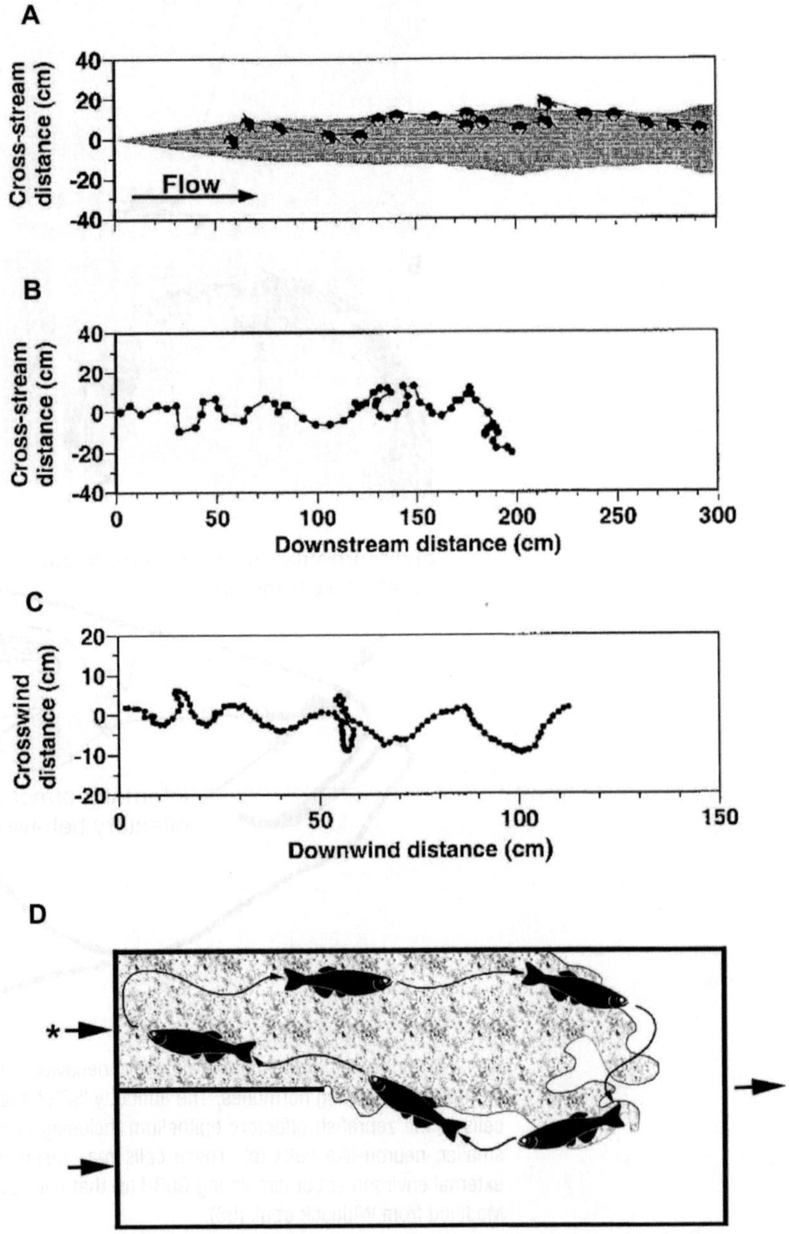

Fig. 2.5. The orientation of animals in odor plumes shows consistent behavioral characteristics. Various animals, crab (**a**), lobster (**b**), moth (**c**) (61), and fish (**d**) (41) show consistent behavior characteristics when orienting into odor plume, moving across the border of the plume gradient to detect concentration differences. **a–c** Modified from Vickers (61); **d** from Vitebsky et al. (38).

of the plume has a weaker gradient (61). When orienting towards an odor source, animals (fish, birds, moths) move into the mean direction of the flow. Strikingly, during this movement the animals do not move in a direct path, rather they cast back and forth across the fluid flow (**Fig. 2.5**). This behavior has been previously described in moths (62), crustaceans (63), and now zebrafish (41). In our analysis of adult zebrafish behavior we used a "Y-maze" system where the two different arms of the maze were baited with a control and PEA, an odor to which the fish had been exposed to as juveniles. The maze had water flowing through it and the fish oriented in this flow. We collected video recordings of the fish during the first 4 min of the trial. The data were analyzed by counting the number of fish in specific regions of the flow over time. Because we were testing groups of fish the analysis was more complicated because the fish moved through space and time. After considering the characteristic behaviors of fish orienting in a plume and sampling across a gradient, the analysis was redesigned to capture changes relative to the odor plume (41). We more recently used Y-mazes to test reactions of native Chilean puye (galaxids) to odors of invasive fish species (brown trout). The Y-maze also employed a flow-through system, only the fish were given a very restricted area in which to move, thus were unable to cast across concentration gradients (see section below).

8. Small Scale Water Movements

Just as fish move through water sampling the odor environment, water must move through the fish, i.e., across the OE, and fishes have evolved a variety of structures that facilitate water flow across the olfactory sensory/respiratory epithelia that line the olfactory chamber. Two principal active sources are the beating of cilia lining the olfactory chamber and the pumping action generated by accessory sacs. In general, each olfactory chamber has two openings through which water enters (anterior) and water leaves (posterior). In contrast, agnathans such as lampreys and hagfish have a single un-paired olfactory chamber with a single nostril (64); some sharks have their olfactory organs located ventrally on the rostrum, and in puffers the olfactory chambers containing the sensory epithelia are stalked, protruding from the head as described by Kleerekoper (65). (It has been proposed that the development of stalked epithelia results from the crowding due to extensive development of jaw muscle in this coral feeding group of fishes (66).)

In order to actively move water around the olfactory chamber fishes have developed a variety of means including beating of cilia of nonsensory cells, the mechanical movement of the olfactory chamber via jaw movements, and expansions of olfactory sacs, which are an expansion of the main olfactory chamber (67). Zebrafish have kinocilia lining their olfactory chamber (one can see this in anesthetized living zebrafish by 2–3 days post-fertilization), although these kinocilia have been proposed to be for the movement of mucus due to their short length (7–8 μm; (68)), not for movement of water (10–20 μm). Due to paucity of data on this subject it is currently not known whether the kinocilia of zebrafish are multipurpose: moving water and mucus (67).

9. Modern Approaches to Behavioral Analysis

Since their inception as a model system for studies in developmental biology, zebrafish have reigned supreme due to their optical clarity and accessibility, which enables us to visualize the first moments of fertilization, gastrulation, neurulation, somitogenesis, and other developmental landmarks. For those of us entering the "fish world" from the "fly world" (*Drosophila*), a constant source of frustration has been the lack of molecular genetic tools essential for manipulating gene expression. In zebrafish, the generation of transgenic lines expressing fluorescent reporters has allowed us to visualize movements of cells in living embryos and identify cells in neuronal circuits. In the past few years there has been an increase in the available transgenic technologies including incorporating the Gal4/UAS system in zebrafish in order to identify and manipulate specific neurons within a circuit of interest (69). The development of transgenic technologies in zebrafish has been coupled with the analysis of behavioral responses to uncover neural circuitry important for the given behavior. Through these types of analyses the motor circuitry underlying the escape response in juvenile zebrafish has been elucidated using a transgenic zebrafish with a neural specific promoter driving a calcium indicator protein (70). The technique of expressing calcium indicator dyes in transgenic fish has also been applied in zebrafish to map odorant responses in the developing olfactory bulb (71). In contrast to the studies of motor circuitry, the mapping of the odor response was not done as the animal was displaying a behavior. Rather, odorants we know fish response to, such as amino acids, were delivered and responses recorded in a tethered preparation. Recently, the circuitry of the primary olfactory sensory system of the zebrafish has been revealed through a combination of fluorescent reporter lines that are expressed in the ciliated sensory neurons (OMP-RFP) and microvillous (TRP-GFP) sensory neurons of the living fish (14) as well as secondary projections from the olfactory bulb (72). This group has made use of the *tol2* transposon-mediated technique combined with the Gal4/UAS system to generate lines of zebrafish with GFP expression in specific subsets of olfactory sensory neurons. They then used these lines to express the tetanus neurotoxin in these neurons to block synaptic transmission and showed that a specific subset of microvillous olfactory sensory neurons are important in behavioral responses to amino acids (73). Thus, this recent study by Koide et al. (73) demonstrates that molecular genetic manipulations in zebrafish are becoming more accessible, and will prove invaluable in the coupling of sensory circuitry with given behavioral responses in developing and adult animals.

10. Water Quality and What Is "Native" to a Zebrafish

When discussing water parameters with zebrafish researchers, there are always a variety of responses as to the conductivity and pH that each individual lab strives for in its fish facility. Zebrafish are tough little creatures, which is why we like them so much as model lab animals. Yet we tend to overlook the fact that while they survive at given water conditions on given diets, these conditions may not be optimal. This is an important consideration when designing olfactory behavior experiments because olfactory physiological responses can vary dependent upon the pH (74, 75) and environmental odors (76). Studies using galaxids have shown that the olfactory response to conspecifics is strong when the fish are tested in tap water yet greatly diminished when tested in stream water (76). Salmon imprint on odors of a specific stream, but we do not know the relative contribution of stream odors versus odors of conspecifics of different year classes resident in the streams. Thus, animals are using multiple, context-dependent olfactory cues when expressing a behavior. This must be kept in mind when performing olfactory behavioral tests with zebrafish: in particular, the source of the water in which the test is being performed and whether it previously contained zebrafish.

11. Future Directions

The earth has undergone great changes during its existence. For the last 10,000 years we have enjoyed a very stable climate, yet now anthropomorphic (man-made) changes are destabilizing the climate resulting in a rate of climatic change never experienced previously by humans. In conjunction with the destabilization of the climate we are experiencing the sixth great extinction of living animals on the planet (the last being 65 million years ago when dinosaurs as well as 75% of species on earth were exterminated), and some predict that by the end of this century 50% of the species will be gone (77, 78). These dramatic changes are caused by human activity and the excesses we generate: the buildup of atmospheric carbon, accumulation of contaminants in natural systems, and overexploitation of natural resources. All of these factors affect fishes and the worlds they live in as evidenced by the potential negative effects on larval fish settlement due to ocean acidification, the destruction of olfactory responses in freshwater fishes exposed to city and agricultural runoff, and the pressures of invasive species competing for the food and resources of the native fish.

11.1. Olfaction and Global Climate Change

Climate change caused by global warming results from the increase in greenhouse gases produced by human activity. The Intergovernmental Panel on Climate Change (IPCC) concluded that this human activity has been and continues to be the dominant cause of the temperature increase since the mid-twentieth century (1.33 ± 0.32 Fahrenheit). A major component of greenhouse gases is carbon dioxide (CO_2), and rising anthropogenic CO_2 levels affect not only the temperature of the planet but also the pH of the oceans, causing the pH to decrease ("ocean acidification") (79, 80). The pH of the oceans has already dropped by 0.1 units in the past century. As we approach the end of the twenty-first century the pH is expected to decline by another 0.3–0.5 units (81). As the oceans become less basic there is much focus on the ability of calcifying animals to adapt, but what happens to animals dependent upon olfactory cues? Only recently have scientists started to address this complex question. It has previously been shown that larval reef fishes such as cardinalfish, damselfish, and clownfish (44, 45), use olfactory cues for the settlement of their pelagic larvae on their natal coral reefs. Thus, the ability of these fishes to find mates and a suitable habitat in which to reproduce is dependent upon their ability to discriminate odors. A recent study by Munday et al. (75) demonstrated that ocean acidification affects the ability of marine fish to discriminate odors. Using the clownfish, researchers tested the behavioral response of the fish to odors at three distinct pH treatments: that of the ocean at their study site (8.15 ± 0.07), and two lower pH values (7.8 ± 0.05 and 7.6 ± 0.05), corresponding to the pH of more acidic oceans as predicted by climate change models. Larvae reared and tested at pH 7.8 displayed altered behavior, where an odor strongly avoided changed to a preferred odor, and the ability to distinguish parents and non-parents was lost. Strikingly, all olfactory response was lost at the lowest pH levels (=7.6) even though their olfactory sensory system was well developed (75). The only possibility of survival would be if the animals were able to adapt to these changing conditions, yet the rapid rate of ocean acidification makes this scenario unlikely.

Because of their rapid development (enabling the creation of 3–4 generations a year), zebrafish are an ideal model system to better understand the ability of the olfactory sensory system to adapt to accelerated climate change. By testing for a specific behavior, such as recognition of conspecifics (26), and selecting for this behavior we can test for physiology (EOG), behavior, and reproductive success. The first step would be to test the olfactory sensory ability under a range of pH values. It is interesting to note that people raise their zebrafish at fairly different ranges of pH. Information collected from the wild (48) suggests that zebrafish are found in streams that tend to be of higher pH (7.8) than that of the average fish facility (7.3–7.4), although the pH of freshwater systems can vary with the seasons.

11.2. Contaminants in Olfactory Environments

The olfactory sensory epithelia is unique in that the dendrites of the sensory neurons come in direct contact with the outside world thus making them extremely sensitive to environmental insults. The ability of the olfactory sensory neurons to regenerate throughout life has been used as an example of the system's adaptability to environmental insult. Contaminants in the aquatic environments arise from multiple sources and can contain a plethora of known and unknown contaminants. Agricultural runoff can contain a variety of neurotoxic pesticides, and copper, one of the most ubiquitous and damaging contaminants (for the olfactory sensory system), has multiple point sources in modern day society. Due to the multiple sources of copper, many watersheds contain this contaminant in environmentally relevant concentrations and copper has now been shown to directly affect the olfactory sensory system of juvenile Coho salmon. Neurotoxic pesticides such as Diazinon have been shown to disrupt olfactory mediated behaviors such as alarm response and homing in Pacific salmon (82). In these studies it was shown that the function of the olfactory sensory system was impaired at both the physiological level and the behavioral level (response to alarm pheromone) (83). These findings about the effects of contaminants on the olfactory sensory system and the behaviors essential to survival are very alarming, yet the picture grows worse as one analyzes the real world. Chemicals do not exist as single contaminants in the environment; rather water samples generally contain multiple contaminants and their potential to interact and intensify toxic effects (84) is poorly understood. Recently toxicology studies have started using zebrafish as a model system to understand the effects of environmental toxicants on the olfactory sensory system. The well-known environmental contaminant cadmium is known to cause cell death in the olfactory sensory system and has now been shown to affect olfactory behaviors in juvenile zebrafish (85). In examining the literature in other fishes it is apparent that the alarm response, a response that zebrafish also display is a robust response amenable for use as a behavioral assay for effects on environmental pollution on the olfactory sensory system.

11.3. Olfaction and Invasive Species

As human beings developed transport and started to cross the seas to explore different continents they brought not only their belongings, but also their livestock and unintended passengers. These invasive species of animals (as well as plants) initiated the wave of introduced species that now wreak havoc in ecosystems around the world. Subsequent purposeful introductions of foreign species for biological control as well as agriculture/aquaculture have only intensified the problems created by these introduced species. In Chile, particularly in the South, the aquaculture industry as well as the sport fishing industry has introduced a variety of non-native fish into the local ecosystems. Some species such as rainbow trout (*Oncorhynchus mykiss*), brown trout (*Salmo trutta*), and Chinook salmon (*Oncorhyncus tshawytscha*); (86) have established reproductive populations within the rivers of Southern Chile. For species such as the intensively farmed Atlantic Salmon (*Salmo salar*) it remains unclear whether the animals are reproducing in the rivers. In order to understand the dynamics of invasive species at the behavioral level we (our lab and the Darwin Initiative, http://www.darwin.defra.gov.uk/project/15020/), have investigated the effects of introduced species on the behavior of native species in the rivers of Chiloé Island in Southern Chile (**Fig. 2.6**). The small streams and rivers are differentially populated by local Chilean Galaxids (including the native "zebra trout," *Aplochiton zebra*, rare). The galaxids are a southern hemisphere fish, and it has been documented that streams containing invasive trout and salmon have negatively impacted their local populations (87). One obvious impact is predation, but what is the response of native population to a newly established population non-native species? In the case of olfactory behaviors essential in the avoidance of predators it may be that the evolutionary adaptive avoidance response to odorants of predators is lacking. To determine whether the native Chilean puye (*Galaxias maculates*), have a fear response to native versus non-native predators we performed behavior tests using odors of brown trout (non-native) and zebra trout (native). To determine whether this response was a learned response from generations of cohabitation in the streams, we compared the puye originating from streams that lacked brown trout with those originating from streams where

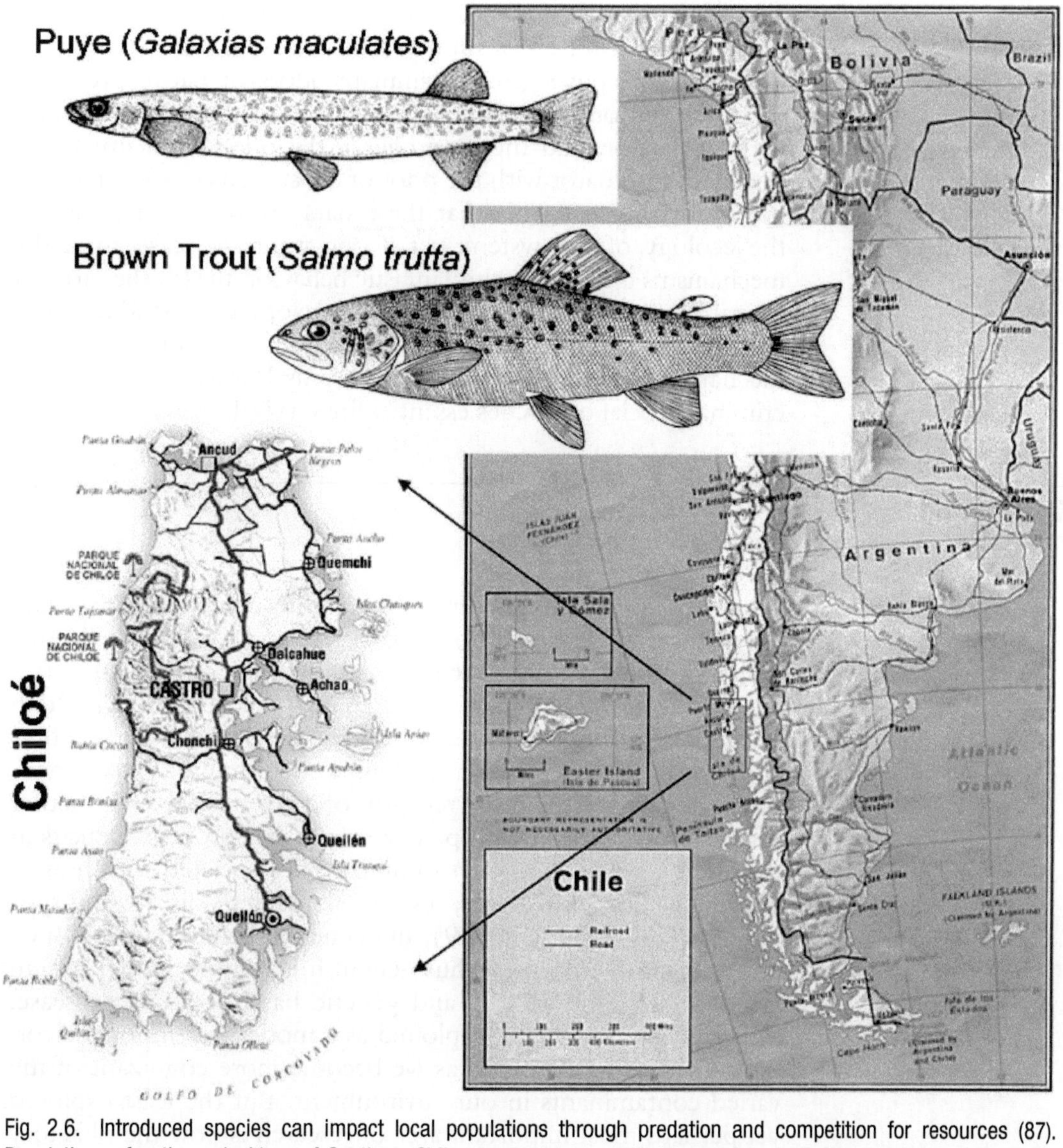

Fig. 2.6. Introduced species can impact local populations through predation and competition for resources (87). Populations of native galaxids **a** of Southern Chile **c** are more severely impacted by introduced species such as the brown trout **b** for they must learn the odor of the predator over generations of cohabitation (88). In the interim their population is impacted by heavy predation.

brown trout were found. The animals were brought to the lab environment, allowed to acclimate, and then tested in a Y-maze. In brief, we found the native puye showed avoidance responses to odors of non-native predators only when they came from sympatric populations (collected from streams where the non-native predator was found). In contrast, the odor of the native predator always evoked an avoidance response although it was stronger in sympatric than allopatric conditions. If the non-native predator was fed puye, then the puye showed an avoidance response regardless of the sympatric/allopatric conditions (88). These studies demonstrate the subtle nuances of olfactory discrimination: that the puye can discriminate odors of native predators that have fed on conspecifics, they can learn the odor of a non-native predator, and they can couple the odor of an unfamiliar non-native predator with the odor of consumed conspecifics. The beauty of this system is that these data are directly relevant to the ecology of the system, yet if one wants to tease apart the mechanisms underlying this fantastic behavior, these fishes are not tractable to molecular, genetic, and developmental studies. Thus the zebrafish is an important model system for teasing apart the mechanisms underlying the innate versus learned ability to discriminate social odor cues essential for survival.

12. Parting Remarks

Where do we go from here? What value does zebrafish olfactory behavioral research hold for society and for our intellectual curiosity about natural systems? In examining the challenges facing society, zebrafish olfactory system is an excellent model system for understanding the crosstalk between basal cells and the differentiating olfactory sensory neurons in determining how a neuron "chooses" to express a receptor of a given type, and whether the olfactory environment plays a role in this process. Deficits in the ability to perceive odors in humans is associated with human neurodegenerative diseases, such as Alzheimer's disease and sporadic Parkinson's disease (89), and genetic-based syndromes such as Kallmann Syndrome. Thus zebrafish is a good model system for dissecting the cellular and genetic basis of human disease. Zebrafish are also being exploited as a model system for toxicology (90), especially useful as we become more cognizant of the varied contaminants in our environment. But the least explored yet perhaps most important area of research is how the olfactory sensory system interfaces with a rapidly changing and degrading environment. We are faced with worldwide changes in climate and a future of diminished availability of natural resources. Studies in fish olfaction will allow us to better understand how rapidly vertebrates can adapt to a changing world, how fish populations will survive in a much depleted ocean, and how resilient the olfactory based behaviors are in the face of environmental contaminants (**Fig. 2.7**). Finally, fish as a group represent 50% of the vertebrate animals on this planet. They are fascinating in their variety of forms and behaviors; knowledge of the important components of their behavior will allow us to preserve these beautiful creatures in the face of increasing environmental stresses.

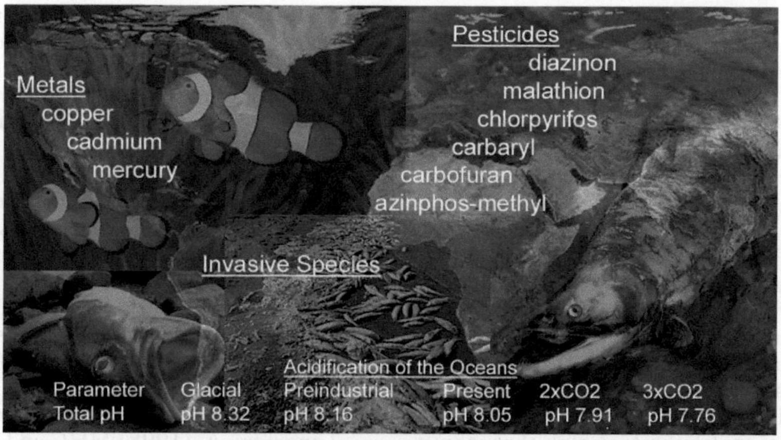

Fig. 2.7. Fish face many challenges in the natural environment in the twenty-first century. There is ample evidence that environmental toxicants (heavy metals, pesticides), as well as invasive species and climate change are adversely affecting the ecosystem in which we live. For fish of both freshwater and marine ecosystems we now know that olfactory behaviors, alarm response, homing, recognition of conspecifics are being negatively impacted by the plethora of environmental contaminants (91). These behaviors, while an intellectual curiosity for us, are essential to maintain reproductive populations of these fascinating and beautiful animals.

Acknowledgments

I would like to thank Kyle Young, Paul Howes, and Jessica Stephensen of the Darwin Initiative for introducing me to the life of puye, J. Ewer for critical reading of the manuscript, and my lab for their interest and support. Our work is supported by FONDECYT 1071071 (KW), MIDEPLAN, Millennium Science Initiative Program (KW), NIH/NIDCD R01 050820 (KW).

References

1. Alioto, T. S. & Ngai, J. (2005) The odorant receptor repertoire of teleost fish. *BMC Genomics* 6, 173.
2. Hashiguchi, Y., Furuta, Y. & Nishida, M. (2008) Evolutionary patterns and selective pressures of odorant/pheromone receptor gene families in teleost fishes. *PLoS One* 3, e4083.
3. Buck, L. & Axel, R. (1991) A novel multigene family may encode odorant receptors: a molecular basis for odor recognition. *Cell* 65, 175–187.
4. Olender, T., Lancet, D. & Nebert, D. W. (2008) Update on the olfactory receptor (OR) gene superfamily. *Hum Genomics* 3, 87–97.
5. Godfrey, P. A., Malnic, B. & Buck, L. B. (2004) The mouse olfactory receptor gene family. *Proc Natl Acad Sci USA* 101, 2156–2161.
6. Gilad, Y., Przeworski, M. & Lancet, D. (2004) Loss of olfactory receptor genes coincides with the acquisition of full trichromatic vision in primates. *PLoS* 2, E5.
7. Liberles, S. D. & Buck, L. B. (2006) A second class of chemosensory receptors in the olfactory epithelium. *Nature* 442, 645–650.
8. Zhang, X., Rodriguez, I., Mombaerts, P. & Firestein, S. (2004) Odorant and vomeronasal receptor genes in two mouse genome assemblies. *Genomics* 83, 802–811.

9. Yang, H., Shi, P., Zhang, Y. P. & Zhang, J. (2005) Composition and evolution of the V2r vomeronasal receptor gene repertoire in mice and rats. *Genomics* 86, 306–315.
10. Luu, P., Acher, F. C., Bertrand, H. O., Fan, J. & Ngai, J. (2004) Molecular determinants of ligand selectivity in a vertebrate odorant receptor. *J Neurosci* 24, 10128–10137.
11. Speca, D. J., Lin, D. M., Sorensen, P. W., Isacoff, E. Y., Ngai, J. & Dittman, A. H. (1999) Functional identification of a goldfish odorant receptor. *Neuron* 23, 487–498.
12. Triballeau, N., van Name, E., Laslier, G., Cai, D., Paillard, G., Sorensen, P. W., Hoffmann, R., Bertrand, H. O., Ngai, J. & Acher, F. C. (2008) High-potency olfactory receptor agonists discovered by virtual high-throughput screening: molecular probes for receptor structure and olfactory function. *Neuron* 60, 767–774.
13. Hansen, A., Anderson, K. T. & Finger, T. E. (2004) Differential distribution of olfactory receptor neurons in goldfish: structural and molecular correlates. *J Comp Neurol* 477, 347–359.
14. Sato, Y., Miyasaka, N. & Yoshihara, Y. (2005) Mutually exclusive glomerular innervation by two distinct types of olfactory sensory neurons revealed in transgenic zebrafish. *J Neurosci* 25, 4889–4897.
15. Hamdani, E. H. & Døving, K. B. (2007) The functional organization of the fish olfactory system. *Prog Neurobiol* 82, 80–86.
16. Pfister, P. & Rodriguez, I. (2005) Olfactory expression of a single and highly variable V1r pheromone receptor-like gene in fish species. *Proc Natl Acad Sci USA* 102, 5489–5494.
17. Lipschitz, D. L. & Michel, W. C. (2002) Amino acid odorants stimulate microvillar sensory neurons. *Chem Senses* 27, 277–286.
18. Schmachtenberg, O. (2006) Histological and electrophysiological properties of crypt cells from the olfactory epithelium of the marine teleost Trachurus symmetricus. *J Comp Neurol* 495, 113–121.
19. Boehm, T. & Zufall, F. (2006) MHC peptides and the sensory evaluation of genotype. *Trends Neurosci* 29, 100–107.
20. Yamazaki, K., Beauchamp, G. K., Kupniewski, D., Bard, J., Thomas, L. & Boyse, E. A. (1988) Familial imprinting determines H-2 selective mating preferences. *Science* 240, 1331–1332.
21. Yamaguchi, M., Yamazaki, K., Beauchamp, G. K., Bard, J., Thomas, L. & Boyse, E. A. (1981) Distinctive urinary odors governed by the major histocompatibility locus of the mouse. *Proc Natl Acad Sci USA* 78, 5817–5820.
22. Milinski, M., Griffiths, S., Wegner, K. M., Reusch, T. B., Haas-Assenbaum, A. & Boehm, T. (2005) Mate choice decisions of stickleback females predictably modified by MHC peptide ligands. *Proc Natl Acad Sci USA* 102, 4414–4418.
23. Reusch, T. B., Haberli, M. A., Aeschlimann, P. B. & Milinski, M. (2001) Female sticklebacks count alleles in a strategy of sexual selection explaining MHC polymorphism. *Nature* 414, 300–302.
24. Olsen, K. H., Grahn, M. & Lohm, J. (2002) Influence of MHC on sibling discrimination in Arctic char, *Salvelinus alpinus* (L.). *J Chem Ecol* 28, 783–795.
25. Olsen, K. H., Grahn, M., Lohm, J. & Langefors, A. (1998) MHC and kin discrimination in juvenile Arctic charr, *Salvelinus alpinus* (L.). *Anim Behav* 56, 319–327.
26. Mann, K. D., Turnell, E. R., Atema, J. & Gerlach, G. (2003) Kin recognition in juvenile zebrafish (Danio rerio) based on olfactory cues. *Biol Bull* 205, 224–225.
27. Skals, N., Anderson, P., Kanneworff, M., Löfstedt, C. & Surlykke, A. (2005) Her odours make him deaf: crossmodal modulation of olfaction and hearing in a male moth. *J Exp Biol* 208, 595–601.
28. Gardiner, J. M. & Atema, J. (2007) Sharks need the lateral line to locate odor sources: rheotaxis and eddy chemotaxis. *J Exp Biol* 210, 1925–1934.
29. Modarressie, R., Rick, I. P. & Bakker, T. C. (2006) UV matters in shoaling decisions. *Proc R Soc B* 273, 849–854.
30. Whitlock, K. E. (2009) Evolution of the terminal nerve. In Binder, M. D., Hirokawa, N. H. & Windhorst, U. (Eds.) *Encyclopedia of Neuroscience*. Berlin, Springer GmbH.
31. Behrens, U. & Wagner, H. J. (2004) Terminal nerve and vision. *Microsc Res Tech* 65, 25–32.
32. Umino, O. & Dowling, J. E. (1991) Dopamine release from interplexiform cells in the retina: effects of GnRH, FMRFamide, bicuculline, and enkephalin on horizontal cell activity. *J Neurosci* 11, 3034–3046.
33. Wirsig-Wiechmann, C. R. & Oka, Y. (2002) The terminal nerve ganglion cells project to the olfactory mucosa in the dwarf gourami. *Neurosci Res* 44, 337–341.
34. Andersen, O. & Døving, K. B. (1991) Gonadotropin releasing hormone (GnRH) – a novel olfactory stimulant in fish. *Neuroreport* 2, 458–460.
35. Behrens, U. D., Douglas, R. H. & Wagner, H. J. (1993) Gonadotropin-releasing hormone, a neuropeptide of efferent projections to the teleost retina induces light-adaptive

spinule formation on horizontal cell dendrites in dark-adapted preparations kept in vitro. *Neurosci Lett* 164, 59–62.
36. Maaswinkel, H. & Li, L. (2003) Olfactory input increases visual sensitivity in zebrafish: a possible function for the terminal nerve and dopaminergic interplexiform cells. *J Exp Biol* 206, 2201–2209.
37. Weiss, O. & Meyer, D. L. (1988) Odor stimuli modulate retinal excitability in fish. *Neurosci Lett* 93, 209–213.
38. Vitebsky, A., Reyes, R., Sanderson, M. J., Michel, W. C. & Whitlock, K. E. (2005) Isolation and characterization of the laure olfactory behavioral mutant in the zebrafish Danio rerio. *Dev Dyn* 234, 229–242.
39. Hasler, A. D. & Scholz, A. T. (1983) *Olfactory Imprinting and Homing in Salmon.* Berlin, Heidelberg, Springer.
40. Nevitt, G. A., Dittman, A. H., Quinn, T. P. & Moody, W. J. J. (1994) Evidence for a peripheral olfactory memory in imprinted salmon. *Proc Natl Acad Sci USA* 91, 4288–4292.
41. Harden, M. R., Newton, L. A., Lloyd, R. & Whitlock, K. E. (2006) Olfactory imprinting is correlated with changes in gene expression in the olfactory epithelia of the zebrafish. *J Neurobiol* 66, 1452–1466.
42. Barbour, J., Neuhaus, E. M., Piechura, H., Stoepel, N., Mashukova, A., Brunert, D., Sitek, B., Stühler, K., Meyer, H. E., Hatt, H. & Warscheid, B. (2008) New insight into stimulus-induced plasticity of the olfactory epithelium in Mus musculus by quantitative proteomics. *J Proteome Res* 7, 1594–1605.
43. McKenzie, M. G., Harden, M. V. & Whitlock, K. E. (2006) Odorant modulation of immediate early gene expression in the zebrafish olfactory epithelia. *Association for Chemoreception Sciences.* Sarasota, FL, Chemical Senses.
44. Gerlach, G., Atema, J., Kingsford, M. J., Black, K. P. & Miller-Sims, V. (2007) Smelling home can prevent dispersal of reef fish larvae. *Proc Natl Acad Sci USA* 104, 858–863. Epub 2007 Jan 9.
45. Dixson, D. L., Jones, G. P., Munday, P. L., Planes, S., Pratchett, M. S., Srinivasan, M., Syms, C. & Thorrold, S. R. (2008) Coral reef fish smell leaves to find island homes. *Proc Biol Sci* 275, 2831–2839.
46. Doving, K. B., Stabell, O. B., Ostlund-Nilsson, S. & Fisher, R. (2006) Site fidelity and homing in tropical coral reef cardinalfish: are they using olfactory cues? *Chem Senses* 31, 265–272. Epub 2006 Jan 25.
47. Engeszer, R. E., Patterson, L. B., Rao, A. A. & Parichy, D. M. (2007) Zebrafish in the wild: a review of natural history and new notes from the field. *Zebrafish* 4, 21–40.
48. McClure, M. (1999) Development and evolution of melanophore patterns in fishes of the genus Danio (Teleostei: Cyprinidae). *J Morphol* 241, 83–105.
49. Darrow, K. O. & Harris, W. A. (2004) Characterization and development of courtship in zebrafish, Danio rerio. *Zebrafish* 1, 40–45.
50. Derby, C. D. & Sorensen, P. W. (2008) Neural processing, perception, and behavioral responses to natural chemical stimuli by fish and crustaceans. *J Chem Ecol* 34, 898–914.
51. Sorensen, P. W., Hara, T. J., Stacey, N. E. & Goetz, F. W. M. (1988) F Prostaglandins Function as Potent Olfactory Stimulants That Comprise the Postovulatory Female Sex Pheromone in Goldfish. *Biol Reprod* 39, 1039–1050.
52. Sorensen, P. W., Scott, A. P., Stacey, N. E. & Bowdin, L. (1995) Sulfated 17,20 beta-dihydroxy-4-pregnen-3-one functions as a potent and specific olfactory stimulant with pheromonal actions in the goldfish. *Gen Comp Endocrinol* 100, 128–142.
53. Guthrie, K. M., Anderson, A. J., Leon, M. & Gall, C. (1993) Odor-induced increases in c-fos mRNA expression reveal an anatomical "unit" for odor processing in olfactory bulb. *Proc Natl Acad Sci USA* 90, 3329–3333.
54. Norlin, E. M., Vedin, V., Bohm, S. & Berghard, A. (2005) Odorant-dependent, spatially restricted induction of c-fos in the olfactory epithelium of the mouse. *J Neurochem* 93, 1594–1602.
55. Hudson, R. & Distel, H. (1998) Induced peripheral sensitivity in the developing vertebrate olfactory system [Review]. *Ann NY Acad Sci* (Nov 30) 855, 109–115.
56. Kleerekoper, H. (1969) *Olfaction in Fishes.* Bloomington, IN, Indiana University Press.
57. Uchida, H., Ogawa, S., Harada, M., Matushita, M., Iwata, M., Sakuma, Y. & Parhar, I. S. (2005) The olfactory organ modulates gonadotropin-releasing hormone types and nest-building behavior in the tilapia *Oreochromis niloticus. J Neurobiol* 65, 1–11.
58. Park, D. & Eisthen, H. L. (2003) Gonadotropin releasing hormone (GnRH) modulates odorant responses in the peripheral olfactory system of axolotls. *J Neurophysiol* 90, 731–738.
59. Whitlock, K. E., Illing, N., Brideau, N. J., Smith, K. M. & Twomey, S. (2006) Development of GnRH cells: setting the stage for puberty. *Mol Cell Endocrinol* 25, 39–50.
60. Flanagan, C. A., Chen, C. C., Coetsee, M., Mamputha, S., Whitlock, K. E.,

Bredenkamp, N., Grosenick, L., Fernald, R. D. & Illing, N. (2007) Expression, structure, function, and evolution of gonadotropin-releasing hormone (GnRH) receptors GnRH-R1SHS and GnRH-R2PEY in the teleost, Astatotilapia burtoni. *Endocrinology* 148, 5060–5071.
61. Gorelick, D. A., Watson, W. & Halpern, M. E. (2008) Androgen receptor gene expression in the developing and adult zebrafish brain. *Dev Dyn* 237, 2987–2995.
62. Vickers, N. J. (2000) Mechanisms of animal navigation in odor plumes. *Biol Bull* 198, 203–212.
63. Vickers, N. J. (2006) Winging it: moth flight behavior and responses of olfactory neurons are shaped by pheromone plume dynamics. *Chem Senses* 31, 155–166.
64. Atema, J. (1995) Chemical signals in the marine environment: dispersal, detection, and temporal signal analysis. *Proc Natl Acad Sci USA* 92, 62–66.
65. Kleerekoper, H. & van Erkel, G. A. (1960) The olfactory apparatus of *Petromyzon marinus*. *Can J Zool* 38, 209–223.
66. Kleerekoper, H. (1969) *Olfaction in Fishes*. Bloomington, Indiana University Press.
67. Cox, J. P. (2008) Hydrodynamic aspects of fish olfaction. *J R Soc Interface* 5, 575–593.
68. Hansen, A. & Zeiske, E. (1998) The peripheral olfactory organ of the zebrafish, Danio rerio: an ultrastructural study. *Chem Senses* 23, 39–48.
69. Scott, E. K. (2009) The Gal4/UAS toolbox in zebrafish: new approaches for defining behavioral circuits. *J Neurochem* 110, 441–456.
70. Higashijima, S., Masino, M. A., Mandel, G. & Fetcho, J. R. (2003) Imaging neuronal activity during zebrafish behavior with a genetically encoded calcium indicator. *J Neurophysiol* 90, 3986–3997.
71. Li, J., Mack, J. A., Souren, M., Yaksi, E., Higashijima, S., Mione, M., Fetcho, J. R. & Friedrich, R. W. (2005) Early development of functional spatial maps in the zebrafish olfactory bulb. *J Neurosci* 25, 5784–5795.
72. Miyasaka, N., Morimoto, K., Tsubokawa, T., Higashijima, S., Okamoto, H. & Yoshihara, Y. (2009) From the olfactory bulb to higher brain centers: genetic visualization of secondary olfactory pathways in zebrafish. *J Neurosci* 29, 4756–4767.
73. Koide, T., Miyasaka, N., Morimoto, K., Asakawa, K., Urasaki, A., Kawakami, K. & Yoshihara, Y. (2009) Olfactory neural circuitry for attraction to amino acids revealed by transposon-mediated gene trap approach in zebrafish. *Proc Natl Acad Sci USA* 106, 9884–9889.
74. Heuschele, J. & Candolin, U. (2007) An increase in pH boosts olfactory communication in sticklebacks. *Biol Lett* 3, 411–413.
75. Munday, P. L., Dixson, D. L., Donelson, J. M., Jones, G. P., Pratchett, M. S., Devitsina, G. V. & Døving, K. B. (2009) Ocean acidification impairs olfactory discrimination and homing ability of a marine fish. *Proc Natl Acad Sci USA* 106, 1848–1852.
76. Hale, R., Swearer, S. E. & Downes, B. J. (2009) Separating natural responses from experimental artefacts: habitat selection by a diadromous fish species using odours from conspecifics and natural stream water. *Oecologia* 159, 679–687.
77. Kingsford, R. T., Watson, J. E., Lundquist, C., Venter, O., Hughes, L., Johnston, E. L., Atherton, J., Gawel, M., Keith, D. A., Mackey, B. G., Morley, C., Possingham, H. P., Raynor, B., Recher, H. F. & Wilson, K. A. (2009) Major conservation policy issues for biodiversity in Oceania. *Conserv Biol* 23, 834–840.
78. Kolbert, E. (2009) *The Sixth Extinction? The Earth's Species in Peril*. The New Yorker, May 25, pp. 53–56.
79. Caldeira, K. & Wickett, M. E. (2003) Oceanography: anthropogenic carbon and ocean pH. *Nature* 425, 365.
80. Orr, J. C., Fabry, V. J., Aumont, O., Bopp, L., Doney, S. C., Feely, R. A., Gnanadesikan, A., Gruber, N., Ishida, A., Joos, F., Key, R. M., Lindsay, K., Maier-Reimer, E., Matear, R., Monfray, P., Mouchet, A., Najjar, R. G., Plattner, G. K., Rodgers, K. B., Sabine, C. L., Sarmiento, J. L., Schlitzer, R., Slater, R. D., Totterdell, I. J., Weirig, M. F., Yamanaka, Y. & Yool, A. (2005) Anthropogenic ocean acidification over the twenty-first century and its impact on calcifying organisms. *Nature* 437, 681–686.
81. Guinotte, J. M. & Fabry, V. J. (2008) Ocean acidification and its potential effects on marine ecosystems. *Ann NY Acad Sci* 1134, 320–342.
82. Scholtz, N. L., Truelove, N. K., French, B. L., Berejikian, B. A., Quinn, T. P., Casillas, E. & Collier, T. K. (2000) Diazinon disrupts antipredator and homing behaviors in chinook salmon (*Oncorhynchus tshawytscha*). *Can J Fish Aquat Sci* 57, 1911–1918.
83. Sandahl, J. F., Baldwin, D. H., Jenkins, J. J. & Scholz, N. L. (2007) A sensory system at the interface between urban stormwater runoff and salmon survival. *Environ Sci Technol* 41, 2998–3004.

84. Laetz, C. A., Baldwin, D. H., Collier, T. K., Heber, T. V., Stark, J. D. & Scholz, N. L. (2009) The synergistic toxicity of pesticide mixtures: implications for risk assessment and the conservation of endangered Pacific salmon. *Environ Health Perspect* 117, 348–353.
85. Matz, C. J. & Krone, P. H. (2007) Cell death, stress-responsive transgene activation, and deficits in the olfactory system of larval zebrafish following cadmium exposure. *Environ Sci Technol* 41, 5143–5148.
86. Tomelleri, J. (2002) *Trout and Salmon of North America*. New York, NY, The Free Press.
87. McDowall, R. M. (2006) Crying wolf, crying foul, or crying shame: alien salmonids and a biodiversity crisis in the southern cool-temperate galaxioid fishes? *Rev Fish Biol Fish* 16, 233–422.
88. Howes, P. N. (2008) Predator recognition of invasive salmonids by native galaxids. *School of the Environment & Society*. Swansea, Wales, Swansea University.
89. Doty, R. L. (2009) The olfactory system and its disorders. *Semin Neurol* 13, S225–S228.
90. Scholz, S., Fischer, S., Gündel, U., Küster, E., Luckenbach, T. & Voelker, D. (2008) The zebrafish embryo model in environmental risk assessment – applications beyond acute toxicity testing [Review]. *Environ Sci Pollut Res Int* (Jul) 15(5), 394–404.
91. Sloman, K. A. & Wilson, R. W. (2006) Anthropogenic impacts upon behavior and physiology. In Sloman, K. A., Wilson, R. W. & Balshine, S. (Eds.) *Behavior and Physiology of Fish*. San Diego, CA, Academic Press.

Chapter 3

Modeling Stress and Anxiety in Zebrafish

Jonathan M. Cachat, Peter R. Canavello, Marco F. Elegante, Brett K. Bartels, Salem I. Elkhayat, Peter C. Hart, Anna K. Tien, David H. Tien, Esther Beeson, Sopan Mohnot, Autumn L. Laffoon, Adam M. Stewart, Siddharth Gaikwad, Keith Wong, Whitlee Haymore, and Allan V. Kalueff

Abstract

While zebrafish (*Danio rerio*) are widely utilized as a model species for neuroscience research, they also possess several qualities that make them particularly useful for studying stress and anxiety-related behaviors. Zebrafish neuroendocrine responses are robust, and correlate strongly with behavioral endpoints. These fish are also highly sensitive to various environmental challenges, including novelty stress, exposure to predators, alarm pheromone, anxiogenic drugs, and drug withdrawal. In addition, varying levels of baseline anxiety can be observed in different strains of zebrafish. Collectively, this supports the validity and efficacy of the adult zebrafish model for studying both acute and chronic anxiety.

Key words: Novel environment, video-aided analysis, stress, anxiety, fear, affective behavior, predator stress, endocrine response, endocrine signaling, behavioral phenotyping, drug withdrawal, novel tank test, genetic differences.

1. Introduction

As summarized in several chapters of this book, the zebrafish is commonly used as a model species in biomedical research (1, 2). A vast array of genetic knowledge and a complete genome sequence is available for zebrafish, placing our genetic understanding of this species on par with the fruit fly and mouse (3). Although these studies have predominantly examined genetic and embryological phenomena (4), zebrafish are increasingly used in

A.V. Kalueff, J.M. Cachat (eds.), *Zebrafish Models in Neurobehavioral Research*, Neuromethods 52,
DOI 10.1007/978-1-60761-922-2_3, © Springer Science+Business Media, LLC 2011

neuroscience research (5–10). Importantly, zebrafish possess all of the "classical" neurotransmitters found in vertebrates (11), suggesting their potential for studying disorders such as Parkinson's, Alzheimer's, anxiety, and depression (12). While complex neuropsychiatric disorders are difficult to reproduce in zebrafish, analogous brain mechanisms may be investigated using such models (11).

Stress and anxiety have been studied extensively using various animal (primarily murine) models (13–15). Recently, zebrafish have emerged as a promising new organism for anxiety research due to their robust cortisol stress response (16), behavioral strain differences (17) and sensitivity to drug treatment (7, 18–20), as well as to various stressors, such as exposure to predators (6) and alarm substance (21). This chapter outlines several aspects of zebrafish behavior that are relevant to the study of fear and anxiety-related states.

2. The Novel Tank Diving Paradigm: A Fish "Open Field"?

Zebrafish behavioral assays are currently used for high-throughput phenotyping and testing various psychotropic drugs (8, 22, 23). A popular method of behavioral analysis in zebrafish is the novel tank diving paradigm (**Fig. 3.1**), conceptually similar to the open field test used for rodents (**Table 3.1**). In the open field test, mice exposed to a novel environment initially exhibit anxiety-like behavior by staying close to the walls (thigmotaxis), but begin to display increased exploration as they become acclimated to the new setting (24). Similarly, exposure of zebrafish to a novel environment evokes a robust anxiety response (8), as the

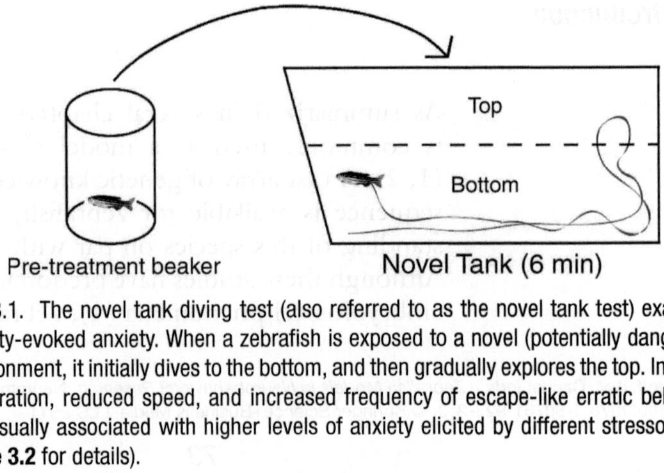

Fig. 3.1. The novel tank diving test (also referred to as the novel tank test) examines novelty-evoked anxiety. When a zebrafish is exposed to a novel (potentially dangerous) environment, it initially dives to the bottom, and then gradually explores the top. Inhibited exploration, reduced speed, and increased frequency of escape-like erratic behaviors are usually associated with higher levels of anxiety elicited by different stressors (see **Table 3.2** for details).

Table 3.1
Comparison of behavioral endpoints in mouse and zebrafish novelty-based paradigms, such as the open field and novel tank tests

Mice		Zebrafish	
Open field	**References**	**Novel tank**	**References**
Thigmotaxis (staying close to the walls)	Heisler et al. (50)	Preference of the bottom of the tank	Levin et al. (22)
Center:periphery ratio	Kalueff et al. (51)	Top:bottom ratio	Own unpublished observations, 2008
Vertical and horizontal exploratory activity	Prut and Belzung (52)	Exploratory swimming	Panula et al. (12) and Egan et al. (17)
Freezing	Heisler et al. (50)	Freezing	Gerlai et al. (19)
Risk assessment	Choleris et al. (53)	Short visits to top of the tank	Own unpublished observations, 2008
Hyperarousal in dangerous situations	Siegmund and Wotjak (54)	Erratic movements	Egan et al. (17) and Gerlai et al. (9)
Locomotor activity (e.g., total distance travelled)	Eilam (55)	Locomotor activity (e.g., total distance swam)	Levin et al. (22)
Vegetative behavior (e.g., defecation, urination)	Fernandez-Teruel et al. (56) and Flint et al. (57)	–	–

animals dive to the bottom and limit exploration until they feel safe to swim in the upper regions of the tank (**Table 3.1**).

Until recently, quantification of zebrafish behavior was mostly performed manually, making it vulnerable to human error and incorrect data interpretation. In contrast, automated video-tracking technologies can analyze animal behavior to provide standardized observation of behavioral endpoints and reduce subjective influence (17). Another advantage of using the video-tracking approach in zebrafish research is the ability to store, replay, and reanalyze videos. Finally, during the novel tank diving test, video-tracking programs can calculate additional behavioral endpoints not available through manual observation, such as distance traveled in top/bottom, velocity, meandering, and angular velocity (**Table 3.2**). Comparison of data produced by the video-tracking system with that recorded manually shows significant (>80–90%) correlation between the two (17), confirming that the video-tracking approach is a reliable method of analysis in zebrafish neurobehavioral research.

3. Analyzing Endocrine Responses to Stress

Physiological phenotypes contribute markedly to the utility of zebrafish models for anxiety research. The zebrafish hypothalamus-pituitary-interrenal (HPI) axis is homologous to the human hypothalamus-pituitary-adrenal (HPA) axis, with cortisol being the primary stress hormone in both species (25, 26) (**Fig. 3.2**). Following animal exposure to stressful stimuli, the hypothalamus secretes corticotropin releasing hormone (CRH), which activates the pituitary gland and signals the pituitary to release andrenocorticotropic hormone (ACTH). Stimulated by ACTH, the adrenal (mammals) or interrenal (zebrafish) glands synthesize glucocorticoid hormones from a cholesterol precursor (26, 27). Increased levels of glucocorticoids initiate metabolic effects that modulate the stress reaction (26, 28), including gluconeogenesis, anti-inflammatory effects, and immune system suppression (29). The effects of the stress reaction are harmful in excess and are alleviated through a negative feedback to the hypothalamus and pituitary, which suppresses CRH and ACTH release (30, 31). This evolutionarily conserved stress response between zebrafish and humans makes zebrafish a valid model to study cortisol-mediated stress responses (16, 32).

Analysis of the physiological (neuroendocrine) responses to stress in zebrafish is a valuable tool complementing behavioral studies. The cortisol assay in zebrafish (5, 17) is relatively simple, inexpensive, can be easily adopted in a variety

Table 3.2
Summary of behavioral endpoints measured in the novel tank diving test

Endpoint (units)	Registration	Definition	Interpretation
Latency to enter the top (s)	m, a	The amount of time to first cross (by the center of mass of the body) from the defined bottom portion to the top of the novel tank	When introduced to a novel environment, zebrafish naturally dive to the bottom of the tank and gradually explore as it habituates to the test apparatus. The longer latency indicates higher anxiety levels
Time spent in top (s)	m, a	Total time spent in the top portion of the novel tank	A longer duration in the top of the tank indicates lower anxiety levels
Time spent top:bottom ratio	c	The ratio of the time spent on top over bottom	Lower ratio indicates higher anxiety level
Number of entries to the top	m, a	The number of crosses from the defined bottom portion to the top of the novel tank	More top entries indicate lower anxiety levels
Entries top:bottom ratio	c	The ratio of the number of entries to the top over bottom	Lower ratio indicates higher anxiety level
Average entry duration (s)	c	The amount of time spent at the top of the novel tank during each crossing	Calculated as time spent in the top divided by the number of entries to the top. Shorter average entry duration indicates higher anxiety level
Distance traveled in the top (m)	a	Total distance traveled in the defined top portion	Zebrafish with high anxiety would travel more distance at the bottom of the tank
Distance traveled top:bottom (m)	c	A ratio of the total distance traveled in the defined top portion versus the defined bottom	A lower top:bottom ratio indicates a higher stressed fish
Total distance traveled (m)	a	Total distance the zebrafish traveled within the novel tank	Reflects general motor/neurological phenotypes. Zebrafish are generally quite sensitive to nonspecific motor impairments and sedative drug effects (see troubleshooting section)

(continued)

Table 3.2 (continued)

Endpoint (units)	Registration	Definition	Interpretation
The number of erratic movements	m, a	Sharp or sudden changes in direction of movement or repeated darting behavior	Indicates increased fear/anxiety, and are generally higher in stressed zebrafish
Average velocity (m/s)	a	Magnitude and direction of zebrafish speed	Reflects motor aspects of zebrafish swimming, may be increased or decreased depending on the nature of behavioral test
Freezing bouts (frequency)	m, a	Total immobility(>1 s), except for the eyes and gills	Indicate increased anxiety and are generally higher in stressed zebrafish
Freezing duration (s)	m, a	Total duration of all freezing bouts	Indicates increased anxiety and is generally higher in stressed zebrafish
Meandering (°/m)	a	The degree of turning (vs. straight locomotion)	Reflects motor aspects of zebrafish swimming, may be increased or decreased depending on the nature of behavioral test
Turning angle (°)	a	Total turning angle	Reflects motor aspects of zebrafish swimming, may be increased or decreased depending on the nature of behavioral test
Angular velocity (°/s)	a	Magnitude and direction of zebrafish angular speed	Reflects motor aspects of zebrafish swimming, may be increased or decreased depending on the nature of behavioral test

a, Automatic observation; m, manual observation; c, calculations based on manually or automatically recorded data.

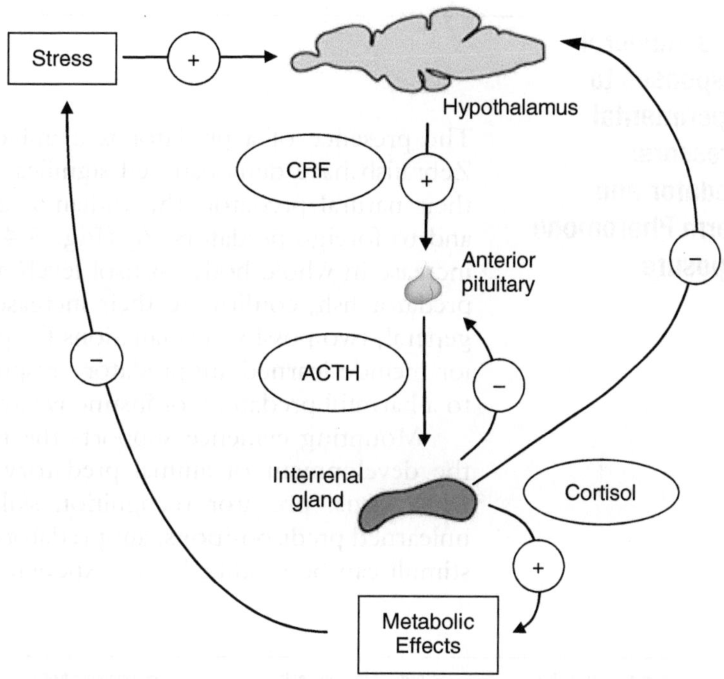

Fig. 3.2. Zebrafish endocrine stress axis. "+" or "−" signs indicate activation or inhibition of activity or secretion, respectively. CRH – corticotropin releasing hormone; ACTH: adrenocorticotropic hormone.

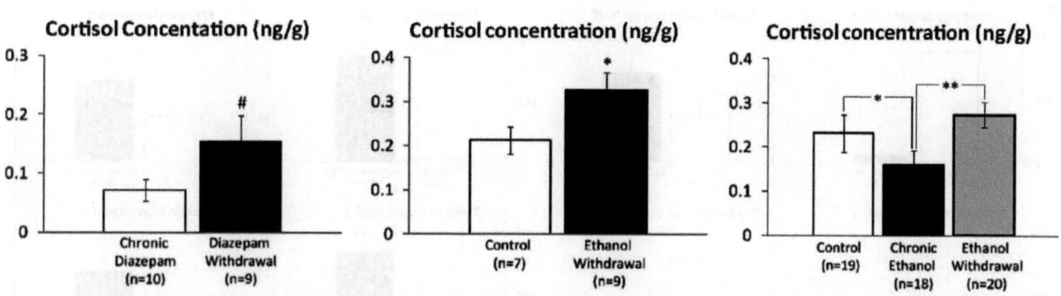

Fig. 3.3. Zebrafish endocrine responses (whole-body cortisol, ng/g fish) to anxiogenic behavioral effects produced by withdrawal from diazepam and ethanol. Data are presented as mean ± SEM, $*p < 0.05$, $**p < 0.01$, #p = 0.05–0.09 (trend) vs. controls, *U*-test (modified from Egan et al. (17)).

of laboratory settings, and strikingly parallels observed anxiety behavior (**Fig. 3.3**). Statistical analysis of correlation between behavior and endocrine response may further assist in data interpretation. For example, the Spearman's rank correlation coefficient, used to assess the relationship between two variables, can determine the level of correlation between behavioral data and cortisol concentration values.

4. Behavioral Responses to Experimental Stressors: Predator and Alarm Pheromone Exposure

The presence of a predator is a universal stressor for animals. Zebrafish have demonstrated significant behavioral responses to their natural predator, the Indian Leaf Fish (*Nandus nandus*), and to foreign predators (6) (**Fig. 3.4**). Zebrafish also show an increase in whole body cortisol levels after visual contact with a predator fish, confirming their increased stress response (5). In general, two possible explanations for predator-avoidance behavior include learned antipredatory responses (following exposure to a harmful predator), or instinctive avoidance behavior.

Mounting evidence supports the importance of learning in the development of animal predatory responses. For example, while visual predator recognition skills seem to be based on unlearned predispositions, antipredatory behavior using olfactory stimuli can be modified with experience, particularly during the

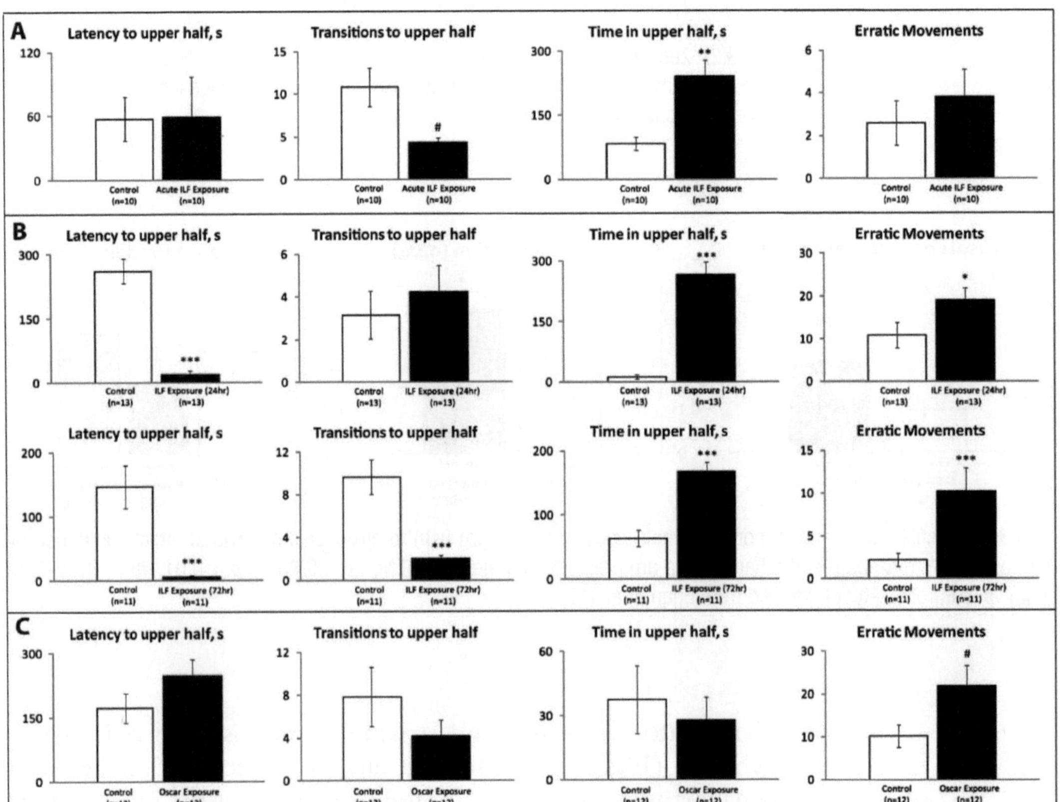

Fig. 3.4. Anxiogenic effects of predator exposure on zebrafish behavior. **a** and **b** Exposure to the sympatric predator Indian Leaf fish (ILF): **a** – acute 5-min exposure; **b** – chronic 24-h (*top*) and 72-h (*bottom*) exposure. **c** Acute 10-min exposure to the allopatric predator Oscar fish. Data are presented as mean ± SEM, $*p < 0.05$, $**p < 0.01$, $***p < 0.005$, $\#p = 0.05–0.09$ (trend) vs. controls, *U*-test.

initial stages of the predator-prey interaction (33). Olfactory cues enable zebrafish to recognize predators following a single exposure to the predator fish (34). However, experimentally naïve zebrafish respond significantly stronger to their natural predator than to an allopatric predator, suggesting a genetically based predator anxiety (6).

Our laboratory has recently examined zebrafish stress responses to the Indian Leaf fish, a natural sympatric predator, and the Oscar fish (*Astronotus ocellatus*), an allopatric predator native to South America. Using experimentally naïve zebrafish, we conducted acute and chronic predator exposure tests using the novel tank diving paradigm. As can be seen in **Fig. 3.4a, b**, both acute and chronic predator exposure produced similar behavioral responses to the Indian Leaf fish. Notably, although the zebrafish displayed a typical response to stress with an increase of erratic movements, they also displayed shorter latency to enter the upper half and more time spent in the upper half, which are not characteristics associated with stress in the novel tank paradigm (**Tables 3.1** and **3.2**). However, as the predator fish spent the majority of the time in the bottom of the tank, it appears that the zebrafish displayed a distinct learned avoidance behavior by moving to the area least likely to be occupied by a predator. In contrast, typical anxiety-like behavior was only significant in the erratic movement endpoint during Oscar fish exposure, indicating weaker responses as compared to the Indian Leaf fish experiment (**Fig. 3.4c**). Although zebrafish were noticeably stressed by the Oscar fish, these findings indicate a greater fear of sympatric (compared to allopatric) predators. This suggests the importance of a genetic, innate influence on the zebrafish fear response.

In line with this, we have also examined the effect of alarm pheromone exposure in zebrafish. As will be mentioned in this book, the zebrafish olfactory system detects alarm pheromone released by injured skin cells, and has been shown to cause behavioral responses. While behavioral alterations in zebrafish could, in theory, be affected by alarm pheromone, the composition of this molecule is not completely understood. Therefore, exact concentrations and dosing cannot be determined when using nonquantifiable extraction from zebrafish skin (35). After extracting alarm pheromone from the epidermal cells of euthanized zebrafish (21), we exposed naïve fish to water containing the alarm pheromone, and measured behavioral responses again through the novel tank paradigm. Acute alarm pheromone exposure (**Fig. 3.5a**) resulted in a robust anxiety-like behavioral response, notably represented through significantly decreased exploration and increased erratic movements and freezing bouts (17). A recent study found that hypoxanthine 3-N-oxide, a molecule common to the alarm substances secreted by several fish species, elicits more erratic movements and jumps when zebrafish were acutely exposed to its

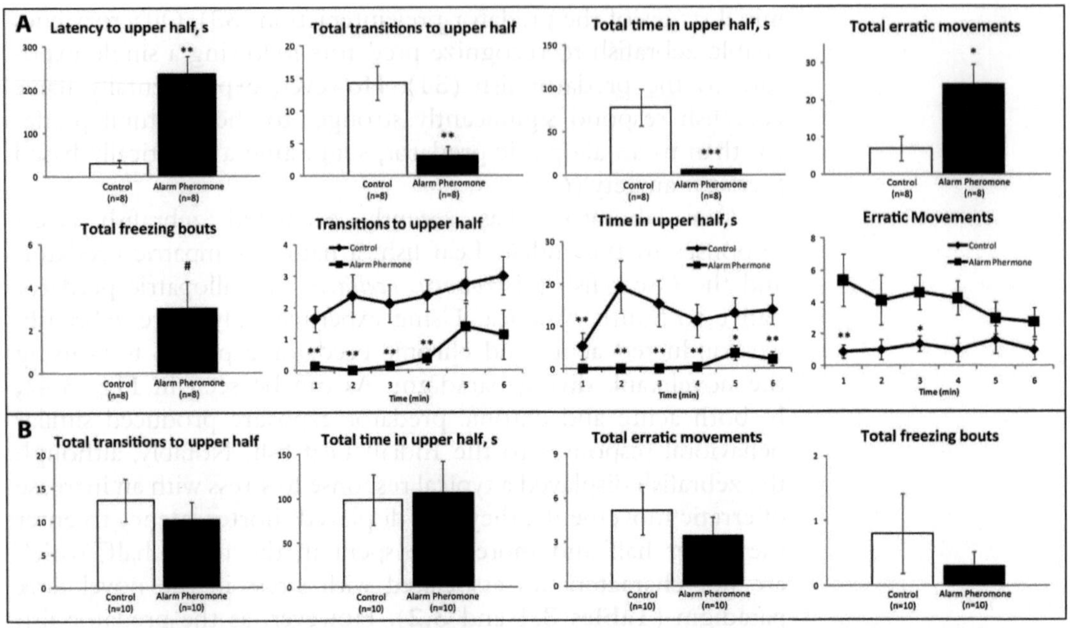

Fig. 3.5. Anxiogenic effects of alarm pheromone on zebrafish behavior in the novel tank diving test: **a** acute alarm pheromone exposure (6 min). **b** Prolonged alarm pheromone (30 min). Data are presented as mean ± SEM, $*P < 0.05$, $**P < 0.01$, $***P < 0.005$, $\#P = 0.05–0.09$ (trend) vs. control, *U*-test (modified from Egan et al. (17)).

increasing doses (35). In contrast, chronic alarm pheromone exposure in our studies produced no significant change from the control cohort (**Fig. 3.5b**), suggesting that alarm pheromone is only effective acutely, most likely reflecting its natural use as a fast-acting danger signal to nearby shoals.

5. Pharmacogenic and Withdrawal-Associated Anxiety

Past zebrafish studies demonstrated robust behavioral phenotypes following acute and chronic exposure to psychotropic agents such as diazepam, caffeine, ethanol, morphine, cocaine, nicotine, barbiturates, and hallucinogens (17, 18, 22, 36–38). The observed predictable bidirectional behavioral responses to known anxiolytic or anxiogenic drugs indicate that zebrafish demonstrate high translation value in stress- and anxiety-related pharmacological research.

Anxiety symptoms are commonly seen in patients withdrawing from chronic drug therapy (39–41). Increasing interest in the underlying neurobiological mechanisms of withdrawal syndrome necessitates the development of appropriate animal models. Robust anxiety phenotypes have been elicited in zebrafish

Table 3.3
Comparison of robustness of zebrafish behavioral phenotypes elicited by different stressors

Type of stress	Stressor	Phenotype
Acute	Alarm pheromone	+++
	Sympatric predator	+++
	Allopatric predator	+
Chronic	Alarm pheromone	0
	Sympatric predator	++
	Strain differences	++
	Drug withdrawal	++

+++, Robust; ++, mild; +, weak effects; 0, no effects; also see Fig. 3.7.

through discontinuation of chronic drug exposure, suggesting the existence of withdrawal syndrome in this species. For example, drug-evoked anxiogenic effects were reported following abrupt cessation of chronic cocaine administration (23), confirming zebrafish as a valid animal model of withdrawal syndrome-associated anxiety (**Table 3.3**).

6. Strain Differences in Zebrafish Behavior

As with other species, genetic differences in zebrafish may lead to varying behavioral phenotypes. One study found that chronic ethanol exposure decreased shoaling behavior in wild-type short-fin zebrafish, but increased shoaling behavior in long-fin striped strain (42). Our group investigated baseline anxiety levels in short-fin and leopard strains, reporting that the leopard zebrafish generally display higher levels of anxiety in the novel tank test (**Fig. 3.6**). Interestingly, using automated video-tracking software, we found no significant differences in swimming velocity or total distance traveled between these two strains (**Fig. 3.6**), indicating that these differences in anxiety were not due to motor/neurological deficits.

Understanding the behavioral differences between zebrafish strains is crucial for expanding this animal model to investigate population differences in humans and their susceptibility to stressors. In addition, selection of a certain strain could optimize data generated in screening of anxiolytic or anxiogenic drugs. For example, due to floor/ceiling effects, choosing a more anxious (e.g., leopard) strain may provide more robust results if examining the behavioral effects of an anxiolytic compound, while the use of

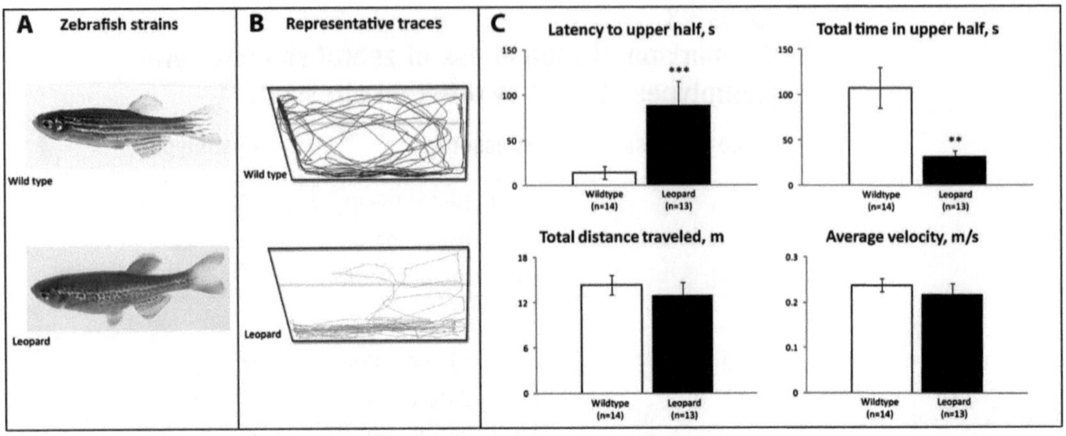

Fig. 3.6. Strain differences in zebrafish novel tank diving test behavior. Two different strains display strain-distinct patterns of their exploratory behavior, as illustrated by representative swimming traces and selected behavioral endpoints analyzed using video-tracking software (CleverSys Inc). Data are presented as mean ± SEM, $^{**}p < 0.01$, $^{***}p < 0.005$ vs. wild type, *U*-test (modified from Egan et al. (17)).

a less anxious strain (such as short-fin zebrafish) could yield more clear-cut phenotypes while testing anxiogenic drugs and manipulations.

7. Mutant and Transgenic Zebrafish

Ease of genetic manipulation, high fecundity, and rapid development make zebrafish a useful tool to study the genetic factors involved in pathogenesis (43). Applied to zebrafish, mutagenesis, transgenesis, and mapping approaches enable the researchers to use invertebrate-style forward genetics on a vertebrate organism (43). There are also certain drawbacks to the use of zebrafish in genetic research, as they have a duplicate genome, and not all duplicated genes have been retained through time (44). For example, it is frequently argued that further comprehension of zebrafish gene function will only uncover invalid redundant and species-specific information (44). However, duplicate genes can also provide significant advantages when zebrafish co-orthologs represent selected expression patterns and developmental functions of mouse orthologs. Thus, restricted expression of zebrafish genes, in comparison to the corresponding mouse orthologs, may lead to an improved comprehension of developmental relations in cell lineage or tissue patterning in mice (44).

Furthermore, several transgenic zebrafish exhibit robust aberrant behavioral phenotypes linked to the knockout of specific target genes. For example, *nevermind* (nev) gene mutant zebrafish display severe disruption of optic nerve innervation (45). While their muscular morphology is normal, *nev* dorsal retinotectal axon projections terminate on both the dorsal and ventral side of the tectum, resulting in atypical locomotion, such as corkscrew swimming, in which zebrafish rotate around their long body axis. Similarly, sphingosylphosphorylcholine knockout zebrafish perform spontaneous erratic movements and escape behaviors (e.g., rapid turning) without provocation from stressful stimuli (46).

Some of the transgenic zebrafish models focus on abnormal developmental patterns that prevent proper innervations between nuclei and in turn disrupt neurophysiology. One example of this is the mutation of the Lhx2 homolog, *bel*, a transcription factor involved in retinotectal axonal growth. In zebrafish, achiasmatic-induced oculomotor deficits generate spontaneous eye oscillations that may model congenital nystagmus in humans, in addition to causing reversed perception of visual stimuli, misappropriated eye movements, and circling swimming behavior (47). It is possible to expect that numerous other zebrafish mutations may lead to interesting motor- and anxiety-related behavioral phenotypes that will be revealed in future studies.

8. Conclusion

Although anxiety-related disorders continue to be one of the most prevalent neuropsychiatric conditions, their pathological mechanisms are poorly understood. One hypothesis stipulates that

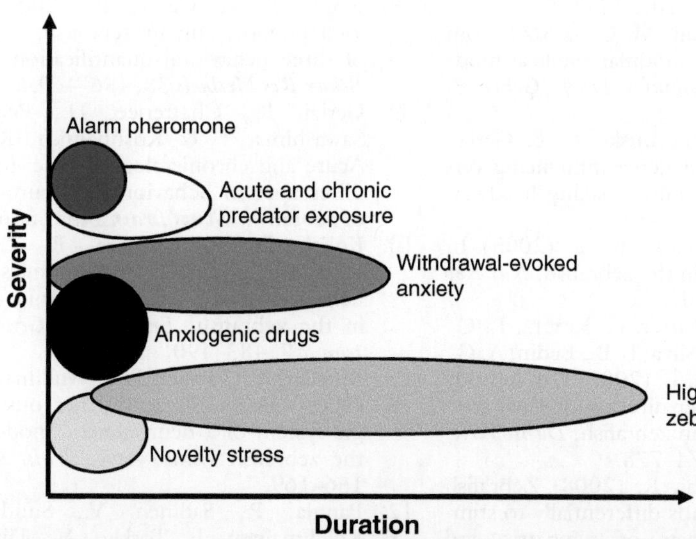

Fig. 3.7. A summary of different forms of stress used in zebrafish neurobehavioral research. Fear-like responses are more likely to occur following alarm pheromone and predator exposure, anticipatory generalized "trait" anxiety is more likely to occur following anxiogenic drug treatment or novelty stress, whereas chronic long-term "state" anxiety can be seen following withdrawal, or in more anxious zebrafish strains (genetic differences).

these disorders are most likely caused by abnormally functioning biological mechanisms of danger avoidance (35). Current challenges to phenotype-based drug discovery include expensive mammalian animal models that require ample physical space and large quantities of compounds for use in experiments. Mammalian animal models also exhibit complex behavioral phenotypes that are sometimes too difficult to characterize and interpret (4). Using zebrafish as an alternative animal model (**Table 3.3, Fig. 3.7**) effectively reduces these limitations, and together with computer-aided video tracking technology, endocrine correlates, and genetic manipulation makes high-throughput behavioral phenotyping and pharmacological screens a promising possibility (2, 4, 17, 48, 49).

Acknowledgments

This work was supported by the Zebrafish Neuroscience Research Consortium (ZNRC), NARSAD YI award (AVK, ALL), Tulane Neuroscience Fellowship (DHT), Tulane LAMP Program (WH), and Tulane University intramural research funds.

References

1. Dooley, K. & Zon, L. I. (2000) Zebrafish: a model system for the study of human disease. *Curr Opin Genet Dev* 10, 252–256.
2. Shin, J. T. & Fishman, M. C. (2002) From zebrafish to human: modular medical models. *Annu Rev Genomics Hum Genet* 3, 311–340.
3. Sison, M., Cawker, J., Buske, C. & Gerlai, R. (2006) Fishing for genes influencing vertebrate behavior: zebrafish making headway. *Lab Anim (NY)* 35, 33–39.
4. Zon, L. I. & Peterson, R. T. (2005) In vivo drug discovery in the zebrafish. *Nat Rev Drug Discov* 4, 35–44.
5. Barcellos, L. J. G., Ritter, F., Kreutz, L. C., Quevedo, R. M., da Silva, L. B., Bedin, A. C., Finco, J. & Cericato, L. (2007) Whole-body cortisol increases after direct and visual contact with a predator in zebrafish, *Danio rerio. Aquaculture* 272, 774–778.
6. Bass, S. L. & Gerlai, R. (2008) Zebrafish (*Danio rerio*) responds differentially to stimulus fish: the effects of sympatric and allopatric predators and harmless fish. *Behav Brain Res* 186, 107–117.
7. Bencan, Z. & Levin, E. D. (2008) The role of alpha7 and alpha4beta2 nicotinic receptors in the nicotine-induced anxiolytic effect in zebrafish. *Physiol Behav* 95, 408–412.
8. Blaser, R. & Gerlai, R. (2006) Behavioral phenotyping in zebrafish: comparison of three behavioral quantification methods. *Behav Res Methods* 38, 456–469.
9. Gerlai, R., Chatterjee, D., Pereira, T., Sawashima, T. & Krishnannair, R. (2009) Acute and chronic alcohol dose: population differences in behavior and neurochemistry of zebrafish. *Genes Brain Behav* 586–599.
10. Kokel, D. & Peterson, R. T. (2008) Chemobehavioural phenomics and behaviour-based psychiatric drug discovery in the zebrafish. *Brief Funct Genomic Proteomic* 7, 483–490.
11. Mueller, T., Vernier, P. & Wullimann, M. F. (2004) The adult central nervous cholinergic system of a neurogenetic model animal, the zebrafish *Danio rerio. Brain Res* 1011, 156–169.
12. Panula, P., Sallinen, V., Sundvik, M., Kolehmainen, J., Torkko, V., Tiittula, A., Moshnyakov, M. & Podlasz, P. (2006) Modulatory neurotransmitter systems and behavior: towards zebrafish models of neurodegenerative diseases. *Zebrafish* 3, 235–247.

13. Jonkman, S., Risbrough, V. B., Geyer, M. A. & Markou, A. (2008) Spontaneous nicotine withdrawal potentiates the effects of stress in rats. *Neuropsychopharmacology* 33, 2131–2138.
14. Teixeira-Silva, F., Antunes, F. D., Silva, P. R. S., Goes, T. C., Dantas, E. C., Santiago, M. F. & Andrade, R. M. D. (2009) The free-exploratory paradigm as a model of trait anxiety in rats: test–retest reliability. *Physiol Behav* 96, 729–734.
15. Zelena, D., Barna, I., Mlynarik, M., Gupta, O. P., Jezova, D. & Makara, G. B. (2005) Stress symptoms induced by repeated morphine withdrawal in comparison to other chronic stress models in mice. *Neuroendocrinology* 81, 205–215.
16. Alsop, D. & Vijayan, M. M. (2008) Development of the corticosteroid stress axis and receptor expression in zebrafish. *Am J Physiol Regul Integr Comp Physiol* 294, R711–R719.
17. Egan, R. J., Bergner, C. L., Hart, P. C., Cachat, J. M., Canavello, P. R., Elegante, M. F., Elkhayat, S. I., Bartels, B. K., Tien, A. T., Tien, D. H., Mohnot, S., Beeson, E., Glasgow, E., Amri, H., Zukowska, Z. & Kalueff, A. V. (2009) Understanding behavioral and physiological phenotypes of stress and anxiety in zebrafish. *Brain Behav Res* 38–44.
18. Braida, D., Limonta, V., Pegorini, S., Zani, A., Guerini-Rocco, C., Gori, E. & Sala, M. (2007) Hallucinatory and rewarding effect of salvinorin A in zebrafish: kappa-opioid and CB1-cannabinoid receptor involvement. *Psychopharmacology (Berl)* 190, 441–448.
19. Gerlai, R., Lee, V. & Blaser, R. (2006) Effects of acute and chronic ethanol exposure on the behavior of adult zebrafish (*Danio rerio*). *Pharmacol Biochem Behav* 85, 752–761.
20. Lopez Patino, M. A., Yu, L., Yamamoto, B. K. & Zhdanova, I. V. (2008) Gender differences in zebrafish responses to cocaine withdrawal. *Physiol Behav* 95, 36–47.
21. Speedie, N. & Gerlai, R. (2008) Alarm substance induced behavioral responses in zebrafish (*Danio rerio*). *Behav Brain Res* 188, 168–177.
22. Levin, E. D., Bencan, Z. & Cerutti, D. T. (2007) Anxiolytic effects of nicotine in zebrafish. *Physiol Behav* 90, 54–58.
23. Lopez-Patino, M. A., Yu, L., Cabral, H. & Zhdanova, I. V. (2008) Anxiogenic effects of cocaine withdrawal in zebrafish. *Physiol Behav* 93, 160–171.
24. Choleris, E., Thomas, A. W., Kavaliers, M. & Prato, F. S. (2001) A detailed ethological analysis of the mouse open field test: effects of diazepam, chlordiazepoxide and an extremely low frequency pulsed magnetic field. *Neurosci Biobehav Rev* 25, 235–260.
25. Alsop, D. & Vijayan, M. (2009) The zebrafish stress axis: molecular fallout from the teleost-specific genome duplication event. *Gen Comp Endocrinol* 161, 62–66.
26. Dedovic, K., Duchesne, A., Andrews, J., Engert, V. & Pruessner, J. C. (2009) The brain and the stress axis: the neural correlates of cortisol regulation in response to stress. *Neuroimage* 864–871.
27. McEwen, B. S. (2007) Physiology and neurobiology of stress and adaptation: central role of the brain. *Physiol Rev* 87, 873–904.
28. Pruessner, J. C., Dedovic, K., Pruessner, M., Lord, C., Buss, C., Collins, L., Dagher, A. & Lupien, S. J. (2009) Stress regulation in the central nervous system: evidence from structural and functional neuroimaging studies in human populations. *Psychoneuroendocrinology* 179–191.
29. Kern, S., Oakes, T. R., Stone, C. K., McAuliff, E. M., Kirschbaum, C. & Davidson, R. J. (2008) Glucose metabolic changes in the prefrontal cortex are associated with HPA axis response to a psychosocial stressor. *Psychoneuroendocrinology* 33, 517–529.
30. Bremner, J. D. (1999) Does stress damage the brain? *Biol Psychiatry* 45, 797–805.
31. Tsigos, C. & Chrousos, G. P. (2002) Hypothalamic-pituitary-adrenal axis, neuroendocrine factors and stress. *J Psychosom Res* 53, 865–871.
32. Winberg, S., Nilsson, A., Hylland, P., Soderstom, V. & Nilsson, G. E. (1997) Serotonin as a regulator of hypothalamic-pituitary-interrenal activity in teleost fish. *Neurosci Lett* 230, 113–116.
33. Kelley, J. L. & Magurran, A. E. (2003) Learned predator recognition and antipredator responses in fishes. *Fish Fish* 4, 216–226.
34. Wisenden, B. D. (2000) Olfactory assessment of predation risk in the aquatic environment. *Philos Trans R Soc Lond B Biol Sci* 355, 1205–1208.
35. Parra, K. V., Adrian, J. C., Jr. & Gerlai, R. (2009) The synthetic substance hypoxanthine 3-N-oxide elicits alarm reactions in zebrafish (*Danio rerio*). *Behav Brain Res* 336–341.
36. Kily, L. J., Cowe, Y. C., Hussain, O., Patel, S., McElwaine, S., Cotter, F. E. & Brennan, C. H. (2008) Gene expression changes in a zebrafish model of drug dependency suggest conservation of neuro-adaptation pathways. *J Exp Biol* 211, 1623–1634.
37. Nikodijevic, O., Jacobson, K. A. & Daly, J. W. (1993) Locomotor activity in mice during chronic treatment with caffeine and

withdrawal. *Pharmacol Biochem Behav* 44, 199–216.
38. Renier, C., Faraco, J. H., Bourgin, P., Motley, T., Bonaventure, P., Rosa, F. & Mignot, E. (2007) Genomic and functional conservation of sedative-hypnotic targets in the zebrafish. *Pharmacogenet Genomics* 17, 237–253.
39. Blum, D., Maldonado, J., Meyer, E. & Lansberg, M. (2008) Delirium following abrupt discontinuation of fluoxetine. *Clin Neurol Neurosurg* 110, 69–70.
40. Hughes, J. R., Higgins, S. T. & Bickel, W. K. (1994) Nicotine withdrawal versus other drug withdrawal syndromes: similarities and dissimilarities. *Addiction* 89, 1461–1470.
41. Kokkinidis, L., Zacharko, R. M. & Anisman, H. (1986) Amphetamine withdrawal: a behavioral evaluation. *Life Sci* 38, 1617–1623.
42. Dlugos, C. A. & Rabin, R. A. (2003) Ethanol effects on three strains of zebrafish: model system for genetic investigations. *Pharmacol Biochem Behav* 74, 471–480.
43. Lieschke, G. J. & Currie, P. D. (2007) Animal models of human disease: zebrafish swim into view. *Nat Rev Genet* 8, 353–367.
44. Key, B. & Devine, C. A. (2003) Zebrafish as an experimental model: strategies for developmental and molecular neurobiology studies. *Methods Cell Sci* 25, 1–6.
45. Solnica-Krezel, L. (1999) Pattern formation in zebrafish–fruitful liaisons between embryology and genetics. *Curr Top Dev Biol* 41, 1–35.
46. Patton, E. E. & Zon, L. I. (2001) The art and design of genetic screens: zebrafish. *Nat Rev Genet* 2, 956–966.
47. Rick, J. M., Horschke, I. & Neuhauss, S. C. (2000) Optokinetic behavior is reversed in achiasmatic mutant zebrafish larvae. *Curr Biol* 10, 595–598.
48. McGrath, P. & Li, C. Q. (2008) Zebrafish: a predictive model for assessing drug-induced toxicity. *Drug Discov Today* 13, 394–401.
49. Rubinstein, A. L. (2006) Zebrafish assays for drug toxicity screening. *Expert Opin Drug Metab Toxicol* 2, 231–240.
50. Heisler, L., Chu, H., Brennan, T., Danao, J., Bajwa, P., Parsons, L. & Tecott, L. (1998) Elevated anxiety and antidepressant-like responses in serotonin 5-HT1A receptor mutant mice. *Proc Natl Acad Sci* 95, 15049–15054.
51. Kalueff, A., Fox, M., Gallagher, P. & Murphy, D. (2007) Hypolocomotion, anxiety and serotonin syndrome-like behavior contribute to the complex phenotype of serotonin transporter knockout mice. *Genes Brain Behav* 6, 389–400.
52. Prut, L. & Belzung, C. (2003) The open field as a paradigm to measure the effects of drugs on anxiety-like behaviors: a review. *Eur J Pharmacol* 463, 3–33.
53. Choleris, E., Thomas, A., Kavaliers, M. & Prato, F. (2001) A detailed ethological analysis of the mouse open field test: effects of diazepam, chlordiazepoxide and an extremely low frequency magnetic field. *Neurosci Biobehav Rev* 25, 235–260.
54. Siegmund, A. & Wotjak, C. (2007) A mouse model of posttraumatic stress disorder that distinguishes between conditioned and sensitized fear. *J Psychiatr Res* 41, 848–860.
55. Eilam, D. (2003) Open-field behavior withstands drastic changes in arena size. *Behav Brain Res* 142, 53–62.
56. Aguilar, R., Gil, L., Flint, J., Gray, J., Dawson, G., Driscoll, P., Gimenez-Llort, L., Escorihuela, R., Fernandez-Teruel, A. & Tobena, A. (2002) Learned fear, emotional reactivity and fear of heights: a factor analytic map from a large F(2) intercross of Roman rat strains. *Brain Res Bull* 57, 17–26.
57. Flint, J., Corley, R., DeFries, J., Fulker, D., Gray, J., Millers, S. & Collins, A. (1995) A simple genetic basis for a complex psychological trait in laboratory mice. *Science* 269, 1432–1435.

Chapter 4

Nicotinic Receptor Systems and Neurobehavioral Function in Zebrafish

Edward D. Levin

Abstract

Nicotinic acetylcholine receptors appear to be quite ancient phylogenetically and are used in the nervous systems of a great number of species across broad parts of the animal kingdom. They play important roles in a variety of neurobehavioral functions from neuromuscular activation to cognitive function. Nicotinic receptor function is an excellent field in which to assess the potential commonalities of neurobehavioral functions across animal species. Nicotinic receptors are remarkably consistent across species and the behavioral effects of nicotinic treatments have been very well determined in mammals. Since zebrafish are an emerging aquatic model for studying neurobehavioral function, we have determined the effects of nicotine, the prototypic nicotinic agonist as well as nicotinic antagonists on cognitive function, exploratory behavior and stress response in a series of behavioral tests we have developed to reliably index these behavioral functions. The overall hypothesis of this line of investigation was that nicotine would have similar behavioral effects in zebrafish as in mammals when analogous tests of behavioral function are used. As with mammalian species, nicotine significantly improves learning and memory at low to moderate doses in zebrafish with an inverted J-shaped dose-effect function. The nicotine-induced learning improvement in zebrafish is reversed with the nicotinic antagonist mecamylamine and is accompanied by increased brain dopamine metabolite levels, an effect which is also reversed with mecamylamine. Also, as in mammals, nicotine has anxiolytic effects in zebrafish. Nicotine significantly reduces bottom dwelling in the novel tank diving task. This effect is reversed by either α7 or α4β2 nicotinic antagonist coadministration. In many respects nicotine has similar effects in zebrafish as in rodents and humans. These studies point to the value of zebrafish as models of human neurobehavioral function.

Key words: Acetylcholine receptor, nicotine, nicotinic receptor antagonist, nicotinic receptor subtypes, spatial learning, spatial alternation, memory, cognitive function, stress response, anxiolytic drug, 3-chambered tank test, novel tank test.

1. Introduction

Zebrafish are emerging as an excellent model for neurobehavioral function. Zebrafish provide useful models for studying the neural

A.V. Kalueff, J.M. Cachat (eds.), *Zebrafish Models in Neurobehavioral Research*, Neuromethods 52,
DOI 10.1007/978-1-60761-922-2_4, © Springer Science+Business Media, LLC 2011

bases of cognitive dysfunction (1, 2) as well as stress response (3). We have conducted a series of studies to determine the biobehavioral effects of nicotine in zebrafish, in particular effects of nicotine on cognitive function, learning, and memory and nicotine effects on stress response.

Nicotinic acetylcholine receptors on neurons are phylogenetically ancient and are used in the brains of a great number of species across the animal kingdom. Nicotine has been shown to improve learning and memory (4) in a variety of species including, humans, monkeys, rats, mice, and zebrafish. Nicotine given acutely or chronically significantly improves spatial learning memory in rats with critical involvement of both α7 and α4β2 nicotinic receptor subtypes (4). In mammals the hippocampus is a crucial region for nicotinic involvement in memory (4). Although zebrafish have a very limited telecephalon and no specific hippocampal formation, the receptor basis for nicotinic effects on cognitive function as described below can be readily seen in the zebrafish model. The functional mechanisms by which neuronal nicotinic receptor actions underlie behavior can be usefully studied in a variety of experimental models. In particular zebrafish provide an excellent model in which to study nicotine effects on cognitive and emotional function.

2. Nicotine Effects on Cognitive Function

Learning in zebrafish has been assessed in experimental studies using a variety of paradigms. Zebrafish have been shown to demonstrate olfactory (5) and visual discrimination learning (6, 7) as well as active avoidance learning (8, 9). We have documented zebrafish of spatial and color discrimination learning and reversal using a three-chamber tank (10).

The three-chamber tank (**Fig. 4.1**) is a useful paradigm for assessing learning and memory in zebrafish. It consists of a center compartment in which the fish is placed to start the test and two side compartments. Gates are opened to the two side compartments and the fish is permitted to make a choice into either side. Once the fish has chosen the side the gates are closed. If the correct side is chosen the fish is permitted to swim freely in the full sized-side compartment. If the incorrect side is chosen the wall separating the central compartment and the side compartment is moved to 1 cm of the side wall restricting the fish to a very narrow space. The contingencies can be managed to assess spatial discrimination and reversal, spatial alternation, and color discrimination and reversal (10–14).

Nicotine ((−) nicotine hydrogen tartrate) significantly improves memory in zebrafish as measured by the delayed spatial

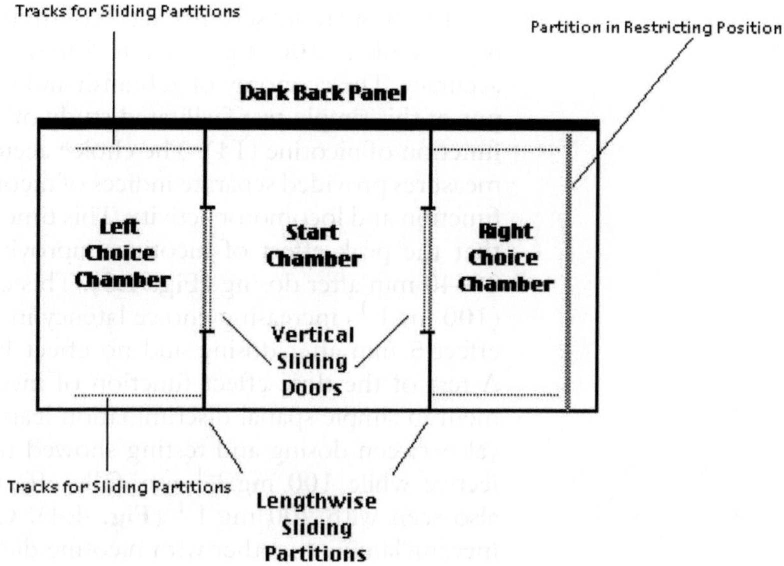

Fig. 4.1. Three-chamber tank for assessing learning and memory in zebrafish (10, 11).

alternation test in the three-chamber tank in which the correct side was alternated between the right and left sides (13). Nicotine exposure was given by immersion with the fish placed in a beaker with fixed doses of nicotine for a 3-min period. The most effective nicotine dose detected for improving delayed spatial alternation was 100 mg l^{-1}. As with mammals, there is an inverted J-shaped dose-effect function (**Fig. 4.2**) with lower doses improving memory accuracy and higher doses impairing it.

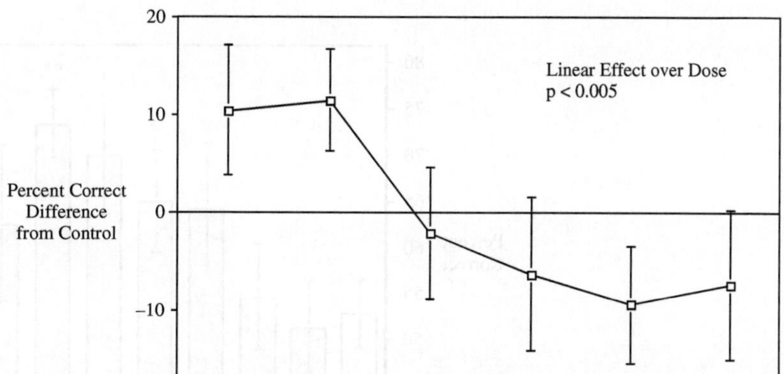

Fig. 4.2. Nicotine effects on memory in zebrafish showing an inverted J-shaped dose-effect function in the delayed spatial alternation task in the three-chamber tank with moderate doses causing improvement in percent correct performance and higher doses having significantly ($p < 0.005$) diminishing effect (13), percent correct difference from control performance (mean ± SEM).

In a more rapid seven-trial test of simple spatial position learning, nicotine (100 mg l^{-1}) was found to significantly improve accuracy. The economy of zebrafish and the more rapid throughput of this simple test facilitated study of an extensive time-effect function of nicotine (14). The choice accuracy and choice latency measures provided separate indices of nicotine effects on cognitive function and locomotor activity. This time-effect function showed that the peak effect of nicotine improving choice accuracy was 20–40 min after dosing (**Fig. 4.3**). This contrasted with nicotine (100 mg l^{-1}) increasing choice latency in the same test with peak effect 5 min after dosing and no effect by 30 min after dosing. A test of the dose-effect function of nicotine-induced improvement in simple spatial discrimination learning with 40-min interval between dosing and testing showed that 50 mg l^{-1} was ineffective while 100 mg l^{-1} was fully effective with improvement also seen with 200 mg l^{-1} (**Fig. 4.4**). Concurrent exposure to mecamylamine together with nicotine did not significantly attenuate the nicotine-induced improvement, suggesting that it was not the activation of nicotinic receptors, which was necessary for the induction of the improvement caused by nicotine (**Fig. 4.5**). Rather, the improvement may have been due to nicotine-induced receptor desensitization, which is not blocked by mecamylamine. Interestingly, mecamylamine given shortly before testing and well after nicotine was effective in eliminating the learning improvement induced by nicotine given earlier. This suggests that recovery from nicotinic receptor desensitization may be key in the expression of nicotine-induced learning improvement.

In a recent study using the more rapid seven-trial test of simple spatial discrimination learning, we replicated the finding

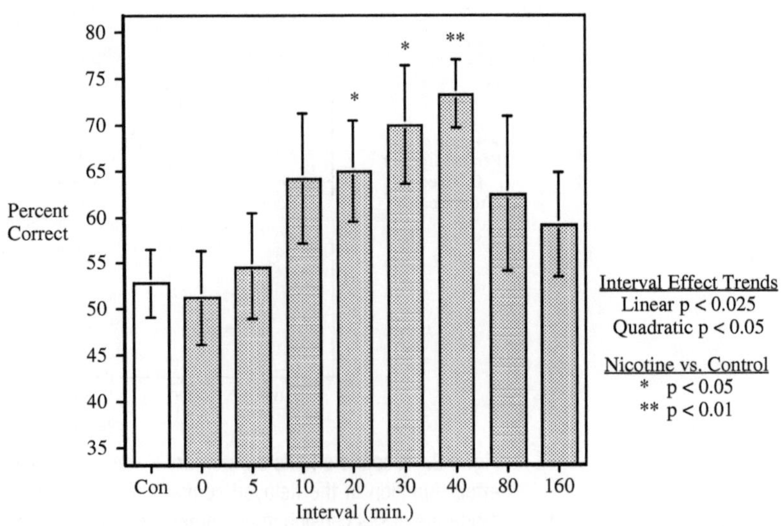

Fig. 4.3. Nicotine (100 mg l^{-1}) improves spatial position learning with the peak effect 20–40 min after dosing (14), percent correct (mean ± SEM).

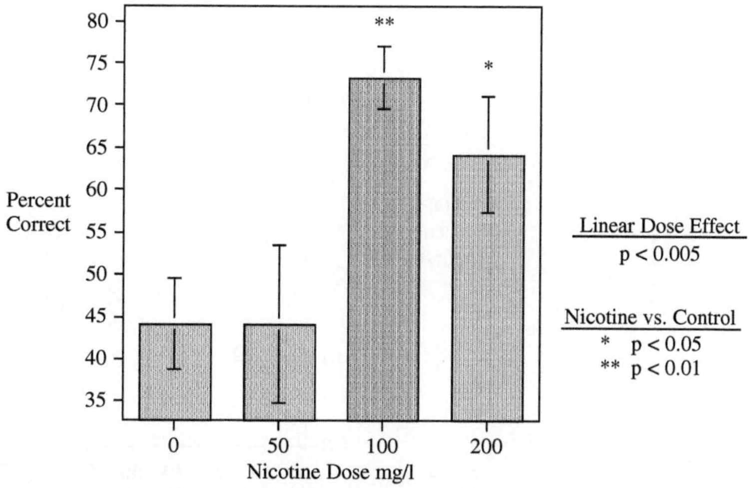

Fig. 4.4. Dose-effect function of nicotine-induced improvement in simple spatial discrimination learning with 40-min interval between dosing and testing (14), percent correct (mean ± SEM).

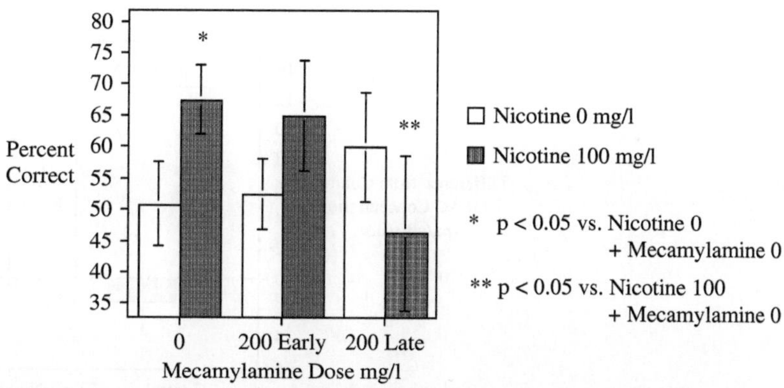

Fig. 4.5. The nicotinic antagonist mecamylamine blocking the expression but not induction of nicotine improvement in spatial discrimination learning (14), percent correct (mean ± SEM).

that nicotine (100 mg l^{-1}) significantly improved learning in zebrafish (15). Overall, spatial discrimination accuracy was significantly correlated with brain levels of 3,4-dihydroxyphenylacetic acid (DOPAC) the principal metabolite of the neurotransmitter dopamine (**Fig. 4.6**). We also found that as in mammals, nicotine treatment significantly increases dopaminergic activity as measured by this metabolite and that the nicotine-induced increase in DOPAC is reversed by the nicotinic antagonist mecamylamine (**Fig. 4.7**).

Thus, as in mammals nicotine significantly improves learning and memory in zebrafish even though the structures in the zebrafish brain with their highly developed optic tectum and much smaller telencephalon are quite different from mammals.

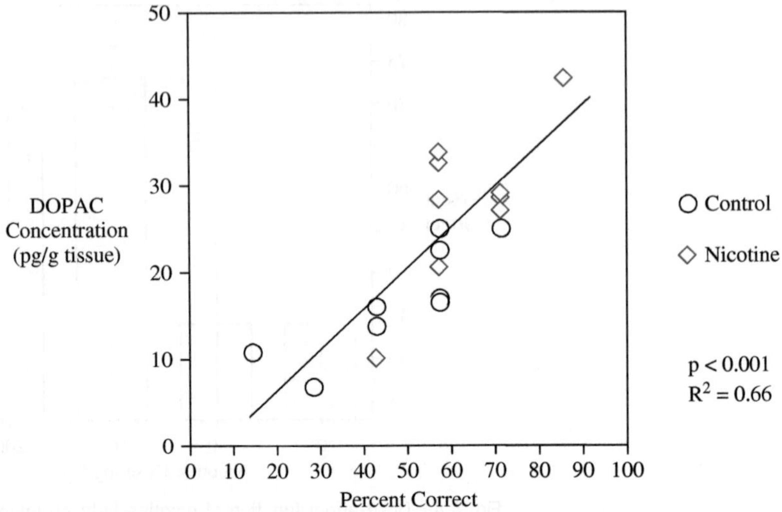

Fig. 4.6. Correlation of percent correct on the simple spatial discrimination test and brain concentration of DOPAC (15).

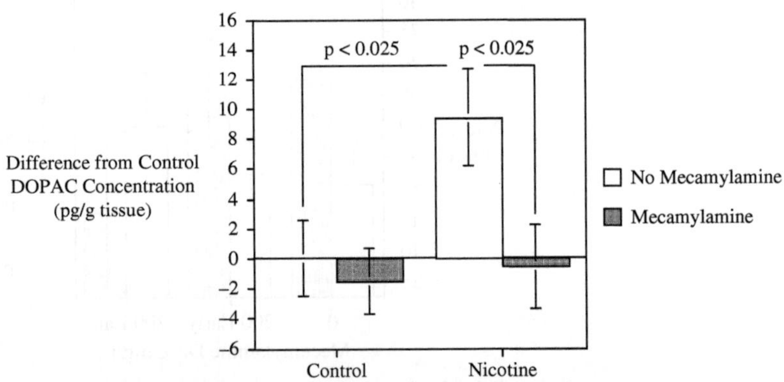

Fig. 4.7. Nicotine (100 mg l^{-1}) significantly increases brain DOPAC concentration in zebrafish, an effect that is reversed by the nicotinic antagonist mecamylamine (15) DOPAC concentration (pg/g tissue) difference from control (mean ± SEM).

Nicotine effects on neuroplasticity may be more similar between fish and mammals. The effect of nicotine improving learning is delayed, possibly involving recovery from receptor desensitization. It appears that nicotine-induced increased dopamine is key to nicotine-induced learning improvement in zebrafish. The effective nicotine dose of 100 mg l^{-1} was the same for the efficacy for improving learning and memory and increasing the dopamine metabolite DOPAC. Further studies will determine the time and dose effect functions of nicotine effects on dopamine systems and how these match the cognitive effects as well as the brain regional specificity of the critical dopamine effects.

3. Nicotine Effects on Stress Response

It is evolutionarily advantageous for an animal to show protective stress responses under potentially dangerous circumstances. Zebrafish have been shown to secrete a specific alarm substance when in stressful situations (16) and their fear responses in the presence of predators have been behaviorally characterized (17). Zebrafish have been found to show anxiety responses induced pharmacologically with cocaine withdrawal (18).

When introduced into a novel environment (**Fig. 4.8**) zebrafish dive to the bottom of the tank and dwell there, and then over time spend less time at the bottom of the tank and swim to higher levels in the tank (**Fig. 4.9**) (3, 19). This diving response in a novel situation is likely a predation avoidance reaction much like thigmotaxis (wall hugging) of rodents in a novel open field. By retreating to be near an impermeable surface an animal can reduce the risk of predatory attack from the direction of the surface. The novel tank diving task is sensitive to the anxiolytic effects of diazepam and buspirone, however the anxiolytic drug chlordiazepoxide was not found to be effective in zebrafish (20). A validation experiment showed that fish with prior experience in a tank identical in size to the test tank did not show the diving response and recovery from it over the course of the 5-min test session compared with fish without such prior experience (20).

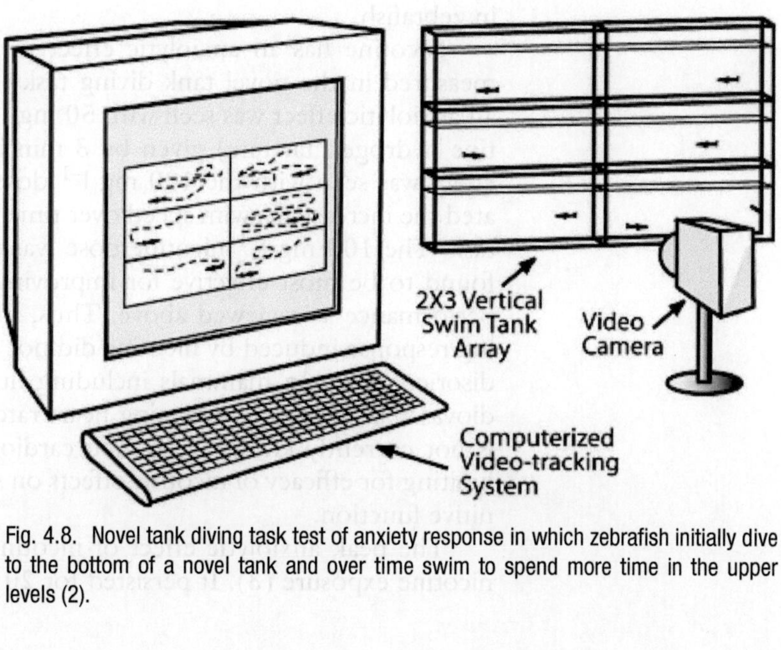

Fig. 4.8. Novel tank diving task test of anxiety response in which zebrafish initially dive to the bottom of a novel tank and over time swim to spend more time in the upper levels (2).

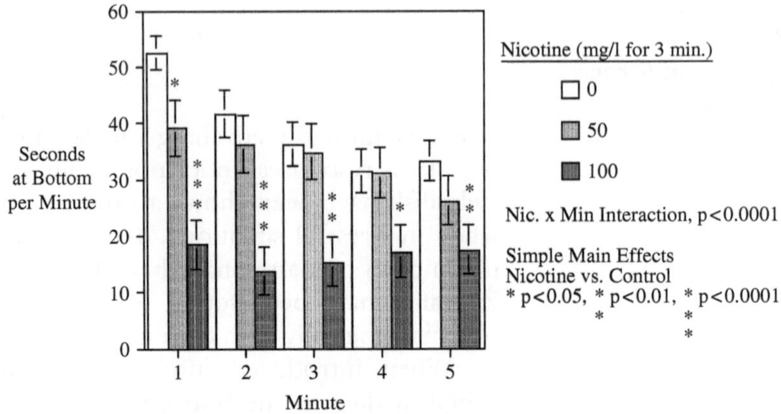

Fig. 4.9. The effect of control zebrafish diving to dwell at the bottom of the novel tank initially with less time in the bottom over the 5-min session and the effect of nicotine attenuating this effect (3), seconds per minute in the bottom third of the tank (mean ± SEM).

To complement our studies of nicotine effects on cognitive function we have conducted a series of studies on nicotinic involvement in stress response in zebrafish. Nicotine has complex effects on anxiety with anxiolytic or anxiogenic effects being seen under different situations (21). Smokers report reduction in anxiety after smoking in a stressful situation despite the fact that their heart rate is elevated (22). Anxiolytic effects of systemically administered nicotine have been seen in rats in the elevated plus maze or social interaction tests of anxiety (23, 24). We investigated whether possible anxiolytic effects of nicotine could be seen in zebrafish.

Nicotine has an anxiolytic effect in zebrafish (**Fig. 4.9**) as measured in the novel tank diving task (3). Initial indication of an anxiolytic effect was seen with 50 mg l^{-1} of nicotine ((–) nicotine hydrogen tartrate) given by 3 min immersion, but the full effect was seen with the 100 mg l^{-1} dose. Nicotine also attenuated the increase in swim speed over time in the novel tank diving task. The 100 mg l^{-1} nicotine dose was the same dose that was found to be most effective for improving learning and memory performance as reviewed above. Thus, the reduction in the diving response induced by nicotine did not seem to be due to mere disorientation. In mammals including humans nicotine has cardiovascular effects of increasing heart rate and blood pressure. It is not currently known how such cardiovascular effects may be limiting for efficacy of nicotine effects on stress response and cognitive function.

The peak anxiolytic effect of nicotine was seen 5 min after nicotine exposure (3). It persisted for 20 min and was no longer

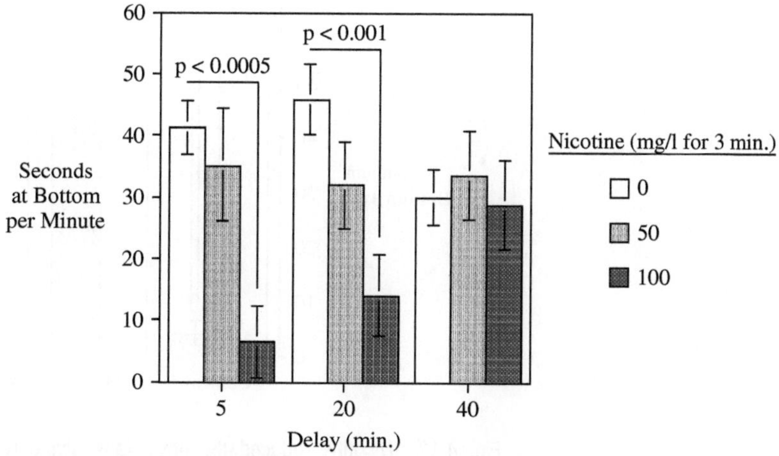

Fig. 4.10. Time-effect function of nicotine in the novel tank diving task. The peak nicotine effect was at 5 min post dosing with no effect seen at 40 min (3), seconds per minute in the bottom third of the tank (mean ± SEM).

evident by 40 min after exposure (**Fig. 4.10**). The nonspecific nicotinic antagonist mecamylamine blocked the expression, but not the induction of nicotine-induced anxiolysis (3) as shown in **Fig. 4.11**. This is a similar pattern of effect as with nicotine-induced learning improvement discussed above. The more specific nicotinic α7 and α4β2 antagonists methyllycaconitine (MLA) and dihydro-β-erythroidine (DHβE) also significantly attenuated the anxiolytic effect of nicotine, as shown in **Fig. 4.12** (25).

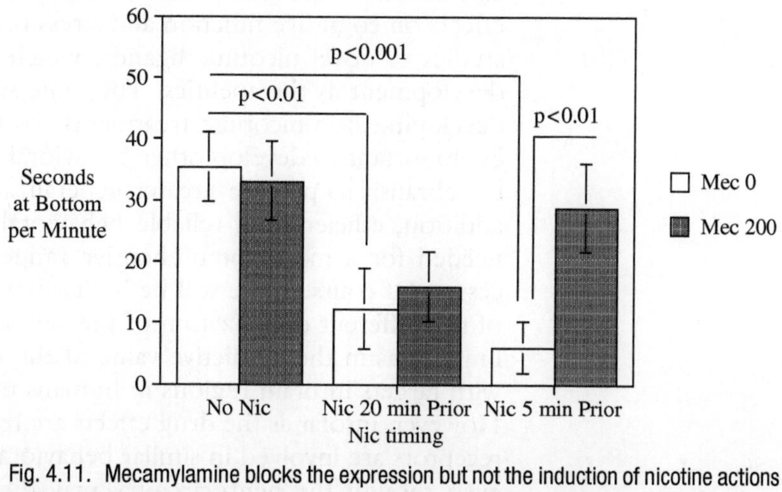

Fig. 4.11. Mecamylamine blocks the expression but not the induction of nicotine actions in the novel tank diving test (3), seconds per minute in the bottom of the tank (mean ± SEM).

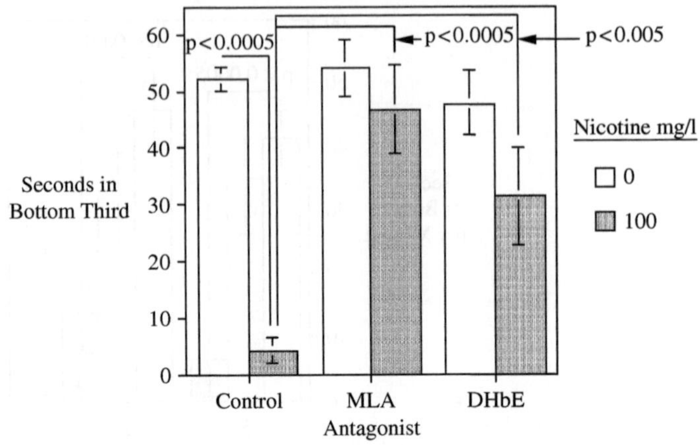

Fig. 4.12. Nicotine reduced the novel tank diving response, an effect, which was reversed by either MLA or DHβE nicotinic α7 and α4β2 antagonists (25).

4. Conclusions

Zebrafish can provide excellent information concerning a variety of neurobehavioral processes. The current research demonstrated that in the case of nicotine, the zebrafish model can be quite useful in determining the neurobehavioral mechanisms pharmacological response in terms of nicotinic involvement in cognitive function as well as anxiety and stress response. In both of these categories of neurobehavioral function zebrafish have qualitatively similar reactions to nicotine as mammals such as rats and humans. This foundation of knowledge concerning nicotine effects on cognitive function and stress response facilitates future studies of novel nicotinic ligands, which may be candidates for development as therapeutics. The same strategy can be used for developing non-nicotinic treatments for these functions. It will be important to develop other behavioral tests of these functions in zebrafish to provide greater generalizability of the results. In addition, efficient and reliable behavioral tests for zebrafish are needed for a more comprehensive range of biobehavioral processes. Of course there will be limitations of the model. Because of the different organization of the zebrafish brain there may be limitations in the predictive value of the model for drug actions with targets in brain regions in humans not present in zebrafish. However, insofar as the drug effects are receptor based and those receptors are involved in similar behavioral functions in zebrafish even though the neurocircuitry may differ from humans, predictive validity may be intact even though the neural structures differ. Important for determining the extent and limitations of the zebrafish model for behavioral pharmacology is to assess in zebrafish the effects of drugs known to be effective in mammals. The current work with nicotine has shown that zebrafish can be very useful complementary models for studying neurobehavioral processes, determining neuronal mechanisms of behavioral function, indicating that it may be useful for screening for new therapeutic drugs.

Acknowledgments

This research was funded by the Duke University Superfund Basic Research Center (NIH ES10356) and an unrestricted grant from Philip Morris-USA.

References

1. Cerutti, D. T. & Levin, E. D. (2006) Cognitive impairment models using complementary species. In Levin, E. D., Buccafusco, J. J. (Eds.) *Animal Models of Cognitive Impairment*. New York, NY, CRC Press, pp. 315–342.
2. Levin, E. D. & Cerutti, D. T. (2008) Behavioral neuroscience of zebrafish. In Buccafusco, J. J. (Ed.) *Methods of Behavior Analysis in Neuroscience*. New York, NY, CRC Press, pp. 293–310.
3. Levin, E. D., Bencan, Z. & Cerutti, D. T. (2007) Anxiolytic effects of nicotine in zebrafish. *Physiol Behav* 90, 54–58.
4. Levin, E. D., McClernon, F. J. & Rezvani, A. H. (2006b) Nicotinic effects on cognitive function: behavioral characterization, pharmacological specification and anatomic localization. *Psychopharmacology* 184, 523–539.
5. Braubach, O. R., Wood, H. D., Gadbois, S., Fine, A. & Croll, R. P. (2009) Olfactory conditioning in the zebrafish (*Danio rerio*). *Behav Brain Res* 198, 190–198.
6. Colwill, R. M., Raymond, M. P., Ferreira, L. & Escudero, H. (2005) Visual discrimination learning in zebrafish (*Danio rerio*). *Behav Processes* 70, 19–31.
7. Gerlai, R., Fernandes, Y. & Pereira, T. (2009) Zebrafish (*Danio rerio*) responds to the animated image of a predator: towards the development of an automated aversive task. *Behav Brain Res* 201, 318–324.
8. Moreira-Santos, M., Donato, C., Lopes, I. & Ribeiro, R. (2008) Avoidance tests with small fish: determination of the median avoidance concentration and of the lowest-observed-effect gradient. *Environ Toxicol Chem* 27, 1576–1582.
9. Xu, X., Scott-Scheiern, T., Kempker, L. & Simons, K. (2007) Active avoidance conditioning in zebrafish (*Danio rerio*). *Neurobiol Learn Mem* 87, 72–77.
10. Arthur, D. & Levin, E. D. (2001) Spatial and non-spatial discrimination learning in zebrafish (*Danio rerio*). *Anim Cogn* 4, 125–131.
11. Levin, E. D., Chrysanthis, E., Yacisin, K. & Linney, E. (2003) Chlorpyrifos exposure of developing zebrafish: effects on survival and long-term effects on response latency and spatial discrimination. *Neurotoxicol Teratol* 25, 51–57.
12. Eddins, D., Petro, A., Williams, P., Cerutti, D. T. & Levin, E. D. (2009a) Nicotine effects on learning in zebrafish: the role of dopaminergic systems. *Psychopharmacology* 202, 103–109.
13. Levin, E. D. & Chen, E. (2004) Nicotinic involvement in memory function in zebrafish. *Neurotoxicol Teratol* 26, 731–735.
14. Levin, E. D., Limpuangthip, J., Rachakonda, T. & Peterson, M. (2006a) Timing of nicotine effects on learning in zebrafish. *Psychopharmacology* 184, 547–552.
15. Eddins, D., Williams, P., Cerutti, D. T. & Levin, E. D. (2009b) Nicotine effects on learning in zebrafish: the role of dopamine systems. *Psychopharmacology* 202, 103–109.
16. Speedie, N. & Gerlai, R. (2008) Alarm substance induced behavioral responses in

zebrafish (*Danio rerio*). *Behav Brain Res* 188, 168–177.
17. Bass, S. L. & Gerlai, R. (2008) Zebrafish (*Danio rerio*) responds differentially to stimulus fish: the effects of sympatric and allopatric predators and harmless fish. *Behav Brain Res* 186, 107–117.
18. Lopez-Patino, M. A., Yu, L., Cabral, H. & Zhdanova, I. V. (2008) Anxiogenic effects of cocaine withdrawal in zebrafish. *Physiol Behav* 93, 160–171.
19. Egan, R. J., Bergner, C. L., Hart, P. C., Cachat, J. M., Canavello, P. R., Elegante, M. F., Elkhayat, S. I., Bartels, B. K., Tien, A. K., Tien, D. H., Mohnot, S., Beeson, E., Glasgow, E., Amri, H., Zukowska, Z. & Kalueff, A. V. (2009) Understanding behavioral and physiological phenotypes of stress and anxiety in zebrafish. *Behav Brain Res* 205, 38–44.
20. Bencan, Z., Sledge, D. & Levin, E. D. (2009) Buspirone, chlordiazepoxide and diazepam effects in a zebrafish model of anxiety. *Pharmacol Biochem Behav* 94, 75–80.
21. Picciotto, M. R., Brunzell, D. H. & Caldarone, B. J. (2002) Effect of nicotine and nicotinic receptors on anxiety and depression. *Neuroreport* 13, 1097–1106.
22. Gilbert, D. G., Robinson, J. H., Chamberlin, C. L. & Spielberger, C. D. (1989) Effects of smoking/nicotine on anxiety, heart rate, and lateralization of EEG during a stressful movie. *Psychophysiology* 26, 311–320.
23. Brioni, J. D., O'Neill, A. B., Kim, D. J. & Decker, M. W. (1993) Nicotinic receptor agonists exhibit anxiolytic-like effects on the elevated plus-maze test. *Eur J Pharmacol* 238, 1–8.
24. Cheeta, S., Irvine, E. E., Tucci, S., Sandhu, J. & File, S. E. (2001) In: adolescence, female rats are more sensitive to the anxiolytic effect of nicotine than are male rats. *Neuropsychopharmacology* 25, 601–607.
25. Bencan, Z. & Levin, E. D. (2008) The role of α7 and α4β2 nicotinic receptors in the nicotine-induced anxiolytic effect in zebrafish. *Physiol Behav* 95, 408–412.

Chapter 5

QTL Mapping of Behaviour in the Zebrafish

Dominic Wright

Abstract

The study of complex traits is one of the greatest current challenges in biology, and the exact mechanism whereby individual genes cause small quantitative variation in any given trait still remains largely unresolved. In the case of behavioural traits, with lower heritabilities and repeatabilities as compared to other character-types, this problem is exacerbated even further. One of the principal forms of genetic analysis for quantitative traits is via QTL (quantitative trait loci) mapping, with the power of this approach even greater in model organisms due to the array of genomic tools available. These tools give a genuine possibility of identifying the actual causative genes or nucleotides responsible for the variation (the quantitative trait nucleotide, or QTN). The zebrafish displays a range of behaviours that are both complex and bear a striking similarity to some of the behavioural measurements performed in other model organisms, notably affecting anxiety and social aggregation. The combination of the behavioural variation present in the zebrafish and the genetic and genomic advantages to QTL mapping available for this species paves the way for its use in generating a new model for the genetic dissection of such trait types. This chapter aims to first discuss the zebrafish as a behavioural model suitable for QTL mapping, focussing in particular on the behaviours of shoaling and predator inspection, before giving an overview of what is contained in a QTL study and the types of crossings, analysis and their relevance to behavioural QTL mapping. Finally two case studies are presented, one of anxiety behaviour in mice, one of shoaling and boldness behaviour in zebrafish.

Key words: Genetic analysis, behavioural genetics, quantitative trait loci mapping, quantitative trait nucleotide, population differences, domestication, anxiety, shoaling behaviour, behaviour variation, single nucleotide polymorphism, oligonucleotide array, predator inspection, environment variation, bioinformatics.

1. Introduction

The study of complex traits is one of the greatest current challenges in biology, and the exact mechanism whereby individual genes cause small quantitative variation in any given trait

A.V. Kalueff, J.M. Cachat (eds.), *Zebrafish Models in Neurobehavioral Research*, Neuromethods 52,
DOI 10.1007/978-1-60761-922-2_5, © Springer Science+Business Media, LLC 2011

still remains largely unresolved (1). Whereas discrete traits have a clearly defined phenotype, a quantitative trait is a continuous or metric character, requiring measurement rather than counting (2). These are generally comprised of numerous small-effect genes, each with primarily additive effects (although even this is now being challenged). The specific regions of the genome that are associated with these characters are termed quantitative trait loci, or QTL.

QTL mapping is based on the same methods that are used for mapping single gene traits, using linkage disequilibrium between alleles at marker loci and at the QTL (2). Information from recombination rates between polymorphic molecular markers is used to estimate a genetic map (the distance between markers in recombination likelihood) and on this map framework QTL are detected, both in terms of their position and effect. The basis of this analysis is conceptually and statistically rather simple. Two populations that differ strongly in the trait of interest and can be separated using polymorphic genetic markers spread throughout the genome are bred together to form a completely heterozygous intercross population. These offspring are heterozygous for every QTL that differs between the parental populations (assuming that QTL are fixed within each parental line in the case of inbred line crosses, though as we will see this is not always the case). This intercross population is often then either bred together to produce a second generation intercross (F_2) or bred back to one of the parental lines (back cross or BC). In either case there will then be a mix of genotypes at each of the QTL and marker positions – both homozygous parental forms as well as heterozygotes in an F_2, or one homozygous and a heterozygous class in a BC. As a complete marker set is available throughout the genome, some QTL should be closely linked to the known polymorphic markers, and the genotypes of which can then be used as a proxy for the underlying QTL genotypes. By doing a simple correlational test between the phenotypes of the mapping population and the genotypes at each marker locus it is then possible to detect differences in genotypic means and hence identify the underlying QTL. Therefore the basic requirements of a QTL study are twofold: two populations that strongly differ in a trait of interest, and a set of markers that are polymorphic between these populations and spread evenly throughout the genome (3). This approach has been used repeatedly in both plants and animals and is amenable to almost any organism with available polymorphic markers. The problem comes, however, when we want to refine these QTL and actually identify the gene or genes causing this variation. In the case of laboratory organisms, with the wealth of genomic and transgenic tools that are available to them, there is a genuine possibility of actually detecting such genes, though even here it is by no means a trivial task (4–6). Given both its physical and genomic properties, the zebrafish therefore represents an excellent, if as yet under-used, candidate for research in such a manner.

The genetic analysis of behaviour is performed in the same way as for other traits, yet the degree of repeatability which may or may not be present and certain ethical reservations at looking at the basis of intelligence and aggression, for example, has meant that the genetics of these trait types was initially restricted to disease phenotypes and major effect genes (7–9). Though the knowledge of such mutations is important, the biggest area for potential gains is in dissecting the genes and genetic architecture involved in small quantitative variation. The knowledge about roughly how many genes are involved in determining these types of traits (termed the number of effective factors), the mode of action (be it mainly additive, dominance or epistatic) and any pleiotropic effects with other behavioural and trait types all require much greater analysis, though lately there have been some large strides in this field using model organisms (5, 10, 11). Despite this proven ability of model organisms to be both amenable to genetic analysis, and particularly QTL mapping, and behavioural analysis, the zebrafish remains largely untapped in this field.

2. The Zebrafish as a Model for Behavioural QTL Mapping

The zebrafish has several advantages that make it excellent for QTL analysis. The generation times in this fish are short – from 3–4 months for laboratory fish, though longer for their wild counterpart – making it ideal for genetic analysis. There currently exist over 60,000 single nucleotide polymorphisms (SNP) markers available for the zebrafish, while a large number of molecular microsatellite markers have been available for some time (12), with more of these being added continually. SNPs are single base alterations, with usually only two or three variants present (due to the limit of four bases) but are the most abundant marker type. Microsatellites in contrast are commonly di- or tri-nucleotide repeats in non-coding regions and are far more diverse in terms of repeat length, giving a large number of polymorphisms, but are less abundant throughout the genome. A zebrafish 16 k oligonucleotide array is already available, as is a 15 k zebrafish Affymetrix expression chip (see zfin.org for details and links for all genomic resources). In addition to these, custom oligo arrays are becoming increasingly cheaper to produce. Although the genome sequence of the zebrafish is not yet fully annotated, the current build (Zv8) has a 6.5-7X coverage, though this sequence still contains misjoins, misassemblies and artificial duplications (details available from the Ensembl website – www.ensembl.org). With any QTL study, the genetic variation that is present between lines is vital, not only as a basis for the phenotypic variation to be selected upon, but also for the presence of informative markers between lines. An initial allozyme analysis using four populations of wild zebrafish from West Bengal in India revealed high levels of genetic variability and weak genetic structure (13). More recently, a comparison of expressed sequence tags (ESTs are subsequences of transcribed cDNA that are generally used to identify genes) to whole-genome shotgun data (the method used to construct the zebrafish genome assembly) predicted more than 50,000 candidate SNPs, based on the divergence between the assembly and the ESTs. Around 65% of these were validated in 16 samples from seven commonly used zebrafish strains (14). This study once again indicated considerable interstrain variation, ranging from 7 to 37% of polymorphic sites being heterozygous, depending on the origin (inbred or wild-derived). The proportion of three allelic variants at a given loci was five times greater than that present in human or mouse. Furthermore, phylogenetic analysis indicated that comparison even between the least divergent strains used in this analysis should provide an informative SNP every 500 nucleotides. This level of variation is not surprising given the ecology of the wild-derived zebrafish lines, with rapid reproduction, large numbers and the monsoon-affected habitat leading to both large-scale mixing and a large and varied environment.

3. Behaviour in the Zebrafish

Though both behavioural and in particular ecological studies in the zebrafish have previously been lacking for some time, these gaps are now being filled (see chapter 1 by Spence, R. in this book). Such increases can be divided into a straight increase in the knowledge of the basic ecology and ethology of the zebrafish (i.e. in the repertoire of behaviours exhibited by the fish) and how these behaviours can vary between populations and to what degree (which is of principal importance in any QTL experiment). The zebrafish themselves perform the basic fish behaviours of shoaling and predator inspection seen in a multitude of fish species and extensively studied (see below), as well as shoaling preferences based on familiarity and kin recognition (15), preferences for the same or opposite sex, depending on age (16, 17) and dominance (18), preference for shoaling partner based on colour phenotype (19–22) and activity levels (23), reactions to potential predators (24) and variation in feeding (25). These behaviours are varied and several are dealt with in detail in other chapters in this book, but certain ones are particularly amenable to QTL analysis, and have also received the most genetic analyses of the behavioural traits exhibited by fish, principally shoaling tendency and predator inspection, though others (notably shoaling preferences and odour-related behaviour) could also be analysed using such techniques.

Shoaling is used to describe a social assembly of fish (26), and includes schooling (a more advanced behaviour involving close coordination of behaviour). Schooling therefore falls into the general definition of shoaling. For ease of terminology, the term shoaling will be used from now on, but will encompass both schooling and shoaling. Broadly speaking, there are three main advantages to shoaling, with these being a hydrodynamic swimming advantage to this type of aggregation, antipredator advantages and certain foraging benefits. Of these, antipredator benefits include the "confusion effect" (where the uniformity of fish serve to reduce hunting efficiency), "many eyes" advantages (with shoals harder to surprise than solitary individuals) and others. See Krause and Ruxton (27), Pavlov and Kasumyan (28) and Wright et al. (29) among others for more in-depth reviews of the benefits (and costs) associated with this behaviour. Another primary form of predator defence in fish is predator inspection. Predator inspection involves either sudden darts at the predator or in random directions followed by quick returns to the school (30), or fish may make slow approaches to the predator to a distance of 4–6 body lengths and then return (31). Seen in numerous fish species, this behaviour detects potentially threatening animals and increases foraging opportunities by exploration at the expense of increased mortality (though other costs and benefits are also apparent). Once again, several reviews are available on this topic, including Krause and Ruxton (27), Pitcher and Parrish (32), Pavlov and Kasumyan (28) and Wright et al. (29).

4. Genetic Basis of Shoaling and Predator Inspection

The genetic components of these traits have been given a limited analysis in a variety of small-scale laboratory studies. These studies can be divided into population studies with controlled breeding and environments and selection studies. Initially, studies on predator inspection in minnows (33) and shoaling in guppies (34) demonstrated that behavioural differences found between different populations must have at least a partial genetic basis. In a study with different laboratory-bred populations of wild-derived zebrafish Wright et al. (35) found population differences in predator inspection, but not shoaling tendency. A potential problem in conducting such studies, where separate populations are bred and reared in different tanks and then a comparison made, is the failure to maintain a population in more than one rearing and holding tank. Maintenance of a population in more than one (and preferably several) tank allows minute differences in rearing environment to be partially accounted for, with these potentially having quite a strong effect depending on the heritability of the trait under investigation. For instance, in the study by Wright et al. (35) significant tank effects were found, with these also being present in a QTL study on shoaling and predator inspection behaviour (36 – see later). These were most apparent in the analysis of shoaling tendency, where without the tank factor significant inter-population differences would have been found (albeit incorrectly). Failure to take such differences into account can therefore lead to spurious population differences that are due to environmental rather than genetic variation. A problem in the literature is that it is often impossible to tell whether such replications have been performed (though if it has been it is rarely formally tested in the analysis, and therefore not statistically controlled for if it is a factor), and for instance several of the earlier studies using stickleback and guppies have not taken this issue into account.

In addition to the above experiments involving the examination of differences between populations, a response to natural or artificial selection leading to variation in a trait is an indicator of additive genetic variation segregating within a population. Magurran et al. (34) performed analysis on two transplanted guppy populations. In a population transplanted 34 years prior to testing, a change in behaviour mirroring the environmental alteration (from high to low predation) was noted. In a population transplanted only 16 years prior to testing, however, though phenotypic modifications were seen these did not appear to be genetically based. In contrast, a laboratory selection experiment on Japanese medaka, *Oryzias latipes*, found changes in aggression and shoaling after only two generations (37, 38), suggesting that substantial additive genetic variation was present in this population.

5. Population Differences in Zebrafish Behaviour

With the behaviours illustrated above and the variety of costs and benefits associated with exhibiting them, it is unsurprising that this has led to a wide variety of population variation, depending on the local environment. Such variation in shoaling tendency and predator inspection between wild populations has been shown in Wright et al. (35) for shoaling and predator inspection in the zebrafish, whilst evidence of within-population variation for shoaling (35) and predator inspection (39) also exists. Feeding behaviour variation is also seen between strains (25). Strong strain variation is vital for QTL mapping (3) – the larger the difference, the fewer individuals that are required to establish a correlation between phenotype and genotype. Similarly, there is also the problem that the smaller the difference the greater the possibility that QTL are not fixed in the line, i.e. individuals are heterozygous rather than homozygous at the QTL of interest, additionally high QTL may exist in the low line and vice versa. This leads to transgressive segregation occurring in the observed QTL and can also be a problem in QTL identification. To combat this, lines often undergo some degree of selection for a few generations to try and fix such differences (40), however there is no guarantee that all such variation will be fixed, especially given that the architecture can be complex and not amenable to such pressure. Perhaps one of the biggest advantages in establishing strong population variation is the use of the domestication paradigm, whereby wild fish are used in combination with their domesticated counterparts to reveal large population differences. This is related to the artificial selection of traits over numerous generations in the domesticated populations. QTL examples exist for several such examples, but include chickens, *Gallus gallus* (41–43), pigs, *Sus scrofa* (6) and others. In the case of the chicken, specific behavioural examples exist (44–46). The differences related to the causation of such a domestication paradigm are briefly outlined below.

6. Domestication

Given the uses of comparing a wild with a domesticated (or in the case of the zebrafish a more accurate term would be 'laboratory') strain, a brief definition of domestication is required, along with a breakdown of the potential forces that are applied in the process. Adaptations to the captive environment are achieved through genetic changes, environmental stimulation and experiences during an animal's lifetime (47). Darwin (48, 49) used the effects of domestication as proof for natural selection and suggested that domestication includes breeding animals in captivity and changes that may occur without conscious effort on the part of man. This idea of unintentional changes (or rather unintentional selection) has also been mirrored by others. Ochieng-Odero (50) defines domestication as habituation and conditioning to environmental stimuli associated with the captive environment.

6.1. Environmental Alterations

During domestication environmental conditions alter greatly from those found in the wild, with Price dividing these differences into four main categories:

1. Space limitations: The degree of space available is almost always limited in domesticated populations, in marked contrast to the territories and range sizes available in the wild.
2. Foraging: With food and water generally provided ad libitum in domestication, one of the basic and most vital of behaviours, foraging, becomes largely obsolete. This leads to a persuasive argument that a relaxation of selection on characters associated with foraging, predator avoidance and exploration (51).
3. Predation: Predation rates on captive-reared animals that have been released into the wild are almost always greater than those on wild-derived individuals. This has been most apparent in the case of captive-reared salmon, *Salmo salar*, that has been used to try and repopulate wild stocks. In this instance the number of successfully returning fish is often low, with these fish suffering very high predation rates (52). When captive-reared fish are subjected to conditioning using both live predators and predator-associated stimuli, an increase in antipredation behaviour is found to occur (53). Using four clonal lines of rainbow trout, *Salmo trutta*, two of which were in captivity for longer, Lucas et al. (54) found significantly different behaviour between the two lines which were captive-bred for longer as compared to the two closer to their wild counterparts. The differences included longer startle responses and decreased visibility of fish in the water column and appeared to indicate less antipredation behaviour in the domesticated strains. Such reduction in antipredation behaviour is also seen in salmonids (55, 56).
4. Social environment: Generally, the age and sex structure present in domesticated populations is more uniform than in natural populations, with the increased density of captive populations also often exacerbating this issue. Though this can have serious effects on behaviour and specifically agonistic behaviour, the direction of effect often appears to be changeable. For instance some studies have found that domestic stocks of salmon (57) and trout (54) display greater aggression compared to wild-derived fish reared in similar conditions, whereas a study on medaka found a decrease in aggression associated with growth, with both linked to domestication (37, 38). Fleming and Einum (58)

analysed the various genotype by environment interactions in salmon, finding that wild-derived fish were more dominant in stream conditions, whereas hatchery-derived fish were more dominant in tank environments.

6.2. Genetic Mechanisms

There are three modes of genetic change that can have an impact on domesticated animals: inbreeding, selection and drift (51). The primary forms of selection are (i) artificial, (ii) natural selection in captivity and (iii) relaxation of natural selection (59).

1. Artificial: This can be intentional or unintentional, with selection often occurring for stress or aggression reduction to allow ease of handling. Fevolden et al. (60) give an example of stress reduction in trout.
2. Natural selection in captivity: All selection that is not artificial in the laboratory must therefore be 'natural' (59). This can be regarded as adaptation to the current environment, i.e. a genetic response to the new space, social structure and foraging requirements. The intensity of any natural selection is dependent on the extent to which the environment allows species-typical biological characteristics (61) and the number of generations in captivity. The separation of this from artificial selection (especially inadvertent artificial selection) is particularly blurred, and the two are arguably indistinguishable.
3. Relaxed selection: As has been seen previously, foraging and antipredation behaviour are often obsolete in a captive environment and can even incur costs in exhibiting them. This can therefore either cause an increase in the genetic variability present in these traits (in the case of no costs) or a reversal of the trait direction where costs are incurred.

7. The Domestication Paradigm in the Zebrafish

In the case of the zebrafish, laboratory strains of fish have undergone a form of domestication for numerous generations (see www.zfin.org for standard strain histories), with marked differences apparent in many laboratory strains in comparison to their wild counterparts. For instance, in the study by Wright et al. (35), a comparison between AB and wik laboratory-strain fish and four different populations collected from Bangladesh revealed differences in predator inspection, shoaling tendency and growth-related traits. Similarly, a study by Robison and Rowland (62) revealed differences in behaviour between a population from India and two laboratory strains. Population differences have also been found with feeding behaviour (25), and given the variety of predation pressures with aerial as well as water-based predators, this is unsurprising. However, further work from Moretz et al. (63) shows that a straight domestic/wild paradigm may be too simplistic, with the environmental stresses associated with a particular wild population leading to a large degree of variation that may on occasion show more extreme phenotypes than certain laboratory populations. What is certain is that the resources present in such wild strains are of paramount importance in dissecting quantitative traits, and efforts to both obtain and maintain such populations should be made.

8. QTL Mapping Overview

Having seen that the behaviour and population differences exhibited by the zebrafish are open to genetic analysis it now remains to give a brief overview of QTL mapping, and in particular in relation to behavioural traits. In addition, fine mapping techniques and potential problems and limitations of QTL mapping are also detailed.

8.1. Types of Cross

There are several different forms of cross that can be utilised in a QTL study. Generally the first step involves the isolation of divergent strains or lines, either through artificial selection or by identification in natural populations. Selective crossbreeding of these strains creates a hybrid F_1 population. F_1 individuals are then either crossed back to the parents (a backcross design) or intercrossed to create an F_2. These offspring are phenotyped and genotyped, with divergence between marker classes taken to indicate a QTL at a basic level. Although recombinant inbred lines can be used, backcrossing and intercrossing have much greater power to detect QTL effects. For instance, with an F_2 study a sample of over 800 animals has over a 99% power to detect a QTL that accounts for 4% of the phenotypic variance, as long as selective genotyping of extremes is not used (with this causing variance estimates to be much larger than those based on the whole data set (9)).

- Backcross: If the aim is to detect at least some of the major QTL present, a backcross design is recommended (64). In this instance the probability of detecting dominance effects is double that of an F_2 study of similar size (64).
- F_2: When a general overview is desired, i.e. the number of QTL segregating and estimates of their additive and dominance effects, an F_2 design is preferred. With an F_2 design,

three genotypes are present at each locus, either homozygous for one of the parental genotypes, or heterozygous. These require 30% fewer offspring than a backcross design to detect additive genetic variance. They were initially thought to be twice as effective (2), however this was due to a failure to adjust the design specific thresholds – the required thresholds for a backcross (BC) are lower due to the reduction in residual genetic variance (65).

- Recombinant inbred lines: Recombinant analysis is performed through brother-sister matings in a line for about 20 generations, producing animals homozygous for recombinant chromosomes. In mice the BXD RI set (66) has been the most popular for behavioural analysis. RI strains tend to only have the power to detect fairly large effect QTL. Even using 100 RI strains the QTL to be identified must still explain at least 12% of the variance in the F_2 to be detectable (67). RI lines are rather used as a technique to detect provisional QTL, which are then confirmed using an independent (e.g. F_2) test. Due to the greater degree of recombination present, they can also be used to give a higher resolution when a QTL is detected. One of the biggest advantages of RIL is the improvement of the mean phenotype of a line, i.e. you can test many individuals from a particular line (with these individuals effectively almost clones of one another), with this serving to provide a very accurate picture of the phenotype (and thereby increase the heritability of the trait measured).
- Advanced intercross lines (AIL) (68): Similar to recombinant inbred lines (RIL) to an extent, these are generated in the same manner as an F_2 cross, only instead of stopping at the second generation, these are continued for a number of additional generations. Unlike the RILs, these additional generations avoid inbreeding by using a large breeding population each time. This results in increasing the number of recombinations present between linked markers, inflating the map size, and decreasing the confidence intervals of detected QTL. For an AIL of t generations, the theoretical decrease in the confidence interval of any given QTL is $t/2 * c$, with c being the QTL confidence interval from the F_2 generation. There are certain complicating issues with an AIL, with probably the greatest being the possibility of non-syntenic association of unlinked markers, due to markers becoming fixed in families in preceding intercross generations (69).

8.2. Linkage Map Construction

Irrespective of the type of cross used, the genotypes of individuals are required to construct a linkage map of the markers spread throughout the genome and then to correlate the phenotypes that are obtained with these markers. In effect the genetic map provides the framework on which to locate the QTL. Genetic map construction is the process whereby these markers are placed in order and the genetic distances between them are calculated. These genetic distances are based on the number of recombinations, with these distances being additive in the best instance and referred to in centi-Morgans (with one centi-Morgan equalling a 1% recombination frequency). The problem here is that these distances are calculated by measuring the number of recombinations present between the markers, with the total number of recombinations required and only odd numbers of crossovers detected. In effect what this means is that if a double recombination has occurred between two markers then it will appear as if none, rather than two, are present and therefore lead to underestimation of the distance between the two markers. The possibility of this occurring is obviously decreased as the two markers are closer together. To correct for both this and the possibility of interference (where the presence of one crossover will suppress the formation of another close by), a variety of mapping functions have been derived. The simplest is Haldane's (70), which assumes that crossovers occur randomly and with no possibility of interference. Others mapping functions allow for a degree of interference, with one very commonly used function being the Kosambi map function (71). As well as distance between markers, the second main problem is the order of the markers themselves. When the distances between markers are small, it can be easy to confuse the correct order of markers. In such instances the problem is akin to minimising the distance between all markers, so in effect getting the smallest possible map length. A variety of different methods are used, all based on a 'multipoint analysis'. These use calculations based on the recombination between chains of markers, rather than considering just two markers at a time, and include the simulated annealing (72), seriation (73), and branch and bound (74) methods, though the problem here is that all can be very computationally intensive. Doerge (75) developed a method to speed up this analysis by starting with a two-point estimation framework to form a preliminary order which is then finalised by resolving local inversions by permuting triplets of markers and using the best fit. One final issue to be aware of during map construction is that of segregation distortion. This problem refers to any deviation from the Mendelian segregation ratio of a marker, with either an excess or a lack of homozygous marker classes from one or both parental genotypes. This can be a serious problem if it is not detected, and can bias the recombination frequencies between markers, reduce the power of QTL identification as well as bias estimates and affect sizes of QTL that are detected. When such distortion is revealed it is still possible to test the distorted region, but a separate threshold detection level is required, and a specific permutation of the affected region should be performed (*see* **Section 8.4**).

8.3. Types of Analysis

- Single marker: At the crudest level of analysis it is possible to detect differences in the means between marker genotypes due to linkage to a nearby QTL (although it must be pointed out that this test only really measures association and not linkage). This will therefore only look at markers on an individual basis and will use none of the information present in the linkage map. As detailed in Falconer and Mackay (2), if there is no linkage of a particular QTL to a marker (i.e. c, the recombination fraction between marker and QTL is 0.5) then there will be no difference between the marker class means derived from the two alternative parental types. Alternatively, a significant difference can be taken to indicate if a QTL is present. Using this non-closed method of QTL detection any effects of the QTL will be diluted by relative proximity to the marker, i.e. a small effect could be a small QTL in tight association or a large QTL in relatively weak association. This type of analysis requires a large number of progeny to overcome these statistical weaknesses (3).
- Interval mapping (IM): First proposed by Lander and Botstein (3), this circumnavigates the problem of QTL effect size and location being confounded by analysing the data in the form of intervals between markers. As the genetic distance is known between markers it is possible to search in increments along this interval. The derived statistic is a maximum likelihood estimate of the proposed QTL, with this being compared to a null model of no QTL present (i.e. no linkage, or 50% recombination between marker and QTL). It is therefore a systematic, linear search through the genome for QTL (76) and as such suffers from problems when a more complicated architecture is present. In such cases it is common for numerous 'ghost QTL' to be identified, where neighbouring QTL can give rise to these false-positive errors (77). Generally, IM calculations are presented in a likelihood map, where the LOD score (log-of-odds) is plotted along every cM of the chromosomal marker map. When this curve exceeds the threshold required for significance a significant QTL has been located. The confidence interval of the detected QTL is usually indicated by the size of the interval indicated by one or two LOD drops i.e. if a peak height has a LOD score of 4.2, then where the curve decreases to either 3.2 or 2.2 on either side is the region considered to contain the QTL (3). The effect size of a QTL can be calculated in terms of the amount of variance explained by it. This is calculated by measuring the proportional decrease in the genetic model with and without the inclusion of that QTL in the

model (i.e. the total variation in the trait minus the residual variation left in the model with the QTL fitted, all divided by the total variation). The problem here is that technically estimates are not additive for each QTL analysed, and as such usually overestimate the variance explained by the QTL (with this being a particular problem in smaller data sets – see the Beavis effect later). As well as using maximum likelihood to calculate the probability of a QTL occurring, at a point along the interval, a simplified regression model can also be used, as proposed by Haley and Knott (78) and Martinez and Curnow (77). This approach makes analysis very straightforward, and also enables a great flexibility in models that can be used (due to the low computational power required; 79). Despite this simplicity, it gives a good approximation to the ML estimation.

- Composite interval mapping (CIM): This is one of several multiple QTL analyses devised, whereby the problem of linked QTL can be distinguished. This arises from looking for QTL individually when several QTL are present. First instigated by Zeng in 1993 (with a similar approach known as multiple QTL mapping by Jansen in the same year), this extends interval mapping by controlling for any other QTL effects in other parts of the genome. It achieves this by isolating a 'window' around the interval in question and then introduces additional markers as cofactors. These cofactors are used to account for the effects of other QTL located in other regions of the chromosome. This should then decrease the residual variance of the model and hence increase the power to detect a QTL. The 'window of analysis' works by blocking out the effects of a region of the genome on either side of the test site; however if a very narrow window size is selected then there will be very little recombination between the markers at each end of the region analysed and hence little gain in power, due to a failure to reduce the residual genetic variation in any appreciable fashion. Conversely, if a very large window is used and many cofactors are added, this will reduce the power of the analysis as well, by decreasing the degrees of freedom that are available (80).
- Multiple interval mapping (MIM): This is one of several methods designed to refine searches for multiple QTL. The basic goal with these types of analyses is to consider every position in the genome simultaneously, locating QTL that act independently, are linked to or interact epistatically with other QTL. Using both these and Bayesian methods it is possible to identify genomic regions that would be missed with a standard one-dimensional, linear search (76). The main limitation here is the vast numbers of interactions that could

be present making calculations prohibitively large. Although it is relatively straightforward to define an equation for a model, selecting the correct model from the potentially millions available is less so. Research has therefore attempted to define techniques that are less computationally intensive and yet are still able to resolve the potentially subtle architecture that may be present in a trait. Kao et al. (81) first suggested the use of multiple interval mapping to address these issues. With this method individual QTL are first identified in a conventional manner (IM or CIM usually), with the model for the multiple interval mapping then constructed around this. Although meant to be simultaneous, the computational power that would be required means that only quasi-simultaneous analysis is possible. This technique is still limited by the initial number of QTL identified, as well as being unable to map QTL that are not individually significant (i.e. any QTL with epistatic effects rendering them invisible during this first-pass mapping will be ignored). Finally, searching through and defining the most applicable model is once again a problem. A further technique, that attempts to deal with the computational problems involved rather than actual QTL-mapping itself, is the use of genetic algorithms (GA) in conjunction with any of these multiple mapping strategies. Using GA, it is possible to search for the optimum multiple QTL genotype by randomly generating models and continually selecting the 'fittest', as defined by a set of parameters. This method can therefore efficiently search through and define the most applicable model for use with any of the other methodologies.

8.4. Threshold Levels

When many different intervals are analysed throughout the genome, a correction needs to be applied for the number of multiple tests performed. To establish thresholds initially, corrections were made for a sparse map case whereby the number of intervals tested was used to correct the overall *p*-value in a standard Bonferroni manner. However for dense map cases, when association occurs between markers, this correction is too stringent (3). Permutation tests are now used to tailor thresholds to individual data sets (82, 83). These tests randomly reassign the phenotypes to genotypes, destroying the relationship between trait values and genotypes, modelling the null hypothesis that no QTL are present in the genome. Once thresholds have been established it is also necessary to correct them for the number of uncorrelated traits being mapped by dividing the significance value required by the number of tests (9). This becomes particularly important in the case of expression QTL data, whereby micro-array chips are used to generate many thousands of gene expression phenotypes for each member of the cross population. In this instance, a standard Bonferroni correction would be far too restrictive, so what is often used in such cases is a False Discovery Rate (FDR) correction (84).

8.5. Marker Density and Power

The power of a study is a combination of the density of the markers used to genotype the cross and the number of individuals used in the study. Given that the lowest power to detect a QTL is at the mid-point between two adjacent markers, where the genotype information of the individuals is at its weakest, the denser the map the less this is an issue. The density required for standard F_2 or backcross population for complete coverage (i.e. no decrease in power) is in fact rather lenient, due to the number of recombinations present in these populations resulting in fairly large recombination blocks (with this also being the reason the confidence intervals for QTL in these populations are rather large). In general, an average marker spacing of 10–15 cM is sufficient for complete coverage (40), with this also holding true for an AIL. However, in the case of an AIL the map size will be inflated due to the increased number of recombinations, therefore more markers will be required to cover the same physical distance. Population sample size is the other obvious principal limiting factor in the power of any QTL study, with the larger the sample size the greater the chance for detecting smaller QTL. This is usually expressed in terms of the probability of detecting a QTL which explains a given percentage of variation in the cross. For the equation for this calculation and its explanation, see Lynch and Walsh (40). Though this is principally related to the sample size and the method of action of the QTL (i.e. principally additive or principally dominant), the degree of difference between the two progenitor inbred strains and the potential number of effective factors can also have a bearing (3). A further issue with using a small sample size is the Beavis effect (85), named after its author. By simulation studies Beavis demonstrated that when QTL studies with small samples sizes do detect a QTL, they tend to greatly overestimate the effect size of the QTL.

8.6. Behavioural QTL Mapping

Though potentially more problematic than QTL mapping on morphological or physiological traits due to the lower heritability and repeatability often associated with them, nevertheless numerous QTL studies have been performed on behavioural traits. As such, some knowledge is available on the number of genes making significant contributions to these traits, as well as their effect size and some of their interactions.

- Effect size: Although the majority of studies identify QTL with around a 10% effect size (9) the actual accuracy of these estimates depend on the power of the study used. For studies with a power estimate of greater than 90% this is reliable, however as most studies have a power of around 50%, this effect estimate can be as much as twice the actual value (85). This estimate is also dependent on the mode of action of the QTL being detected as well as the type of cross used in the study – for example a back cross design will not detect certain dominant QTL. LOD score and effect size are linked in a complex manner – small effect values of 10% or less tend to be overestimated, while larger effect sizes tend to be underestimated, with larger LOD scores being more accurate (9). A further problem can be related back to selective genotyping of extremes – with these studies there is a tendency to not include the phenotypic data that has no corresponding genotypic information – which can lead to a large overestimation of QTL effect (see Blizard study – (9)). In general, where large studies of behaviour are performed, small effects tend to be the norm. The Belknap and Atkins (86) study found only 4 QTL, all with effect size less than 10%, similarly Turri et al. (87) and Hitzemann et al. (88) found one QTL of 10%, with the average effect size being around 5%. In terms of a desirable sample size, Beavis noted that the effects of the bias he discovered were greatly reduced when $n = 500$ or more, so an experiment of this size or greater is ideal, though this may be more constrained by genotyping and other costs. Obviously, smaller sample sizes can be used, but the effects of inflated effect sizes and the potential for missing small-effect loci must be borne in mind.

- Number of factors: Studies in humans often indicate a complex basis for behaviour (89), and often many QTL appear to influence behaviour. Crabbe et al. (90) found that over 24 QTL for drug abuse had been detected over a number of studies, although individual studies tend to find fairly small numbers of QTL. Similarly, with selected traits the architecture appears to be relatively simple with only a small number of QTL contributing to the bulk of genetic variance (91, 92). There are several reasons for this. On the one hand the studies used do not often have the power to detect small effects (in a study that cannot detect factors contributing to less than 2% of the variance, Otto and Jones (93) argue that only half the QTL have been identified). Conversely, selection experiments tend to capture a few loci in opposing lines, dependent on the number of generations over which selection has occurred and the initial population size of the experiment. This has been shown by the Turri (87) study, which was performed in two parts, with the same 6 QTL identified in each experiment. When the two halves were combined, with a commensurate increase in resolution power, no further QTL were identified. Finally, as the confidence intervals in studies tend to be very large, the detected QTL could well in fact resolve to be several linked loci, rather than one.

- Epistasis: These interactions are the hardest component of any quantitative trait to measure and are therefore often taken to have little effect, or at any rate effects that are not quantifiable. For example when quantifying heritability such epistatic terms that exist are considered equally distributed throughout the additive and dominance components and effectively ignored in standard analysis (40). The differences in heritability of the various trait types, and in particular the low narrow-sense heritability of certain traits, have been used in the past to indicate that significant non-additive interactions are present. In a study by Mousseau and Roff (94), morphological traits were shown on average to have the highest heritability, followed by behavioural and then life history traits. Hill et al. (95) recently demonstrated using twin studies that even when such heritabilities are low, they are often close to the maximum, broad-sense heritability for the trait in question, therefore leaving little variation remaining for epistasis. By modelling, it appears that even if epistatic interactions are possible, the actual influence in natural populations may be low due to the low allele frequency required in such circumstances. In QTL studies epistatic interactions were initially assumed to be rare (9, 40), although some studies have found effects (96–98). More recently, QTL studies can actually be considered to be one of the better methods for analysing epistasis. However, despite more studies analysing for such a phenomenon, the extent of observed epistasis is often highly variable, especially being dependent on the types of trait under investigation. In some cases large effects are present (see a review by Carlborg and Haley (99)), though how representative such effects are remains equivocal. Morphological examples of epistasis provide perhaps the most striking examples (100–102), though fitness traits also contain some (103, 104). Generally, where present, epistatic pairs appear to be in similar numbers to QTL for additive effects (105). Behavioural epistasis has perhaps the least evidence for epistatic interactions, possibly due to the problems associated with obtaining repeatable measures, with examples including drug tolerance in mice (106), odour-guided behaviour in *Drosophila* (107–109), foraging behaviour in the honeybee (110) and antipredator behaviour in the zebrafish (111). However, the study with the largest sample size to look at epistasis in behaviour, anxiety behaviour in selected mice (112), failed to find any major epistasis, and in total there are relatively few studies that have looked at epistasis in relation to behaviour. The main problem with the analysis of epistasis in this context is the sample sizes required for accurate assessment, with the large number of tests leading to a very stringent threshold of significance that is required. Though epistasis has been found in smaller samples (103, 111), it may not be possible to rule out the possibility of epistasis existing in datasets where it is not found. Although currently such analyses are mainly finding two-way allelic effects this is not to say that larger combinations are not possible or even probable, rather that higher order interactions cannot be feasibly tested using conventional QTL studies due to insufficient sample size (113).

- Sex limited effects: Sex limitation occurs when a QTL is expressed in one sex but not the other, however detection of sex limitation effects can be somewhat controversial (9). The first study to robustly provide evidence for such effects was in alcohol preference in mice (114). Often detection is based on a significant QTL segregating in one sex at a particular locus but not in the other. However this in itself is not sufficient – simply put, just because a study does not find effects in one sex at a locus does not mean that none exist there. Flint (9) gives a formula that can be used to calculate significant sex effects, calculating likelihoods for each sex independently as well as in combination and placing them in the following formula: $2(L_c - (L_f + L_m))$, which is chi-distributed with 2 degrees of freedom.

9. Repeatability/ Robustness of Assay

One of the first problems when attempting to assay behaviour is to ensure it is both robust and repeatable. Repeatability is the degree to which the behaviour is reproducible between tests, or the correlation between two measures on the same individual (2). The variance among repeated measures, i.e. within individual variation, can only be due to environmental variance, which in the case of a behaviour is far more severe than many other trait types. The repeatability of a trait gives an upper limit on the heritability (40). Therefore with the analysis of behaviour, it is highly desirable to have both a robust test, which minimises environmental noise, as well as repeated measures. Repeated measures allow the repeatability to be ascertained, as well as allow one to use a mean value per individual which should more accurately reflect the actual underlying trait. For instance, one of the greatest attractions about RIL is that multiple individuals can be used per line, and as individuals within a RIL are essentially clonal, this will serve to increase the heritability of the trait through repeated measures over repeated individuals.

10. Genetic Definition of Behaviour

An important question when quantifying a behavioural trait is how relevant the test devised is at defining the behaviour in question. It is possible that tests of drug preference, emotion, learning, etc., are actually mapping different, if related, behaviours. The fact that significant QTL are found tends to dissuade closer examination of exactly what has been measured. There is a remarkable diversity in how people use behavioural tests and how they interpret the results. For instance, in the open field arena using the mouse, measures have been made of anxiety, hyperactivity, latency to move relating to locomotion and emotional response (92, 115). Crabbe et al. (116) argue that behavioural tests can be hard to standardise and that it is difficult to control fully for variations in test performance. It can be difficult to ascertain if animals are emotionally different or whether some other factor, irrelevant to anxiety, explains the variation. In this instance, differences in locomotor activity could masquerade as differences in fearfulness. In the case of the elevated plus maze, drugs that reduce anxiety in humans increase entries into the open arms of the maze (117), however anxiolytic effects are reduced or abolished after a single exposure (although prior test experience increases base-line open-arm avoidance (118)). Dawson et al. (119) give evidence that the anxiolytic effects of chlordiazepoxide are confounded with increases in motor activity; in fact stimulants have an anxiolytic effect in the apparatus. Flint (9) argues that problems of interpretation can be dealt with genetically by choosing tests that map the same underlying trait from different perspectives; it is then possible to apply QTL mapping to multiple measures. Identified QTL that act pleiotropically and effect more than one measure in a way consistent with predictions are then assumed to be those that actually influence the trait of interest.

As well as the potential for mapping an incorrect behaviour, the characterisation of a trait generally is also an issue. This is especially the case with the study of behaviour, as any one behaviour may contain many different components or facets. Mather and Jinks (120) defined 'super' and 'sub' characters to address this issue. An overall behavioural trait or super-character can often be better analysed by breaking it down into a series of sub-characters. Whereas the overall trait may be highly complex with a large number of genes and numerous interactions, some of the component parts may be less intricate. By first obtaining information on some of these underlying characters the overall trait can thus be more easily quantified. Generally speaking in almost all behaviours analysed to date it is these initial components that are first identified and analysed with further mechanisms leading on from these then coming under scrutiny.

11. Further Work – Fine Mapping, Selective Sweeps and Bioinformatics Approaches

11.1. Fine Mapping

To be able to actually get down to a resolution suitable for gene isolation techniques it is necessary to identify an interval of around 1 cM (121, 122). Inbred strains have a poor mapping resolution (123), with confidence intervals often encompassing the entire chromosome (86). With fine mapping, it now becomes necessary to shift searching from a genome-wide scale down to single-QTL detection methods. These tend to be based around genetic chromosome dissection or GCD, as pioneered in *Drosophila* (124, 125). These methods are summarised more fully in Darvasi (64).

- In recombinant progeny testing individuals are screened for recombinants over the QTL region of interest. These are then backcrossed to one of the parental strains, to help determine the QTL genotype, with the position of the breakpoint then narrowing the region accordingly. Reduction of the interval from y to x cM will require y/x recombinant individuals. This method does have certain limitations, especially if the QTL under examination is of rather small effect making it easy to lose in the background 'noise' (and therefore requiring a larger number of progeny to ascertain the QTL genotype).
- Interval-specific congenic strains (ISCS) is a similar process whereby individuals with recombinants in the region of interest are backcrossed for several generations with the background parental strain, removing alleles from the donor parental strains at all other QTL affecting the trait. These animals are then intercrossed, with homozygotes being used to establish one ISCS (126). This method has the advantage over the recombinant progeny testing method of decreasing the background 'noise' greatly, though it is obviously more laborious to perform.

- The recombinant inbred segregation test (RIST) uses the theoretical high mapping resolution present in RI strains in QTL mapping. RI strains are selected with recombinations in the region of interest. These are then crossed with both parental strains to produce two separate F_1 populations, which can then be crossed or backcrossed. The QTL will then by necessity be segregating in one population but not the other, analysis revealing whether the QTL is above or below the recombination point.
- A further technique that has been used in mice is through outbred animals such as the genetically heterogeneous (HS) mice derived from known progenitor strains (122). These have been randomly intercrossed for more than 30 generations and can be used to map small-effect QTL to under a cM. These lines are derived from crossings involving around eight different progenitor strains. As each different progenitor set is known, using these strains as the background, regions can be finely isolated according to how the QTL varies in each progenitor strain. It is also theoretically possible to use a cross between HS mice and an inbred strain to screen the genome and map a QTL to within a cM relatively cheaply (121). Proof of the efficacy of this approach comes from a major QTL mapping experiment using such a mouse HS line: 1,904 mice were phenotyped for 94 different complex traits and genotyped for 13,459 markers, revealing a total of 843 QTL with an average 95% confidence interval of 2.8 Mb (127).

11.2. Population Disequilibrium and Selective Sweeps

A further method of fine mapping, that is particularly amenable to domestic populations (128), is the use of linkage disequilibrium and the identification of selective sweeps in populations that have actively undergone selection. In many ways this is similar to association mapping (129), whereby haplotype blocks which are associated with the trait in question are identified in a population and used to provide far narrower regions than would otherwise be possible with standard linkage. In the case of association the issue becomes how likely a given SNP under selection is both associated with the phenotype and fixed within the target population. Though a straight association mapping could be performed in zebrafish (given enough individuals from a wild population that were not too closely related), the use of domestic (or more correctly laboratory) lines in the zebrafish opens up the possibility of identity by descent (IBD) mapping with selective sweeps. In the case of the zebrafish the high within-population variation even in such laboratory or domestic strains is actually of great benefit here. The basis of IBD mapping is that when a mutation occurs in an individual from a population, it will be in LD with all the other surrounding loci, i.e. it will have a discrete haplotype.

In subsequent generations this haplotype will be eroded. If one considers a domestic population that has undergone intense selection for a trait, the initial haplotype that gave rise to the mutation that is to be selected upon (be it high growth or reproduction, decreased aggression, etc.) should eventually become fixed in the population. This will then cause all polymorphic SNPs around the mutation site to decrease in heterozygosity, due to the 'hitchhiking' effect (130), depending on the number of generations that have elapsed since the mutation, the time taken until fixation, the degree of recombination in the region and the strength of selection acting upon the locus (with this also being dependent on the population size). Obviously, in the case of domestic populations the strength of selection is often extremely high, therefore these homozygous haplotype blocks will initially start as being quite large and then be slowly eroded. This approach has been used successfully to identify both discrete mutations in domestic animals – for instance the pea comb mutation in chickens – (131) as well as to identify QTNs for pig fatness (6), cattle twinning rate (132) and milk production (133). The main problem with the identification of selective sweeps to date was the fine-scale resolution of SNPs that was required to detect what was often still very small regions (for instance selective sweeps mapping in dogs, considered to have large haplotype blocks within breeds, still only found blocks of around 106 kb in size for rod-cone degeneration (134)). In stickleback a set of natural populations showed evidence of a selective sweep between 20 and 90 kb in size (135). This problem can now largely be ameliorated using the latest next generation sequencing technology, allowing representatives from strains to be fully sequenced at a relatively low cost, with the detection of the selective sweeps within the QTL regions of interest then being excellent candidates for further analysis.

11.3. Bioinformatic Approaches

With the wide array of bioinformatic tools now available, a variety of additional methods are now also available to help fine-map a region and go from hundreds of potential candidate genes to far fewer. Burgess-Herbert et al. (136) detail these techniques in the fine mapping of HDL cholesterol in mice, but in brief they involve using combined cross analysis, comparative genomics and haplotype block analysis. Depending on the trait being analysed and the extent to which it has already been analysed in other populations or species, all of these may or may not be applicable to a particular cross, but the use of one or more can still significantly reduce a region. Using combined cross analysis (137), several different QTL crosses between different strains can actually be combined, resulting in greater power to narrow a region, though only if the crosses share mutual QTL (138). Comparative genomics involves the analysis of conserved regions between different species, with such resources allowing the comparison of orthologous regions becoming commoner. For instance, in the Burgess-Herbert study they use the concordance between QTL for plasma lipids (139) in humans and mice to assume that the genes for HDL cholesterol may also be the same in both humans and mice, so narrowed the regions containing HDL QTL in mice to regions homologous to concordant human HDL QTL. In the case of zebrafish behaviour, the large numbers of mouse studies looking at behaviour could well be an excellent source for comparison (127), though obviously this method relies on the same QTL affecting the trait across species, of which there is still only limited evidence (140). Haplotype block analysis uses the linkage disequilibrium evidenced by linkage 'blocks' of haplotypes present in different strains and is another name for the IBD mapping mentioned above. In this case rather than identifying signatures of selective sweeps by homozygosity, mutually shared haplotype regions between populations are identified and used to narrow down an interval.

12. Problems of QTL Analysis and Alternative Strategies

The main problem once a QTL has been identified is making the step from QTL to the actual nucleotide variation (QTN) (123). The size of the regions isolated by standard QTL mapping means a great deal of work is required to make this step. Even in the case of standard linkage with discrete Mendelian traits, this is by no means straightforward; so in the case of a behavioural QTL, with their rather small effect sizes, this can be extremely challenging, though successful examples even for behaviour do exist (5, 141). Initial QTL studies tend to find surprisingly simple genetic architecture due to the large regions that are identified, but when these are more finely dissected, far more complex architecture is revealed (142). As QTL mapping detects a region, rather than an actual gene (though see eQTL studies below), and the regions tend to be extremely large using a standard analysis (on average between 20 and 30 cM), there are potentially several actual genes that are involved in affecting the trait of interest in that particular region. If these linked genes together act in the same direction then they will appear to be one large-effect gene, whereas if they act in opposite directions they may be missed altogether. A further issue is to what extent an inbred cross actually represents the variation seen in the original population. Especially when we are trying to unlock the architecture from a general or wild-derived population, it is quite possible that many of the important genes contributing to variation in other populations may be missed. However, this does add to the corroborating effect of identifying the same QTL using different strains (and can then also add additional weight to attempts to replicate QTL between species – (136)). A further problem is that of higher order epistasis. Though as has been seen the question of epistatic interactions can divide the community, it is a certainty that higher order interactions (such as three and four locus interactions) will be entirely missed in these studies simply due to the massive computing requirements and stringent significance thresholds, though these may still be important (see (100) for an example of a four locus interaction affecting body weight in the chicken).

When studies are attempted to find the QTN responsible, the more molecular and genomic tools that are available for the species used, the better. One of the main advantages with the zebrafish in terms of QTN detection and most importantly verification is the use of transgenics, which is often considered the 'purest' method of proving a given genomic mutation influences a trait (though there are still numerous difficulties attached to transgenics see (123) for a brief review of different approaches). Given that a QTN is a very subtle effect, it is more likely to be a non-coding mutation in a regulatory region (143), and could well be in either cis- or trans-effect (i.e. close to or far from the gene it regulates), even transgenic analysis is by no means simple. Complementation studies can be used, whereby the gene of interest is inserted into a transgenic animal, and then both expression and phenotype are measured. If the effect is additive, then both should be affected (144). The problem here is that if the QTL is not confined to a suitably narrow region, it is impossible to rule out other small effect loci also contributing to the phenotype from within that region. Even then, the effects of exogenous DNA may not be analogous to a genuine diploid mutation. A more refined approach can be made using quantitative complementation (see (106) for a review), though this requires deficiency stocks and balancer chromosomes for the regions of interest and are therefore restricted to *Drosophila* and mouse studies at present. Similarly, the Cre-*lox* engineering system in mice is also very powerful, allowing the expression of your transgene at the specific time and in the specific tissue of your choice (145). This approach is now also being used in zebrafish (146), though even here the problem of when and where to express the gene can be problematic.

One method of circumnavigating the technical difficulties of transgenics is the use of expression profiling in conjunction with QTL analysis. Although this will not identify the causative mutation, it should in theory indicate the gene of interest and is far easier to perform. It is possible to assay cDNA from genes in the region of interest and check their correlation with the QTL phenotype. This approach has been used successfully in the case of 5-lipoxygenase affecting obesity and bone traits in the mouse (4),

Cd36 in hypertensive rats (147) and *Usp46* in mouse immobility behaviour (5). Although this is becoming increasingly commonplace to perform, it still has associated problems. These are related principally to identifying the correct tissue and developmental stage on which to run the analysis – it is highly possible that expression differences may be restricted to very narrow developmental time windows in specific regions, and even then may only have modest expression differences. Finally it is even possible that the QTN may not even be regulated at the transcription level. As the cost of expression profiling decreases, a further possible approach is to map expression QTL or eQTL (148). In this instance an expression array is run on cDNA from each individual in the study from a tissue of interest, with these expression levels then being used as a series of thousands of phenotypes. By regressing these onto the QTL markers this opens up the possibility to detect cis- and trans-effects, as not only is the gene position obviously known but also the region causing the variation. With such studies it was initially thought that many pleiotropic QTL hotspots would be found, though this has largely turned out not to be the case (149).

Mutagenesis screens and the study of single gene mutants is a viable alternative to QTL mapping to disentangle the genetic effects of behaviour in the zebrafish (150). Such an approach has been successfully used to identify behavioural mutations in several other model organisms, including the neural bases of learning, courtship and circadian rhythms in *Drosophila* (151, 152), social aggregation in *C. elegans* (10, 153, 154), and larval foraging in *Drosophila* (11, 155, 156). Advantages of mutagenesis are principally related to the ability to identify the causative gene immediately. This means the myriad of problems associated with going from a QTL to a QTN no longer apply in this case. Additionally, it also creates strong mutations having large effects, improving the signal-to-noise ratio and also has the ability to target any of the genes involved with the trait due to the resulting mutations being at random in the genome. As complex traits are made up of intricate developmental and physiological pathways, in theory this should increase the probability of disrupting a gene critical for the trait of interest as compared to a single gene trait. The problems of mutagenesis are initially similar to other complex trait analysis methods – namely the reliability on the assay itself and whether it is repeatable and a true measure of the actual trait of interest (150) and the large numbers of individuals required for the actual screen itself. However, one of the main issues with mutagenesis is that it actually examines an entirely different genetic architecture to that involved in a QTL study. In the latter, numerous units responsible for small quantitative change are identified, whereas the former looks more at the basic machinery involved in the trait, with effects on these genes causing such massive and fundamental changes it is highly unlikely that such genes are also responsible for subtle variation (156).

This zebrafish itself is also highly amenable to genetic screens; in fact the first vertebrate genetic screens were performed on the zebrafish, identifying over 2,000 mutated genes that control development (157, 158). It is possible to perform both haploid and homozygous diploid screens with the zebrafish, through the use of early pressure (EP) or heat shock (HS) on eggs fertilised with UV sterilised sperm. This allows easy analysis of recessive mutations without the need to generate large numbers of F_2 individuals (in the case of haploid screens), or ensures that 50% of F_2 individuals created are homozygous for the recessive mutation (in the case of homozygous diploid screens). Of the mutagens available, most are effective in the zebrafish. ENU (ethyl nitrosourea) induces mainly point mutations, and can generate mutations in most genes although the mutation frequency can vary widely between loci. EMS (ethyl methane sulfonate) is a mild mutagen in zebrafish (as opposed to having potent mutagenic effects in *Drosophila*), while radiation can cause a varied degree of mutations, ranging from point through large-scale genomic deletions and translocations (159). A chemical mutagen, trimethylpsoralen, is also available that can produce more modest deletions (from 100 bp to 1.5 kb). With all these induced mutations, the positional cloning is made far easier by the knowledge of the zebrafish sequencing project. Though principally used to examine disease phenotypes (160), this by no means precludes behavioural phenotypes. Drug tolerance (161), visual behaviour-response (162, 163), defective tail-flick response (164, 165) and resistance to epileptic seizures (166) have all been analysed using this technique.

13. Case Study: QTL Mapping Anxiety in Mice

Though QTL mapping has been more commonly used for other, more easily measured, traits there are nevertheless still numerous examples of it being applied to behavioural trait types. Such cases can highlight how these behavioural analyses are performed, and also what can be obtained using the technique. One of the largest sets of studies on animal behaviour via QTL mapping comes from the Flint laboratory in Oxford. These studies include a series of conventional F_2 crosses using large populations of mice, and more recently a heterogeneous strain (HS) cross comprising of a series of heterogeneous intercrossed populations. Although this latter study has mapped a huge number of QTL to very narrow regions, the initial studies will be presented here to demonstrate behavioural QTL mapping, as they are an excellent example of a standard cross type that is more likely to be seen, at least initially, in the zebrafish. The crosses in question are set up between two inbred mouse strains that have been assayed for different measures of anxiety. These measures are made using numerous apparatuses, including an open field arena, elevated plus maze and a light-dark box. The open field arena consists of a 60-cm square box that is brightly lit. Animals placed here are considered to be in a stressful environment with differences at both the level of the individual and between populations observed in the extent of movement within this apparatus. An elevated plus-shaped maze consists of four different arms suspended over a drop, two of which are closed and considered less anxiogenic, two of which are open and considered more anxiogenic. A light-dark box consists of two joined rooms – one lit, the other dark. Latency to emerge to the lit side as well as activity in either side is measured. It is possible to select for extremes in this anxiety-based behaviour; if the experiment results in animals that show consistent heritable differences in the apparatus, selection is assumed to have produced a stable model of the behavioural trait (87). Using the open-field arena to measure anxiogenic behaviour in mice, only 3 loci were revealed in mice selected for high and low activity (91). A larger-scale study, using all of the aforementioned apparatus, was then performed by Turri et al. (87), negating the problems of small effects being missed due to insufficient sample sizes (85) and also demonstrating replication of results. These two new studies of 815 and 821 individuals each genotyped for 79 markers, both identified the same six QTL. By combining the two studies there is a much greater power to detect small effect QTL, yet no more were detected, enabling researchers to be confident that no further small-effect QTL had been omitted. Although it is possible that within the large chromosomal regions isolated by the study different molecular variants are segregating within it, this region was reduced via composite interval mapping and analysis of chromosomal structure of the four strains (with this region agreeing with actual QTL positions of the common chromosomal structure) indicating that the different effects may be due to the same QTL, especially regarding activity in the open field arena. Elevated plus maze and light-dark box trials also found a similar pattern, with small numbers of loci influencing the phenotypes in both crosses, suggesting the QTL are acting in a pleiotropic manner, as indicated by the close map positions. Assuming such pleiotropy, as little as 12 loci could account for the variation in all of the phenotypes measured. As well as standard analysis of the data by combining groups of related characteristics within tests, factor analysis was used to try and obtain composite values relating to several tests at one time (167). Although these were found to be weaker than standard within-test analysis, several broad measures of anxiety were identified in this study (87, 167), including (i) tendency to avoid the more anxiogenic areas of the test environments and (ii) hesitancy to enter novel areas of the test environment, with all of the QTL identified through the analysis influencing at least two of the five general anxiety measures. A standard concern with activity-based and anxiety-based behavioural phenotypes is the possibility that these definitions can be misleading, with exploration in the low anxiogenic regions considered to be an assessment of general activity level whereas activity in the more anxiogenic areas is considered to be a measure of anxiety level. Henderson et al. (167) point out that out-of-cage test environments range from moderately to severely anxiogenic; hence all behaviours are in fact a measure of anxiety.

14. Case Study: QTL Mapping for Behaviour and Growth in the Zebrafish Using the Wild × Domestic Paradigm

The only recent QTL study using zebrafish was performed by Wright et al. (111, 168), and utilised the wild × domestic paradigm detailed earlier. The cross was made using a wild-derived population obtained from Bangladesh (named Santal after the village nearest to the collection site) and the AB laboratory strain.

14.1. Population Differences

Shoaling tendency and boldness (as determined by the approach to a novel object) had previously been shown to vary between wild populations and, given the strong differences between wild and domestic zebrafish, these two behaviours were used as the starting point in a behavioural analysis between the AB and Santal populations mentioned above (35). Tendency to shoal was measured using a preference tank consisting of a central arena, flanked on either side with two outer compartments. Two 40-mm stimulus zones were demarked in front of both outer compartments. A stimulus shoal of six fish was placed randomly in one of the two outer compartments. These compartments were separated by one-way glass dividers so only visual cues were given to the focal fish, while the stimulus shoal was unaware of the focal individual. Focal fish were first acclimatised in a beaker for 10 min, prior to being gently poured in to a release tube in the arena, where they acclimatised for a further 5 min. The release tube was then remotely raised, and once the stimulus fish had entered the stimulus zone adjacent to the stimulus shoal the test began, lasting for 10 min, with the amount of time spent associating with the shoal recorded. Trials were conducted twice for each fish, with fish being housed in a tank subdivided into mesh compartments (allowing fish not only to see and smell conspecifics, but also to be identified again, see below). Boldness measures were performed using a 600-mm long tank, with a 140 mm, long, roughly cylindrical 'novel' object, suspended at one end of the tank. The focal fish was placed in a release tube at the opposite end of the tank and given 5 min to acclimatise prior to remote release. The measures for boldness were time spent within 1.5 standard body-lengths of the novel object, the time the novel object was first approached and the number of times the zone around the novel object was entered. Once again, all trials were performed twice per fish. An analysis of 20 Santal and 19 AB fish showed differences between the two population means ranging from 1–2 standard deviations for measurements taken from these two tests (see **Fig. 5.1**). Wild-derived (Santal) fish appeared less 'bold', taking more time to enter the zone around the novel object and spending less time in the zone, and more 'social', spending more time associating with the stimulus shoal of conspecifics. This appears to be broadly consistent with heightened predator-avoidance behaviour in the wild fish, certainly in terms of an increased tendency to shoal, though the desire to approach the novel object can be seen in a number of ways (e.g. potentially the domesticated laboratory fish showed more fear of anything novel, conversely the wild fish could also be displaying greater predator inspection behaviour). Other measures taken also included standard body-length, weight and percent body fat, with these all measured post-mortem following the

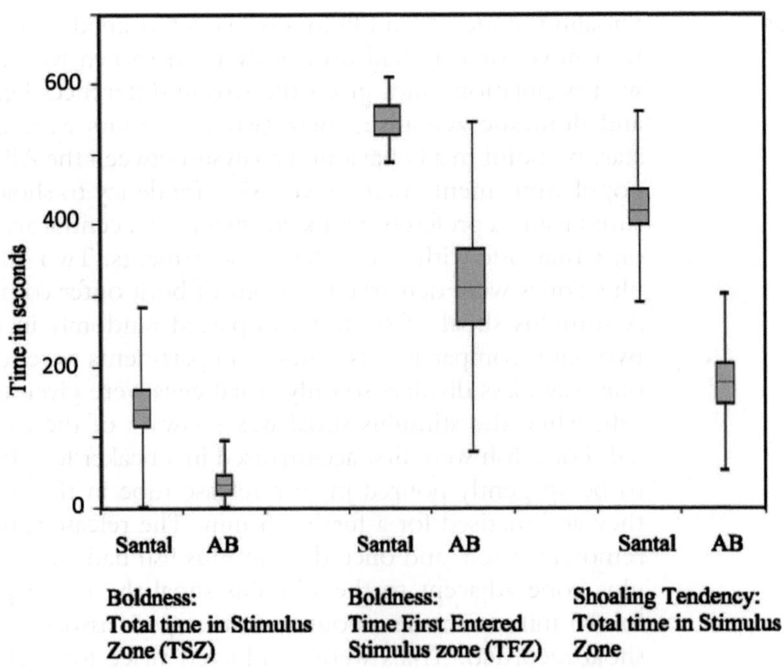

Fig. 5.1. Box and whisker plots of population means, S.E. and S.D. of wild-derived (Santal) and laboratory (AB) strains for boldness and shoaling tendency.

final behavioural tests. In terms of growth rates relating to food consumption, all fish were fed ad libitum, to ensure food was not a limiting factor for growth.

14.2. Repeatability of Tests

As mentioned previously, with the analysis of any behaviour the repeatability of individual performances is an obvious concern, as is the robustness of the assay. There is therefore a strong need to conduct multiple replicates per fish, however distinguishing unmarked fish in a tank containing even a small number of individuals is virtually impossible. Similarly, the isolation of individuals (i.e. in individual containers) has been shown to affect behaviour (21) and in the case of the experiment described here was found to increase stress, resulting in a strongly increased tendency to shoal between the first and subsequent trials. To navigate this issue and allow repeated measures of the same individuals, Wright et al. used a holding tank consisting of several compartments separated by a thin gauze mesh to house fish between trials. This enabled fish housed within to see, smell and even, to a limited degree, touch conspecifics. Fish treated in such a way displayed no adverse reactions to this apparatus, though other methods may also be used. In the case of the zebrafish PIT tags (passive integrated transponders, used routinely to distinguish fish) are too large, though coloured polymer injected under the skin has been used repeatedly on guppies (169) and could be a possibility. In this case, some thought has to be given to the potential effect on the behaviour being analysed, though at least the same treatment is given to all fish. In terms of the robustness of the assay described here, the primary uncontrolled environmental stress acting on fish in these instances is that of netting and handling stress. The stress in this instance can easily serve to modify the desire to aggregate or explore, and as such conditions should be as standardised as possible, with handling kept to a minimum. In the case of Wright et al. (168), the authors first netted the focal fish from the holding tank and placed into a beaker of water for 10 min to acclimatise. Before the start of the experiment, fish were then poured into the release tube of the test apparatus and left to acclimatise for a further 5 min. Though rather laborious, this decrease in stress does result in repeatabilities for the behavioural measures recorded to be between 58 and 75% (depending on whether extreme outliers are removed from the dataset (35)). Further details of the holding tank construction and the shoaling protocol are available in Wright and Krause (36).

14.3. Mapping Density, Sample Size, Power of Study and Map Construction

A total of 66 markers were used, spread out over the 25 linkage groups, with 2–4 markers per chromosome, an average marker separation of 29 cM, and an average distance between the distal marker and the telomere of 17 cM. The F_2 population consisted of 166 fish derived from one F_1 pairing and 18 from a second F_2 pair. As can be seen in comparison to previous QTL experiments, these sample sizes are therefore rather small, with this being reflected in the relatively low power of the study. Using a critical p value of 5%, this experiment had approximately a 90% probability of detecting a QTL which accounted for a minimum of 5% of the differences between populations, where this QTL is fully associated with a marker. At the midpoint between markers or at the telomere, where the power of detection is at its lowest, there was approximately a 90% probability of detecting a QTL with a 10% effect size.

14.4. Analysis – LOD Scores, GA Approach

Analysis of the data was performed with two separate techniques – interval mapping and a genetic algorithm. As is common in these experiments, rearing tank was found to have some effect on the traits in question and therefore included as a fixed factor in the IM model. Similarly, though body-length fell just short of significance, this was also included as a covariate. The threshold for significance was determined by permutation testing, as detailed previously, while the threshold was also additionally modified by the number of uncorrelated tests performed. Here the traits broadly fell into two separate suites of antipredator and growth traits, leading to a Bonferroni correction of 2, giving a final threshold of 3.3. Using this threshold a total of three behavioural and three morphological QTL were obtained. See **Fig. 5.2** for an LOD score graph of one of the boldness QTL. In addition to a standard regression-based interval mapping analysis, a genetic algorithm was also used to search for loci (170). This approach fits multiple loci with models selected using Akaike's information criterion (AIC), and as such should allow a global analysis of the genetic architecture. The problems with this approach is that first missing data is a problem to

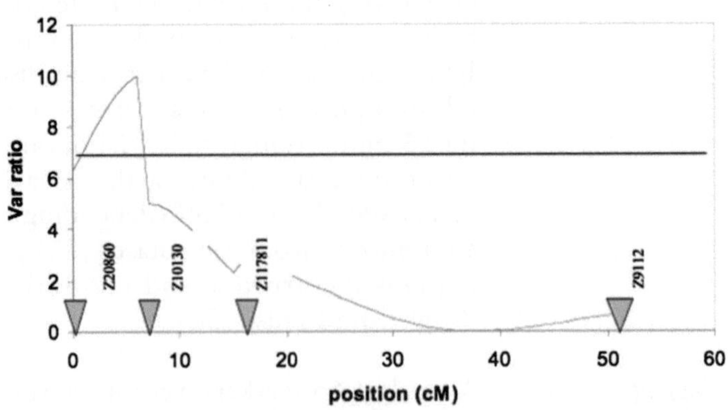

Fig. 5.2. QTL LOD graph for a boldness measure (time first entered stimulus zone) on chromosome 9, as presented in Wright and Krause (36). Positional markers are indicated with grey triangles, while the horizontal bar represents the significance threshold cut-off. The graph was calculated using interval mapping (IM) analysis.

the approach and second false QTL may be identified as long as their inclusion serves to decrease the AIC, due to the AIC being slightly more liberal than the likelihood tests used by IM. This approach detected all those QTL found by IM, as well as revealing several additional loci for all traits considered, leading to a total of 20 QTL over all traits. Of particular note is that two of the QTL for boldness (on chromosomes 9 and 16) were detected in multiple measures of the boldness assay, increasing the likelihood that the loci were actually involved in boldness in general.

14.5. Epistatic Analysis

Two-locus epistatic interactions were analysed throughout the genome for this cross. This was performed by taking every position of the linkage map and analysing it in combination with every other position, using a regression method (171). Though statistically simple, the problem is the large number of calculations that are required, and the resulting significance threshold calculation. In this case, a genetic algorithm (172) was used to help permute the data to find the required threshold for significance and suggestivity (where QTL may fall short of actual significance, but a potential QTL is suggested at a given location). This analysis revealed two suggestive and one significant set of pair-wise epistatic interactions for boldness, involving four different loci that created a small network of loci. The effects of these are shown in **Fig. 5.3**, with a strong degree of dominance present. One of the loci (on chromosome 21) was dominant over two of the other loci, while the loci on chromosomes 9 and 12 had a multiplicative effect (so, the presence of alleles increasing the trait at these two loci caused an even greater increase in the phenotype than would be expected). Though this does give possible indications of what may be occurring in the architecture of a behavioural trait, the effect sizes in this instance must be treated with caution due to the small sample size used (186 F_2 fish). In terms of the ideal sample size for the analysis of epistatic interactions, it is generally a case of the bigger the better, with even two-way interactions serving to decrease the number of double homozygotes (with these providing the most power for additive × additive interactions) by half, as compared to a single QTL analysis. That is, in a standard F_2 QTL analysis, there are 50% homozygotes (25% of each parental class) and 50% heterozygotes at any given loci, however when looking at two loci simultaneously, only 25% will be double homozygotes and of these only half will be double homozygotes of the same parental class (so possessing for instance two wild-type alleles at each of the two loci analysed). Therefore a corresponding doubling or even quadrupling of sample size would be ideal to accurately measure epistatic interactions, though often not practical.

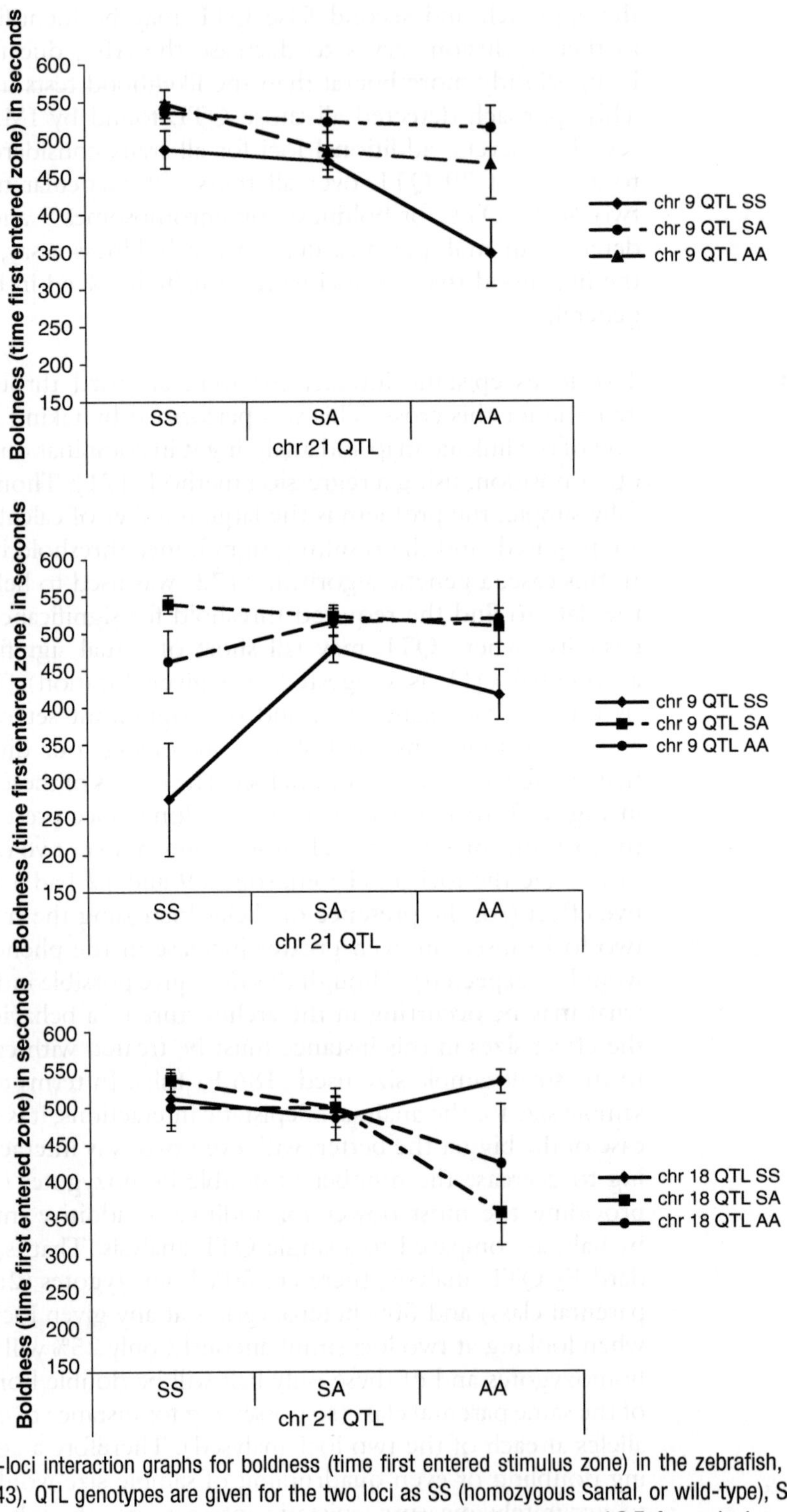

Fig. 5.3. Two-loci interaction graphs for boldness (time first entered stimulus zone) in the zebrafish, as presented in Wright et al. (43). QTL genotypes are given for the two loci as SS (homozygous Santal, or wild-type), SA (heterozygous and AA) (homozygous AB, or laboratory) for each of the two QTL, with the mean and S.E. for each class.

15. Summary

Overall, this line of research serves as an important proof-of-principle (were one required) that behavioural QTL mapping in zebrafish is not only possible, but has great potential. Though the study does indicate the potential of epistasis in such a system, the issue of the small sample size limits what it is possible to conclude about the genetic architecture of shoaling and boldness in this animal, and further studies are definitely required if these behaviours are to be more completely mapped and used in comparison with some of the more established anxiety and aggregation-related behaviour measures performed in other laboratory organisms. More generally, the further use of the zebrafish in behavioural QTL mapping could add considerably to the understanding of behaviour in general, and more fully open up the use of another model organism in the genetic dissection of such traits with a complexity level above that of the *Drosophila* but below that of the mouse.

References

1. Barton, N. H. & Keightley, P. D. (2002) Understanding quantitative genetic variation. *Nat Rev Genet* 3, 11–21.
2. Falconer, D. S. & Mackay, T. F. C. (1996) *Introduction to Quantitative Genetics.* Upper Saddle River, NJ, Prentice Hall.
3. Lander, E. S. & Botstein, D. (1989) Mapping mendelian factors underlying quantitative traits using RFLP linkage maps. *Genetics* 121, 185–199.
4. Mehrabian, M., Allayee, H., Stockton, J., Lum, P. Y., Drake, T. A., Castellani, L. W., Suh, M., Armour, C., Edwards, S., Lamb, J., Lusis, A. J. & Schadt, E. E. (2005) Integrating genotypic and expression data in a segregating mouse population to identify 5-lipoxygenase as a susceptibility gene for obesity and bone traits. *Nat Genet* 37, 1224–1233.
5. Tomida, S., Mamiya, T., Sakamaki, H., Miura, M., Aosaki, T., Masuda, M., Niwa, M., Kameyama, T., Kobayashi, J., Iwaki, Y., Imai, S., Ishikawa, A., Abe, K., Yoshimura, T., Nabeshima, T. & Ebihara, S. (2009) Usp46 is a quantitative trait gene regulating mouse immobile behavior in the tail suspension and forced swimming tests. *Nat Genet* 41, 688–695.
6. van Laere, A. S., Nguyen, M., Braunschweig, M., Nezer, C., Collette, C., Moreau, L., Archibald, A., Haley, C. S., Buys, N., Tally, M., Andersson, G., Georges, M. & Andersson, L. (2003) A regulatory mutation in IGF2 causes a major QTL effect on muscle growth in pigs. *Nature* 425, 832–836.
7. Bolivar, V., Cook, M. & Flaherty, L. (2000) List of transgenic and knockout mice: behavioural profiles. *Mammal Genome* 11, 260–274.
8. Flint, J. (1999) The genetic basis of cognition. *Brain* 122, 2015–2032.
9. Flint, J. (2003) Analysis of quantitative trait loci that influence animal behaviour. *J Neurobiol* 54, 46–77.
10. de Bono, M. & Bargmann, C. I. (1998) Natural variation in a neuropeptide y receptor homolog modifies social behavior and food response in *C. elegans. Cell* 94, 679–689.
11. Osborne, K. A., Robichon, A., Burgess, E., Butland, S., Shaw, R. A., Coulthard, A., Pereira, H. S., Greenspan, R. J. & Sokolowksi, M. B. (1997) Natural behavior polymorphism due to a cGMP-dependent protein kinase of *Drosophila. Science* 277, 834–836.
12. Shimoda, N., Knapik, E. W., Ziniti, J., Sim, C., Yamada, E., Kaplan, S., Jackson, D., de Sauvage, F., Jacob, H. & Fishman, M. C. (1999) Zebrafish genetic map with 2000 microsatellite markers. *Genomics* 58, 219–232.

13. Gratton, P. et al. (2004) Allozyme and microsatellite genetic variation in natural samples of zebrafish, Danio rerio. *J Zoolog Syst Evol Res* 42, 54–62.
14. Guryev, V. et al. (2006) Genetic variation in the zebrafish. *Genome Res* 16, 491–496.
15. Gerlach, G. & Lysiaki, N. (2006) Kin recognition and inbreeding avoidance in the zebrafish, *Danio rerio*, is based on phenotype matching. *Anim Behav* 71, 1371–1377.
16. Bloom, H. D. & Perlmutter, A. (1977) A sexual aggregating pheromone system in the zebrafish *Brachydanio rerio*. *J Exp Zool* 199, 215–226.
17. Bloom, H. D. & Perlmutter, A. (1978) Possible pheromone mediated reproductive isolation in two species of cyprinid fish of the genus, *brachydanio*. *J Fish Biol* 13, 47–50.
18. Spence, R. & Smith, C. (2006) Mating preference of female zebrafish, *Danio rerio*, in relation to male dominance. *Behav Ecol* 17, 779–783.
19. Engeszer, R. & Ryan, M. J. (2004) Learned social preference in zebrafish. *Curr Biol* 14, 881–884.
20. McCann, L. I., Koehn, D. J. & Kline, N. J. (1971) The effects of body size and body markings on nonpolarized schooling behaviour of zebrafish (*Brachydanio rerio*). *J Psychol* 79, 71–75.
21. McCann, L. I. & Matthews, J. J. (1974) The effects of lifelong isolation on species identification in zebrafish (*Brachydanio rerio*). *Dev Psychobiol* 7, 159–163.
22. Rosenthal, G. G. & Ryan, M. J. (2005) Assortative preference for stripes in danios. *Anim Behav* 70, 1063–1066.
23. Pritchard, V. L., Lawrence, J., Butlin, R. K. & Krause, J. (2001) Shoal choice in zebrafish (*Danio rerio*): the influence of shoal size and activity. *Anim Behav* 62.
24. Bass, S. & Gerlach, R. (2008) Zebrafish (*Danio rerio*) responds differentially to stimulus fish: the effects of sympatric and allopatric predators and harmless fish. *Behav Brain Res* 186, 107–117.
25. Oswald, M. & Robison, B. D. (2008) Strain-specific alteration of zebrafish feeding behavior in response to aversive stimuli. *Can J Zool* 86, 1085–1094.
26. Pitcher, T. J. (1983) Heuristic definitions of fish shoaling behaviour. *Anim Behav* 31, 611–612.
27. Krause, J. & Ruxton, G. (2002) *Living in Groups*. Oxford, Oxford University Press.
28. Pavlov, D. S. & Kasumyan, A. O. (2000) Patterns and mechanisms of schooling behavior in fish: a review. *J Ichthyol* 40, 163–231.
29. Wright, D., Ward, A. J. W., Croft, D. P. & Krause, J. (2006d) Social organization, grouping, and domestication in fish. *Zebrafish* 3, 141–156.
30. Pitcher, T. J. (1992) Who dares wins: the function and evolution of predator inspection behaviour in fish shoals. *Neth J Zool* 42, 371–391.
31. Magurran, A. E. & Pitcher, T. J. (1987) Provenance, shoal size and the sociobiology of predator-evasion behavior in minnow shoals. *Proc R Soc Lond B* 229, 439–465.
32. Pitcher, T. J. & Parrish, J. K. (1993) Functions of shoaling behaviour in teleosts. In Pitcher, T. J. (Ed.) *Behaviour of Teleost Fishes*. 2nd ed. London, Chapman and Hall.
33. Magurran, A. E. (1990) The inheritance and development of minnow antipredator behavior. *Anim Behav* 39, 834–842.
34. Magurran, A. E., Seghers, B. H., Shaw, P. W. & Carvalho, G. R. (1995) The behavioural diversity and evolution of guppy. *Poecilia reticulata*. Populations in Trinidad. *Advances in the Study of Behavior*. San Diego, CA, Academic Press Inc.
35. Wright, D., Rimmer, L. B., Pritchard, V. L., Krause, J. & Butlin, R. K. (2003) Inter and intra-population variation in shoaling and boldness in the zebrafish (*Danio rerio*). *Naturwissenschaften* 90, 374–377.
36. Wright, D. & Krause, J. (2006) Repeated measures of shoaling tendency in zebrafish (*Danio rerio*) and other small teleost fishes. *Nat Protocols* 1, 1828–1831.
37. Ruzzante, D. E. & Doyle, R. W. (1991) Rapid behavioral-changes in medaka (*oryzias-latipes*) caused by selection for competitive and noncompetitive growth. *Evolution* 45, 1936–1946.
38. Ruzzante, D. E. & Doyle, R. W. (1993) Evolution of social-behavior in a resource-rich, structured environment – election experiments with medaka (*Oryzias-Latipes*). *Evolution* 47, 456–470.
39. Dugatkin, L. A., McCall, M. A., Gregg, R. G., Cavanaugh, A., Christensen, C. & Unseld, M. (2005) Zebrafish (*Danio rerio*) exhibit individual differences in risk-taking behavior during predator inspection. *Ethol Ecol Evol* 17, 77–81.
40. Lynch, M. & Walsh, B. (1998) *Genetics and Analysis of Quantitative Traits*. Sunderland, MA, Sunderland Sinauer Associates.
41. Kerje, S., Carlborg, O., Jacobsson, L., Schutz, K., Hartmann, C., Jensen, P. & Andersson, L. (2003) The twofold difference in adult size between the red junglefowl and White Leghorn chickens is largely explained

by a limited number of QTLs. *Anim Genet* 34, 264–274.
42. Wright, D., Kerje, S., Brändström, H., Schütz, K., Kindmark, A., Andersson, L., Jensen, P. & Pizzari, T. (2008) The genetic architecture of a female sexual ornament. *Evolution* 62, 86–98.
43. Wright, D., Kerje, S., Lundström, K., Babol, J., Schutz, K., Jensen, P. & Andersson, L. (2006b) Quantitative trait loci analysis of egg and meat production traits in a red junglefowl × white leghorn cross. *Anim Genet* 37, 529–534.
44. Buitenhuis, A. J., Rodenburgb, T. B., Siweka, M., Cornelissena, S. J. B., Nieuwland, M. G. B., Crooijmansa, R. P. M. A., Groenen, M. A. M., Koene, P., Bovenhuis, H. & van der Poel, J. J. (2004) Quantitative trait loci for behavioural traits in chickens. *Livestock Product Sci* 93, 95–103.
45. Schutz, K., Kerje, S., Carlborg, O., Jacobsson, L., Andersson, L. & Jensen, P. (2002) QTL analysis of a red junglefowl × white leghorn intercross reveals trade-off in resource allocation between behavior and production traits. *Behav Genet* 32, 423–433.
46. Schutz, K. E., Kerje, S., Jacobsson, L., Forkman, B., Carlborg, O., Andersson, L. & Jensen, P. (2004) Major growth QTLs in fowl are related to fearful behavior: possible genetic links between fear responses and production traits in a red junglefowl × white leghorn intercross. *Behav Genet* 34, 121–130.
47. Price, E. O. (1984) Behavioral aspects of animal domestication. *Quart Rev Biol* 59, 1–32.
48. Darwin, C. (1859) *The Origin of Species*. New York, NY, Mentor.
49. Darwin, C. (1868) *The Variation of Animals and Plants Under Domestication*. London, John Murray.
50. Ochieng-Odero, J. P. R. (1994) Does adaptation occur in insect rearing systems, or is it a case of selection, acclimatization and domestication? *Insect Sci Appl* 15, 1–7.
51. Price, E. O. (1999) Behavioral development in animals undergoing domestication. *Appl Anim Behav Sci* 65, 245–271.
52. Kallio-Nyberg, I., Salminen, M., Saloniemi, I. & Kannala-Fisk, L. (2009) Marine survival of reared Atlantic salmon in the Baltic Sea: the effect of smolt traits and annual factors. *Fish Res* 96, 289–295.
53. Olla, B. L. & Davis, M. W. (1989) The role of learning and stress in predator avoidance of hatchery and wild steelhead trout. *Aquaculture* 76, 209–214.
54. Lucas, M. D., Drew, R. E., Wheeler, P. A., Verrell, P. A. & Thorgaard, G. H. (2004) Behavioral differences among rainbow trout clonal lines. *Behav Genet* 34, 355–365.
55. Berejikian, B. A. (1995) The effects of hatchery and wild ancestry and experience on the relative ability of steelhead trout fry (*oncorhynchus mykiss*) to avoid a benthic predator. *Can J Fish Aquat Sci* 52, 2476–2482.
56. Johnsson, J., Petersson, E., Jonsson, E., Bjornsson, B. & Jarvi, T. (1996) Domestication and growth hormone alter antipredator behaviour and growth patterns in juvenile brown trout, *Salmo trutta*. *Can J Fish Aquat Sci* 53, 1546–1554.
57. Swain, D. P. & Riddell, B. E. (1990) Variation in agonistic behavior between newly emerged juveniles from hatchery and wild populations of coho salmon. *Can J Fish Aquat Sci* 47, 566–571.
58. Fleming, I. A. & Einum, S. (1997) Experimental tests of genetic divergence of farmed from wild Atlantic salmon due to domestication. *ICES J Marine Sci* 54, 1051–1063.
59. Price, E. O. & King, J. A. (1968) Domestication and adaptation. In Hafez, E. S. E. (Ed.) *Adaptation of Domestic Animals*. Philadelphia, PA, Lea and Febiger.
60. Fevolden, S. E., Refstie, T. & Roed, K. H. (1991) Selection for high and low cortisol stress response in Atlantic salmon (*Salmo salar*) and rainbow-trout (*Oncorhynchus mykiss*). *Aquaculture* 95, 53–65.
61. Spurway, H. (1955) The causes of domestication: an attempt to integrate some ideas of Konrad Lorenz with evolution theory. *J Genet* 53, 325–362.
62. Robison, B. D. & Rowland, W. (2005) A potential model system for studying the genetics of domestication: behavioral variation among wild and domesticated strains of zebrafish (*Danio rerio*). *Can J Fish Aquat Sci* 62, 2046–2054.
63. Moretz, J. A., Martins, E. P. & Robison, B. D. (2007) Behavioral syndromes and the evolution of correlated behavior in zebrafish. *Behav Ecol* 18, 556–562.
64. Darvasi, A. (1998) Experimental strategies for the genetic dissection of complex traits in animal models. *Nat Genet* 18, 19–24.
65. Lander, E. S. & Kruglyak, L. (1995) Genetic dissection of complex traits: guidelines for interpreting and reporting linkage results. *Nat Genet* 11, 241–247.
66. Williams, R. W., Gu, J., Qi, S. & Lu, L. (2001) The genetic structure of recombinant inbred mice: high-resolution consensus maps for complex trait analysis. *Genome Biol* 2, 1–18.

67. Belknap, J. K., Mitchel, S. R. & Crabbe, J. C. (1996) Type I and type ii error rates for quantitative trait loci (QTL) mapping studies using inbred mouse strains: computer simulation and empirical results. *Behav Genet* 26, 149–160.
68. Darvasi, A. & Soller, M. (1995) Advanced intercross lines, an experimental population for fine genetic-mapping. *Genetics* 141, 1199–1207.
69. Peirce, J. L., Broman, K. W., Lu, L., Chesler, E. J., Zhou, G., Airey, D. C., Birmingham, A. E. & Williams, R. W. (2008) Genome reshuffling for advanced intercross permutation (graip): simulation and permutation for advanced intercross population analysis. *PLoS One* 3, e1977.
70. Haldane, J. B. S. (1919) The combination of linkage values, and the calculation of distance between the loci of linked factors. *J Genet* 8, 299–309.
71. Kosambi, D. D. (1944) The estimation of map distances from recombination values. *Ann Eugen* 12, 172–175.
72. Weeks, D. E. & Lange, K. (1987) Preliminary ranking procedures for multilocus ordering. *Genomics* 1, 236–242.
73. Buetow, K. H. & Chakravarti, A. (1987) Multipoint gene mapping using seriation. I. General methods. *Am J Hum Genet* 41, 180–188.
74. Thompson, E. A. (1987) Crossover counts and likelihood in multipoint linkage analysis. *Math Med Biol* 4, 93–108.
75. Doerge, R. W. (1996) Constructing genetic maps by rapid chain delineation. *J Quant Trait Loci* 2, Article 6.
76. Doerge, R. W. (2002) Mapping and analysis of quantitative trait loci in experimental populations. *Nat Rev Genet* 3, 43–52.
77. Martinez, O. & Curnow, R. N. (1992) Estimating the locations and size of effects of quantitative trait loci using flanking markers. *Theor Appl Genet* 85, 480–488.
78. Haley, C. S. & Knott, S. A. (1992) A simple regression method for mapping quantitative trait loci in line crosses using flanking markers. *Heredity* 69, 315–324.
79. Haley, C. S., Knott, S. A. & Elsen, J. M. (1994) Mapping quantitative trait loci in crosses between outbred lines using least squares. *Genetics* 136, 1195–1207.
80. Piepho, H. P. & Gauch, H. G. (2001) Marker pair selection for mapping quantitative trait loci. *Genetics* 157, 433–444.
81. Kao, C. -H., Zeng, Z. -B. & Teasdale, R. D. (1999) Multiple interval mapping for quantitative trait loci. *Genetics* 152, 1203–1216.
82. Churchill, G. A. & Doerge, R. W. (1994) Empirical threshold values for quantitative trait mapping. *Genetics* 138, 964–971.
83. Doerge, R. W. & Churchill, G. A. (1996) Permutation tests for multiple loci affecting a quantitative character. *Genetics* 142, 285–294.
84. Benjamini, Y. & Yekutieli, D. (2005) Quantitative trait loci analysis using the false discovery rate. *Genetics* 171, 783–790.
85. Beavis, W. D. (1998) QTL analyses: power, precision and accuracy. *Proceedings of the 49th Annual Corn and Sorghum Industry Research Conference*. Washington, DC, ASTA.
86. Belknap, J. K. & Atkins, A. L. (2001) The replicabilitity of QTLs for murine alcohol preference drinking behavior across eight independent studies. *Mammal Genome* 12, 893–899.
87. Turri, M. G., Henderson, N. D., Defries, J. C. & Flint, J. (2001) Quantitative trait locus mapping in laboratory mice derived from a replicated selection experiment for open-field activity. *Genetics* 158, 1217–1226.
88. Hitzemann, R., Cipp, L., Demarest, K., Mahjubi, E. & McCaughran, J. J. (1998) Genetics of ethanol-induced locomotor activation: detection of QTLs in a C57BL/6 J × DBA/2 J F2 intercross. *Mammal Genome* 9, 956–962.
89. Risch, N. & Botstein, D. (1996) A manic depressive history. *Nat Genet* 12, 351–353.
90. Crabbe, J. C., Philips, T. J., Buck, K. J., Cunningham, C. L. & Belknap, J. K. (1999a) Identifying genes for alcohol and drug sensitivity: recent progress and future directions. *Trends Neurosci* 22, 173–179.
91. Flint, J., Corley, J. C., Defries, J. C., Fulker, D. W. & Gray, J. A. (1995) A simple genetic basis for a complex psychological trait in laboratory mice. *Science* 269, 1432–1435.
92. Moisan, M. P., Courvoisier, H., Bihoreau, M. T., Bgaugier, D., Hendley, E. D., Lathrop, M., James, M. R. & Mormede, P. (1996) A major quantitative trait locus influences hyperactivity in the Wkha rat. *Nat Genet* 14, 471–473.
93. Otto, S. P. & Jones, C. D. (2000) Detecting the undetected: estimating the total number of loci underlying a quantitative trait. *Genetics* 156, 2093–2107.
94. Mousseau, T. A. & Roff, D. A. (1987) Natural-selection and the heritability of fitness components. *Heredity* 59, 181–197.
95. Hill, W. G., Goddard, M. E. & Visscher, P. M. (2008) Data and theory point mainly to additive genetic variance for complex traits. *PLoS Genet* 4(2), e1000008.

96. Cockerham, C. C. & Zeng, Z. -B. (1996) Design III with marker loci. *Genetics* 143, 1437–1456.
97. Damerval, C., Maurice, A., Josse, J. M. & de Vienne, D. (1994) Quantitative trait loci underlying gene product variation: a novel perspective by analyzing regulation of genome expression. *Genetics* 137, 289–301.
98. Doebley, J., Stec, A. & Gustus, C. (1995) *Teosinte branched1* and the origin of maze: evidence for epistasis and the evolution of dominance. *Genetics* 141, 333–346.
99. Carlborg, O. & Haley, C. S. (2004) Epistasis: too often neglected in complex trait studies? *Nat Rev Genet* 5, 618–625.
100. Carlborg, O., Jacobsson, L., Ahgren, P., Siegel, P. & Andersson, L. (2006) Epistasis and the release of genetic variation in response to selection. *Nat Genet* 38, 418–420.
101. Carlborg, O., Kerje, S., Schutz, K., Jacobsson, L., Jensen, P. & Andersson, L. (2003) A global search reveals epistatic interaction between QTL for early growth in the chicken. *Genome Res* 13, 413–421.
102. Routman, E. J. & Cheverud, J. M. (1997) Gene effects on a quantitative trait: two-locu epistatic effects measured at microsatellite markers and at estimated QTL. *Evolution* 51, 1654–1662.
103. Peripato, A. C., de Brito, R. A., Matioloi, S. R., Pletscher, L. S., Vaughn, T. T. & Cheverud, J. M. (2004) Epistasis affecting litter size in mice. *J Evol Biol* 17, 593–602.
104. Shook, D. R. & Johnson, T. E. (1999) Quantitative trait loci affecting survival and fertility-related traits in *Caenorhabditis elegans* show genotype-environment interactions, pleiotropy and epistasis. *Genetics* 153, 1233–1243.
105. Malmberg, R. L. & Mauricio, R. (2005) QTL-based evidence for the role of epistasis in evolution. *Genet Res Camb* 86, 89–95.
106. Hood, H. M., Belknap, J. K., Crabbe, J. C. & Buck, K. J. (2001) Genomewide search for epistasis in a complex trait: pentobarbital withdrawal convulsions in mice. *Behav Genet* 31, 93–100.
107. Anholt, R. R. H., Dilda, C. L., Chang, S., Fanara, J. -J., Kulkarni, N. H., Ganguly, I., Rollman, S. M., Kamdar, K. P. & Mackay, T. F. C. (2003) The genetic architecture of odor-guided behaviour in *Drosophila*: epistasis and the transcriptome. *Nat Genet* 35, 180–184.
108. Anholt, R. R. H., Lyman, R. F. & Mackay, T. F. C. (1996) Effects of single P-element insertions on olfactory behavior in *Drosophila melanogaster*. *Genetics* 143, 293–301.
109. Fedorowicz, G. M., Fry, J. D., Anholt, R. R. H. & Mackay, T. F. C. (1998) Epistatic interactions between *smell-impaired* loci in *Drosophila melanogaster*. *Genetics* 148, 1885–1891.
110. Ruppell, O., Pankiw, T. & Page, R. E., Jr. (2004) Pleiotropy, epistasis and new QTL: the genetic architecture of honey bee foraging behavior. *J Heredity* 96, 481–491.
111. Wright, D., Butlin, R. K. & Carlborg, Ö. (2006a) Epistatic regulation of behavioural and morphological traits in the zebrafish (*Danio rerio*). *Behav Genet* 36, 914–922.
112. Flint, J., Defries, J. C. & Henderson, N. D. (2004) Little epistasis for anxiety-related measures in the DeFries strains of laboratory mice. *Mammal Genome* 15, 77–82.
113. Sen, S. & Churchill, G. A. (2001) A statistical framework for quantitative trait mapping. *Genetics* 159, 371–387.
114. Melo, J. A., Sjendure, J., Pociask, K. & Silver, L. M. (1996) Identification of sex-specific quantitative trait loci controlling alcohol preference in C57BL/6 mice. *Nat Genet* 13, 147–153.
115. Gershenfeld, H. K. & Paul, S. M. (1997) Mapping quantitative trait loci for fear-like behaviors in mice. *Genomics* 46, 1–8.
116. Crabbe, J. C., Wahlsten, D. & Dudek, B. C. (1999b) Genetics of mouse behavior: interactions with laboratory environment. *Science* 284, 1670–1672.
117. Pellow, S., Chopin, P., File, S. & Briley, M. (1985) Validation of open:closed arms entries in an elevated plus maze as a measure of anxiety in the rat. *J Neurosci Methods* 14, 149–167.
118. File, S. E., Mabbutt, P. S. & Hitchcott, P. K. (1990) Characterisations of the phenomenon of "one-trial tolerance" to the anxiolytic effect of chlordiazepoxide in the elevated plus maze. *Psychopharmacology* 102, 98–101.
119. Dawson, G. R., Crawford, S. P., Collinson, N. & Iversen, S. D. (1995) Evidence that the anxiolytic-like effects of chlordiazepoxide on the elevated-plus maze are confounded by increases in locomotor-activity. *Psychopharmacology* 118, 316–323.
120. Mather, K. & Jinks, J. L. (1982) *Biometrical Genetics*. Cambridge, Chapman and Hall.
121. Mott, R. & Flint, J. (2002) Simultaneous detection and fine mapping of quantitative trait loci in mice using heterogeneous stocks. *Genetics* 160, 1609–1618.
122. Talbot, C. J., Radcliffe, R. A., Fullerton, J., Hitzemann, R., Wehner, J. M. & Flint, J. (2003) Fine scale mapping of a genetic locus

for conditioned fear. *Mammal Genome* 14, 223–230.
123. Flint, J. & Mott, R. (2001) Finding the molecular basis of quantitative traits: successes and pitfalls. *Nat Rev Genetics* 2(6), 437–445.
124. Breese, E. L. & Mather, K. (1957) The organization of polygenic activity within a chromosome in *Drosophila*: 1. Hair characters. *Heredity* 11, 373–395.
125. Shrimpton, A. E. & Robertson, A. (1988) The isolation of factors controlling bristle score in *Drosophila melanogaster*: I. allocation of third chromosome sternopleural bristle effects to chromosome sections. *Genetics* 118, 437–443.
126. Darvasi, A. (1997) Interval-specific congenic strains (iscs): an experimental design for mapping a QTL into a 1-centimorgan interval. *Mammal Genome* 8, 163–167.
127. Valdar, W., Solberg, L. C., Gauguier, D., Burnett, S., Klenerman, P., Cookson, W. O., Taylor, M. S., Rawlins, J. N. P., Mott, R. & Flint, J. (2006) Genome-wide genetic association of complex traits in heterogeneous stock mice. *Nat Genet* 38, 879–887.
128. Andersson, L. & Georges, M. (2004) Domestic animal genomics: deciphering the genetics of complex traits. *Nat Rev Genet* 5, 202–212.
129. Carlson, C. S., Eberle, M. A., Kruglyak, L. & Nickerson, D. A. (2004) Mapping complex disease loci in whole-genome association studies. *Nature* 429, 446–452.
130. Smith, J. M. & Haigh, J. (1974) The hitchhiking effect of a favourable gene. *Genet Res* 23, 23–35.
131. Wright, D., Boije, H., Meadows, J. R. S., Bed'hom, B., Gourichon, D., Vieaud, A., Tixier-Boichard, M., Rubin, C. -J., Imsland, F., Hallböök, F. & Andersson, L. (2009) Transient ectopic expression of SOX5 during embryonic development causes the Pea-comb phenotype in chickens. *PloS Genet* 5(6), e1000512.
132. Meuwissen, T. H., Karlsen, A., Lien, S., Olsaker, I. & Goddard, M. E. (2002) Fine mapping of a quantitative trait locus for twinning rate using combined linkage and linkage disequilibrium mapping. *Genetics* 161, 373–379.
133. Riquet, J., Coppieters, W., Cambisano, N., Arranz, J. -J., Berzi, P., Davis, S. K., Grisart, B., Farnir, F. D. R., Karim, L., Mni, M., Simon, P., Taylor, J. F., Vanmanshoven, P., Wagenaar, D., Womack, J. E. & Georges, M. (1999) Fine-mapping of quantitative trait loci by identity by descent in outbred populations: application to milk production in dairy cattle. *Proc Natl Acad Sci USA* 96, 9252–9257.
134. Goldstein, O., Zangerl, B., Pearce-Kelling, S., Sidjanin, D. J., Kijas, J. W., Felix, J., Acland, G. M. & Aguirre, G. D. (2006) Linkage disequilibrium mapping in domestic dog breeds narrows the progressive rod-cone degeneration interval and identifies ancestral disease-transmitting chromosome. *Genomics* 88, 541–550.
135. Makinen, H. S., Shikano, T., Cano, J. M. & Merila, J. (2008) Hitchhiking mapping reveals a candidate genomic region for natural selection in three-spined stickleback chromosome VIII. *Genetics* 178, 453–465.
136. Burgess-Herbert, S. L., Cox, A., Tsaih, S.-W. & Paigen, B. (2008) Practical applications of the bioinformatics toolbox for narrowing quantitative trait loci. *Genetics* 180, 2227–2235.
137. Li, R., Lyons, M. A., Wittenburg, H., Paigen, B. & Churchill, G. A. (2005) Combining data from multiple inbred line crosses improves the power and resolution of quantitative trait loci mapping. *Genetics* 169, 1699–1709.
138. Dipetrillo, K., Wang, X., Stylianou, I. M. & Paigen, B. (2005) Bioinformatics toolbox for narrowing rodent quantitative trait loci. *Trends Genet* 21, 683–691.
139. Wang, X. & Paigen, B. (2005) Genetics of variation in HDL cholesterol in humans and mice. *Circ Res* 96, 27–42.
140. Flint, J. & Mackay, T. F. C. (2009) Genetic architecture of quantitative traits in mice, flies, and humans. *Genome Res* 19, 723–733.
141. Yalcin, B., Flint, J. & Mott, R. (2005) Using progenitor strain information to identify quantitative trait nucleotides in outbred mice. *Genetics* 171, 673–681.
142. Legare, M. E., Bartlett, F. S. & Frankel, W. N. (2000) A major effect QTL determined by multiple genes in epileptic el mice. *Genome Res* 10, 42–48.
143. Lyman, R. F., Lai, C. & Mackay, T. F. (1999) Linkage disequilibrium mapping of molecular polymorphisms at the scabrous locus associated with naturally occurring variation in bristle number in *Drosophila melanogaster*. *Genet Res* 74, 303–311.
144. Symula, D. J., Frazer, K. A., Ueda, Y., Denefle, P., Stevens, M. E., Wang, Z. E., Locksley, R. & Rubin, E. M. (1999) Functional screening of an asthma QTL in YAC transgenic mice. *Nat Genet* 23, 241–244.
145. Sauer, B. & Henderson, N. (1988) Site-specific DNA recombination in mammalian cells by the Cre recombinase of bacteriophage P1. *PNAS* 85, 5166–5170.

146. Langenau, D. M., Feng, H., Berghmans, S., Kanki, J. P., Kutok, J. L. & Look, A. T. (2005) Cre/lox-regulated transgenic zebrafish model with conditional myc-induced T cell acute lymphoblastic leukemia. *Proc Natl Acad Sci USA* 102, 6068–6073.
147. Aitman, T. J., Glazier, A. M., Wallace, C. A., Cooper, L. D., Norsworthy, P. J., Wahid, F. N., Al-Majali, K. M., Trembling, P. M., Mann, C. J., Shoulders, C. C., Graf, D., St Lezin, E. M., Kurtz, T. W., Kren, V., Pravenec, M., Ibrahimi, A., Abumrad, N. A., Stanton, L. W. & Scott, J. (1999) Identification of Cd36 (Fat) as an insulin-resistance gene causing defective fatty acid and glucose metabolism in hypertensive rats. *Nat Genet* 21, 76–83.
148. Gibson, G. & Weir, B. (2005) The quantitative genetics of transcription. *Trends Genet* 21, 616–623.
149. Breitling, R., Li, Y., Tesson, B. M., Fu, J., Wu, C., Wiltshire, T., Gerrits, A., Bystrykh, L. V., de Haan, G., Su, A. I. & Jansen, R. C. (2008) Genetical genomics: spotlight on QTL hotspots. *PLoS Genet* 4, e1000232.
150. Nadeau, J. H. & Frankel, W. N. (2000) The roads from phenotypic variation to gene discovery: mutagenesis versus QTLs. *Nat Genet* 25, 381–384.
151. Sawyer, L. A., Hennessy, J. M., Pexioto, A. A. & Kyriacou, P. (1997) Natural variation in a *Drosophila* clock gene and temperature compensation. *Science* 278, 2117–2120.
152. Tully, T. (1996) Discovery of genes involved with learning and memory: an experimental synthesis of Hirschian and Benzerian perspectives. *Proc Natl Acad Sci USA* 93, 13460–13467.
153. Coates, J. C. & de Bono, M. (2002) Antagonistic pathways in neurons exposed to body fluid regulate social feeding in *C. elegans*. *Nature* 419, 925–929.
154. de Bono, M., Tobin, D. M., Davis, M. W., Avery, L. & Bargmann, C. I. (2002) Social feeding in *C. elegans* is induced by neurons that detect aversive stimuli. *Nature* 419, 899–903.
155. de Belle, J. S. & Sokolowksi, M. B. (1989) Genetic localization of *foraging* (*for*): a major gene for larval behavior in *Drosophila melanogaster*. *Genetics* 123, 157–163.
156. Sokolowski, M. B. (1998) Genes for normal behavioural variation: recent clues from flies and worms. *Neuron* 21, 463–466.
157. Driever, W. & Al, E. (1996) A genetic screen for mutations affecting embryogenesis in the zebrafish. *Development* 123, 37–46.
158. Haffter, P. (1996) The identification of genes with unique and essential function in the development of the zebrafish, *Danio rerio*. *Development* 123, 1–36.
159. Walker, C. & Streisinger, G. (1983) Induction of mutations by gamma-ras in pregonial germ cells of zebrafish embryos. *Genetics* 103, 125–136.
160. Amsterdam, A. & Hopkins, N. (2006) Mutagenesis strategies in zebrafish for identifying genes involved in development and disease. *Trends Genet* 22, 473–478.
161. Darland, T. & Dowling, J. E. (2001) Behavioral screening for cocaine sensitivity in mutagenized zebrafish. *Proc Natl Acad Sci USA* 98, 11691–11696.
162. Baier, H. (2000) Zebrafish on the move: towards a behavior-genetic analysis of vertebrate vision. *Curr Neurobiol* 10, 451–455.
163. Neuhass, S. C. (1999) Genetic disorders of vision revealed by a behavioural screen of 400 essential loci in zebrafish. *J Neurosci* 19, 8603–8615.
164. Granato, M. (1996) Genes controlling and mediating locomotion behaviour of the zebrafish embryo and larvae. *Development* 123, 399–413.
165. Lorent, K. (2001) The zebrafish space cadet gene controls axonal pathfinding of neurons that modulate fast turning movements. *Development* 128, 2131–2142.
166. Baraban, S. C., Dinday, M. T., Castro, P. A., Chege, S., Guyenet, S. & Taylor, M. R. (2007) A large-scale mutagenesis screen to identify seizure-resistant zebrafish. *Epilepsia* 48, 1151–1157.
167. Henderson, N. D., Turri, M. G., Defries, J. C. & Flint, J. (2004) QTL analysis of multiple behavioral measures of anxiety in mice. *Behav Genet* 34, 267–293.
168. Wright, D., Nakamichi, R., Krause, J. & Butlin, R. K. (2006c) QTL analysis of behavioural and morphological differentiation between wild and laboratory zebrafish (*Danio rerio*). *Behav Genet* 36, 271–284.
169. Croft, D. P., Krause, J. & James, R. (2004) Social networks in the guppy (*Poecilia reticulata*). *Proc R Soc Lond Biol Lett* 271, 516–519.
170. Nakamichi, R., Ukai, Y. & Kishino, H. (2001) Detection of closely linked multiplre quantitative trait loci using a genetic algorithm. *Genetics* 158, 463–475.
171. Carlborg, O. & Andersson, L. (2002) Use of randomization testing to detect multiple epistatic QTLs. *Genet Res* 79, 175–184.
172. Carlborg, O., Andersson, L. & Kinghorn, B. (2000) The use of a genetic algorithm for simultaneous mapping of multiple interacting quantitative trait loci. *Genetics* 155, 2003–2010.

Chapter 6

Genetics of Ethanol-Related Behaviors

Cynthia A. Dlugos

Abstract

Alcoholism is a disorder that affects human beings during every stage of the lifespan. Many animal models have been developed to study alcoholism, including those used to assess alcohol preference, the effects of alcohol withdrawal, and the development of tolerance. Knowledge gained from multiple studies on twins has supported a strong genetic basis for predisposition to alcohol. Our laboratory has chosen to investigate the use of the zebrafish, a vertebrate with an accessible and 75% sequenced genome as a possible model for ethanol research. We have used a simple, noninvasive evaluation of swimming behavior in which we measured the distance between each fish and its nearest neighbor to gage the response of the central nervous system to pharmacologically relevant doses of acute and chronic ethanol. In the acute studies, we have shown that WT (wild type) zebrafish show a dose dependent increase in nearest neighbor distance. Conversely, another strain, the LFS (long-fin striped) zebrafish demonstrated a biphasic response to acute alcohol exposure in that change from baseline was larger at the 0.5 than at the 1.0% (v/v) ethanol concentration. A third strain, the BLF (blue longfin) zebrafish, showed no apparent response to acute alcohol exposure. Subsequent studies showed that behavioral response to ethanol in BLF zebrafish was age dependent, as nearest neighbor distance was increased in juvenile but not in adult fish. Investigations using chronic ethanol exposure in zebrafish also support differential strain sensitivity to ethanol and the capacity to develop tolerance. Ethanol-induced alterations in gender were also investigated. Gender does not appear to be a factor in acute sensitivity to ethanol. Chronic ethanol treatment demonstrated that female WT zebrafish are preferentially affected compared to males of the WT strain. The results of chronic studies suggest that the zebrafish may be a useful model for dissecting the rather complex differential effects of ethanol on gender. Taken together, these studies demonstrate with a simple noninvasive behavioral test that zebrafish of three strains demonstrate differential sensitivity to ethanol and suggest that zebrafish are useful models in sorting out the genetic factors concerning the mechanisms of ethanol's actions.

Key words: Alcoholism, alcohol withdrawal, tolerance, ethanol, genetic differences, genetic long fin striped, blue longfin, nearest neighbor distance, strain sensitivity, chronic treatment.

A.V. Kalueff, J.M. Cachat (eds.), *Zebrafish Models in Neurobehavioral Research*, Neuromethods 52,
DOI 10.1007/978-1-60761-922-2_6, © Springer Science+Business Media, LLC 2011

1. Introduction

The National Longitudinal Epidemiologic Survey indicates that alcohol abuse in the US population is on the rise, having increased from 3.03 to 4.65% in the decade between 1991 and 2001 (1). Alcohol abuse exerts its effects on US citizens during every phase of the life cycle. For example, 1 in 100 children in the United States are born with FASD (fetal alcohol spectrum disorders) as a consequence of maternal drinking during pregnancy (2). In addition, adolescent drinkers suffer deficits in learning and memory due to ethanol-related disruption of developing neural pathways while young adult drinkers (18–29 years of age) suffer from more accident-related deaths than their nondrinking peers (3). Heavy drinking in midlife (30–59 years of age) is associated with alcoholic liver disease, pancreatitis, cancer, heart or circulatory problems, and brain disorders. Alcohol abuse in seniors, moreover, may exacerbate other health conditions, complicate drug side effects, and result in increased falls and accidents in a population already susceptible to these conditions (3).

Genetics plays a major role in susceptibility to alcohol, as demonstrated in numerous studies of identical twins. From these studies, it has been estimated that 50–60% of the risk of developing alcoholism is genetic (4–7). Gender differences in alcohol sensitivity are also well recognized and many of these are pharmacokinetic. For example, it is well known that more body fat and greater bioavailability to ethanol in females compared to males predisposes women to ethanol's effects. These effects occur at lower doses than in males of equal size (8, 9). It is likely, however, that in addition to pharmacokinetics, there are other unidentified factors that increase female sensitivity to ethanol and predispose females to ethanol-related diseases such as gastric ulcers, fatty liver, and hypertension. These conditions occur following shorter durations of exposure to lesser concentrations of ethanol than in males (10, 11). Studies in animal models have reiterated that female subjects may have increased sensitivity to alcohol under a variety of experimental paradigms (12–14). The importance of gender in ethanol withdrawal has also been shown with microarrays (15). Age can also alter genetic expression and sensitivity to alcohol, resulting in different predispositions to ethanol-induced damage throughout the lifespan. For example, the effects of chronic ethanol treatment on motor coordination and memory are more lasting in adolescent compared to adult rats (16, 17). The elderly may be predisposed to ethanol effects on motor functions, nutrition, and cognitive status as these factors may already be compromised by their advanced age (18).

Rodents comprise the majority of animals used for ethanol research. Their proliferative behavior and understandable genetics have been extensively employed to develop models with specific ethanol-related characteristics. For example, rats and mice have been developed that express specific characteristics pertaining to alcohol sensitivity, alcohol preference, the development of tolerance, withdrawal, organ damage, and ablation or overexpression of specific genes (19, 20). One disadvantage of the rodent model is that generation of transgenic and mutant models in rodents is costly and time consuming compared to the efficiency of genetic manipulations in models with shorter developmental periods and easier access for genetic manipulations (21). Invertebrate models such as *Drosophila melanogaster* (fruit fly) and *Caenorhabditis elegans* (nematode) are useful models for genetic manipulations and have provided important insights into molecules with potential roles in alcohol intoxication (22, 23), acute ethanol sensitivity (24), and the molecular targets for ethanol's actions (25). It has been recognized, however, that despite the advantages of invertebrate models, the dissimilarities between the vertebrate and invertebrate central nervous systems is a major drawback in extrapolating the results of invertebrate studies to human beings (19). Additional drawbacks include the crucial metabolic differences between *C. elegans* and vertebrates and the time consuming strain construction that is required in Drosophila prior to each investigation (21). The fact that the signaling pathways that occur during development in Drosophila are conserved in vertebrates is an apparent advantage until it is realized that these pathways code for body parts that differ both in structure and in function, thus creating difficulties in relating a particular pathway to the structure produced. For example, some common signaling pathways result in somite formation in vertebrates and segmentation in Drosophila (26).

Danio rerio (the zebrafish), the little teleost described as the canonical vertebrate, is bridging the gap in many disciplines between the efficiency, low cost, and mutagenic capacity of invertebrate models and the obvious similarities between rodent models and humans (27). Zebrafish are low-cost models with high fecundity, producing 70–300 eggs per clutch (28). Zebrafish share marked organ similarities with other vertebrates. Genetic manipulations, including the production of mutants and transgenics, are much easier in zebrafish than in rodent models. In zebrafish, fertilization occurs externally and the development from fertilized egg to free-swimming larvae takes only 72 h. In addition, 75% of the zebrafish genome is sequenced by the Sanger Institute (http://www.sanger.ac.uk/Projects/D_rerio/). Zebrafish are excellent models for drug studies as both embryos and adults are extremely permeable to small molecules, providing potential for identification and validation of gene targets for drugs as well as for drug screening and toxicological studies (29). Zebrafish have been used to evaluate a variety of drugs such as the antipsychotic clozapine (30, 31), nicotine (32), roscovitine (33), cocaine (34), atorvastatin (Lipitor) (35), and alcohol (36–42). Many of the alcohol studies focus on larval development (36–42), but the remaining studies focus on ethanol-related adult behaviors. There are also disadvantages to the use of the zebrafish. The study of alcohol preference or non-preference is irrelevant in zebrafish and is better accomplished in mammals with a cerebral cortex. The duplicity of about 30% of zebrafish genes is a disadvantage that may confound interpretation of mutant characteristics if a single member of a duplicated pair is altered. Finally, alcohol is a drug that affects all organ systems, some of which, such as the cerebral cortex, are present in humans but not in the zebrafish.

The first adult teleost to be used for alcohol studies was not the zebrafish but the goldfish. In these early studies, the benefits of teleost use in studying alcohol were enumerated. For example, at room temperature (20°C), it was demonstrated that the metabolism of the carp more closely corresponds to human metabolism than does the much higher (10x) metabolism of the rat (43). In addition, in the goldfish, constant absorption of ethanol through gills and skin allows ethanol levels in the fish to come into equilibrium with the tank water (44). A similar time course has since been confirmed in the zebrafish, showing that brain alcohol levels rise rapidly for the first 2 h and plateau at 6 h. Constant ethanol levels are maintained for at least 24 h (45). The ability to survive long periods in ethanol has also been determined in goldfish for the study of the biochemical, physiological, and pathological basis of alcohol effects (43). Ethanol-related deficits in learning (64) and enhancement of aversive learning (46) have been demonstrated in the goldfish.

Additional teleost species have been used to show ethanol-induced alterations in aggressive behaviors. For example, ethanol treatment increased aggression in Siamese fighting fish (*Betta splendans*) (47). Convict cichlids were used to show increases in aggressive behavior after acute exposure to low levels of ethanol (0.15%), although the converse was true at higher concentrations (0.30%) of ethanol (48). This ethanol-induced response to aggression was determined in adult zebrafish at 0.25 and 1.0% (v/v) ethanol in a study that also showed that high acute doses of ethanol (1.0%) resulted in decreased locomotor activity and preference of the zebrafish for the bottom of the tank (41). Subsequent studies have shown ethanol-induced alteration in response to a predator with fish of different strains (49, 50).

Efforts in our laboratory have focused on assessing ethanol-related alterations in swimming behavior in several zebrafish strains as an index of the response of the central nervous system to ethanol (45). Zebrafish swim together in shoals, where the fish benefit from enhanced access to mates, food, and reduced risks from predators (51). We have not analyzed shoaling per se, but rather used the shoaling behavior to assess ethanol-induced increases in distance between each zebrafish and its nearest neighbor within the shoal. A surveillance camera was used to observe the normal swimming behavior of the fish so that habituation or human manipulation did not confound the analysis. Other studies have also used this behavioral test to correlate with the effects of ethanol-induced changes in proteins such as VDAC1 (voltage-dependent anion channel 1) and Hsp70 (heat shock protein 70), as assessed with proteomics (52). We have also determined ethanol-related alterations on the startle reaction, a reaction often used to assess alcohol-induced effects in human and animal studies (53, 54). The overall goal of the studies presented here was to determine whether the zebrafish, with its accessible genome, was an appropriate model to study alcoholism, a disease with a very strong genetic predisposition. Results of our analyses are presented below.

2. Methods/Results/Discussion

2.1. Methods

Approximately 700 zebrafish were used in these studies. In our initial study (45), all fish were obtained from a local zebrafish supplier (The Fish Place; Tonawanda, NY) except for the blue longfin fish (BLF) that were obtained from Markeim Tropical Fish and Pet Store (Amherst, NY). Subsequent studies used offspring from the original fish raised in our laboratory. All zebrafish were housed in an Aquatic Ecosystems Benchtop Systems Habitat (Aquatic Habitats^{TM}, Apopoka, FL) in 3 l tanks (25 fish/tank). Ammonium levels and pH of the water were monitored bi-weekly, temperature was maintained at 24°C, and lighting was regulated to a 14 h light/10 h dark cycle. Fish were fed Tetra Flake fish food once daily and their diet supplemented 3 times/week with live brine shrimp nauplii. Three strains of zebrafish were used in these experiments. They included WT, LFS , and BLF zebrafish. The WT fish had steel blue body stripes and were about 3 cm in length. The LFS were slightly longer than the WT (3.25 cm in length) with markings identical to the WT except for the presence of very long pectoral and dorsal fins. BLF fish were the longest (3.8 cm) with a single stripe composed of many small spots. The spots are evidence of derivation of the BLF zebrafish from the leopard danio strain of zebrafish (http://animalworld.com/encyclo/fresh/cyprinids/LongfinBlueDanio.php). Females of the WT and LFS strains can be readily distinguished by their slightly distended gray to white abdomens compared to the slimmer, yellow abdomens of the males (55).

For both swimming and startle studies, the behavioral apparatus used was a 20-cm diameter bowl, containing 700 ml of 24°C tank water or ethanol solution. A 324 cm^2 grid, subdivided into 36 blocks (9 cm^2/block) was aligned under the bowl. Fish were acclimatized to the bowl for 10 min prior to testing. Single fish were added to the bowl for the startle study and 8 fish were added to the bowl for the swimming behavior study. Observation of swimming behavior was noninvasive with the fish isolated in a quiet room with no humans or distractions. In this test, the swimming behavior of the eight fish in each bowl was recorded every 30 s over a 30-min period with an Intel digital camera and Intel Create and Share software (Intel, Hudson, MA). The startle reaction of the fish was captured by video in the first seconds after a glass bead was dropped about 12 cm away from the fish's head. Fish always faced the bead when tested. The response measured the number of squares the fish traversed on the grid in the rapid swimming phase following the fast turn away component of the fast start response to startle (56). All animal care and experimental procedures were approved by the Institutional Animal Care and Use Committee at the State University of New York at Buffalo.

Captured frames were analyzed for swimming behavior by measuring the distance between each fish and its nearest neighbor. The distance between the most cranial points on each nearest neighbor pair was determined with the Image Tool Program (http://ddsdx.uthscsa.edu/dig/itdesc.html). Means (± S.E.M.) were determined for each frame and treatment group. Videos of the startle reaction were analyzed by quantitating the mean (± S.E.M.) number of squares that each fish traversed in the fast-swimming phase of the startle response. ANOVA (analysis of variance), repeated measures ANOVA, Student's *t*-test, and Tukey post-hoc tests using the SPSS program were used to analyze the data. An alpha level of 0.05 was accepted as significant.

2.2. Acute Studies

The purpose of this study (45) was to determine whether swimming behavior in zebrafish was sensitive to acute ethanol treatment and whether there is differential strain sensitivity in the ethanol response. The WT, LFS, and BLF strains and alcohol concentrations of 0.25, 0.5, and 1.0% (v/v) ethanol were used. All fish were ethanol naive until they were exposed to ethanol 2 h prior to testing. The 2 h period was selected on the basis of a time course which demonstrated that after 2 h in ethanol solution, brain alcohol levels were near maximum (45). Statistical analyses of the data showed that nearest neighbor distance in LFS and WT fish was increased at 0.50, and 1.0% (v/v) ethanol ($p < 0.001$) relative to baseline. The LFS fish also showed significant ethanol-induced increases in nearest neighbor distance at the 0.25% (v/v) ethanol concentration. In contrast, the BLF fish proved insensitive to acute ethanol treatment at either 0.50 or 1.0% (v/v) ethanol concentrations, and thus, were not tested with lower ethanol concentrations. A 2 (concentration) × 3 (strain) ANOVA, holding baseline as a covariant, revealed effects of strain [$F(2, 191) = 10.598$, $p < 0.001$] and a significant strain × concentration interaction [$F(2,191) = 7.882$, $p = 0.001$] (45). For presentation here, data were converted to percent change from baseline (**Fig. 6.1**) and tested again to confirm the effect on strain. In the WT strain, percent change from baseline increased with increased dose. In the LFS fish, there was a biphasic response to dose in that there was an increased change from baseline at 0.25 and 0.50% (v/v) ethanol, but a decrease at 1.0% (v/v) ethanol. LFS performance at 1.0% (v/v) ethanol may be atypical as baseline values in this group were significantly larger than baseline values at the other two ethanol concentrations ($p < 0.001$).

The lack of ethanol-induced changes in nearest neighbor distance in the BLF fish was not due to pharmacokinetic differences between the BLF and other strains, as brain alcohol levels measured at 7 time points over a 24-h period did not differ when the three strains were compared (45). Genetic differences between the three strains may be responsible for their differential sensitivity

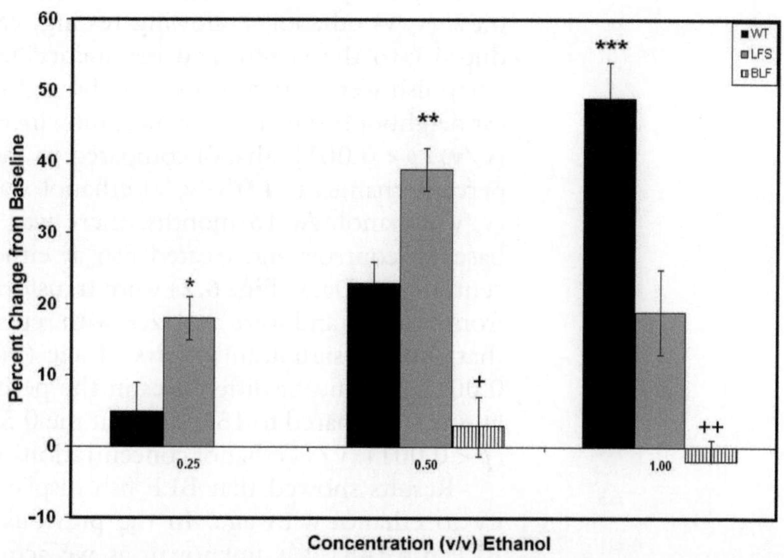

Fig. 6.1. Mean (±S.E.M.) percent change from baseline in WT, LFS, and BLF fish acutely exposed to ethanol. Data were tested with ANOVA and showed significant effects of strain [$F(2, 191) = 10.598$, $p < 0.001$]. Pair-wise comparisons showed that at 0.25% ethanol percent change from baseline was larger in LFS than in WT fish (*, $p = 0.012$). Post-hoc analysis showed that a similar effect was noted at 0.50% (v/v) between LFS and WT (**, $p = 0.006$). At 1.00%, the WT showed a larger percent change than the LFS (***, $p = 0.001$). Percent change in the BLF fish was significantly less at 0.5% (v/v) (+, $p < 0.001$) and 1.0% ethanol (++, $p < 0.005$) than in WT and LFS strains.

to ethanol. The fact that the WT and LFS fish responded similarly to acute alcohol, albeit with slightly different dose response curves, taken with the similar phenotype between the two strains, is suggestive of strong similarities of the genomes of WT and the LFS zebrafish. The only gross phenotypic difference between them, in fact, is in the length of their fins. The gene for long fins (*lof*) was developed within the tropical fish trade (57) to add beauty to the stalwart and hardy zebrafish. The *lof* gene has since been mapped in inbred strains and shown to be a single point mutation (58). In contrast, the BLF strain has several phenotypic dissimilarities in coloration and patterning compared to the WT and LFS strains that may be attributed to its origin from the leopard danio. These dissimilarities are suggestive of more genetic dissimilarities between the BLF and WT strain than between the WT and LFS strain.

To further our understanding of ethanol effects on the BLF fish, we then raised our own from adults in our lab. In this study, 64 BLF zebrafish, approximately 3 months of age and approximately 2 cm in length (compared to 3.8 cm; 45) were used. Fish were divided into eight groups ($n = 8$/group). Groups 1 and 2 were treated acutely with 0.5% (v/v) ethanol for 2 h and assessed for nearest neighbor analysis. Groups 3 and 4 served as their baselines or controls. Groups 5–6 and 7–8 were similarly treated, except that groups 5 and 6 received 1.0% rather than 0.5% (v/v) ethanol. Following testing, each group was reintroduced into the system and maintained until 15 months of age when fish were retested. Results showed that at 3 months, nearest neighbor distance was significantly increased at 0.50 and 1.0% (v/v) ($p < 0.001$) ethanol compared to the baseline controls and percent change at 1.0% (v/v) ethanol was greater than at 0.5% (v/v) ethanol. At 15 months, there was no difference between baseline controls and treated fish at either of the ethanol concentrations. Data (**Fig. 6.2**) were transformed to percent change from baseline and were analyzed with repeated measures ANOVA that showed significant effects of age ($F(1, 64) = 58.984$, $p < 0.001$). Significant differences in the percent change from baseline at 3 compared to 15 months at the 0.5% ($p < 0.001$) and 1.0% ($p < 0.001$) (v/v) ethanol concentrations were also observed.

Results showed that BLF fish display alterations in sensitivity to ethanol with age. In the previous study, the age of our BLF fish (45) was unknown, as we acquired them from a pet supplier. Zebrafish do, however, become larger in adulthood. Reviewing the digitized frames used for swimming behavior and measuring the size of the fish with the Image Tool program, we did find a marked correspondence in size, and perhaps age, in 15-month-old BLF fish from the current study and those from the previous study (45), suggesting that the two groups were of the same age. This observation taken with the current find-

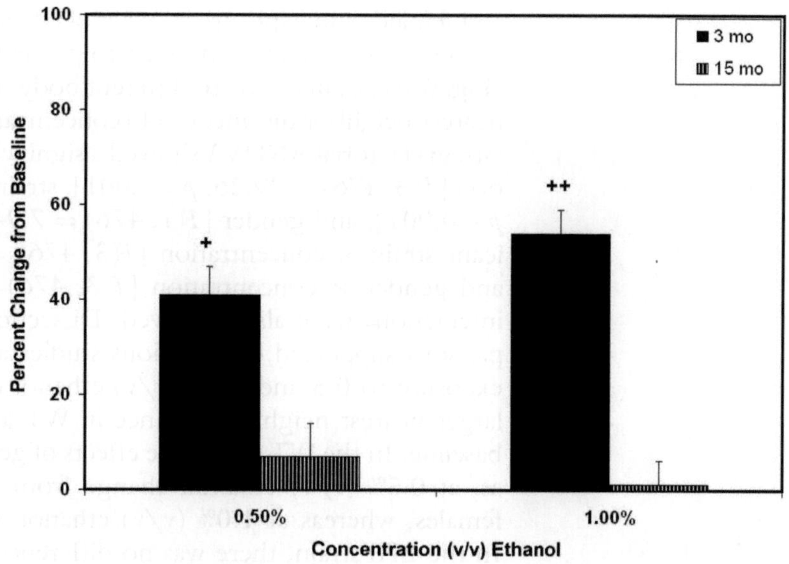

Fig. 6.2. Mean (±S.E.M.) percent change from baseline in two separate groups of BLF fish acutely exposed to ethanol at 3 months and a year later as adult fish, 15 months of age. At 3 months of age, ethanol-treated fish had significantly larger nearest neighbor distances compared to baseline at both 0.5 and 1.0% (v/v) ethanol ($p < 0.001$). There were no significant changes relative to baseline at 15 months. Repeated measures ANOVA showed a significant effect of age on percent change from baseline [$F(2, 191) = 10.598$, $p < 0.001$]. Pair-wise comparisons showed that at 3 months of age changes relative to baseline were significantly greater at 0.5% (+, $p < 0.001$) and 1.0% (++, $p < 0.001$) (v/v) ethanol than similar values at 15 months of age.

ing of decreased acute ethanol sensitivity in the 15-month-old compared to the 3-month-old zebrafish suggests that additional studies with BLF fish of different ages should be undertaken to confirm age-related alterations in acute ethanol sensitivity. Size should also be addressed although it may be a secondary factor that correlates strongly with increased age. Age-related alterations in gene expression are not unlikely as they have been confirmed in Drosophila (59) and acute ethanol exposure has also been shown to alter gene expression in the zebrafish brain (60). The rather short lifespan of the zebrafish (5 years) makes it a good candidate for studies of ethanol-related alterations in gene expression across the lifetime and the BLF zebrafish may prove an important model for these explorations.

In the next study (61), ethanol-induced effects on gender were addressed in 64 young adult (32/gender) WT and 64 young adult LFS zebrafish (32/gender). Fish were separated by gender and exposed to 0.125, 0.25, 0.5, or 1.0% ethanol. Baseline measures of nearest neighbor distance were obtained in the same animals that were subsequently treated with ethanol, the day prior to ethanol exposure. There were 16 test groups (8/ethanol concentration and 8/gender) of ethanol-naive males and females and exposure to ethanol occurred two hours prior to testing. Results were converted to percent change from baseline (**Fig. 6.3**) because, due to different body size, the LFS have larger nearest neighbor distances. A4 (concentration) × 2 (gender) × 2 (strain) factorial ANOVA showed a significant effect of concentration [$F(3, 476) = 99.26$, $p < 0.001$], strain [$F(1, 476) = 64.627$, $p < 0.001$], and gender [$F(1, 476) = 7.942$, $p = 0.005$]. Significant strain × concentration [$F(3, 476) = 17.644$, $p = 0.001$] and gender × concentration [$F(3, 476) = 7.811$, $p = 0.001$] interactions were also observed. Dissection with pair-wise comparisons supported our previous studies as it showed that acute exposure to 0.5 and 1.0% (v/v) ethanol resulted in consistently larger nearest neighbor distance in WT and LFS fish relative to baseline. In the WT strain, the effects of gender, if any, are unclear as, at 0.5% (v/v) ethanol, change from baseline was higher in females, whereas at 1.0% (v/v) ethanol, the converse was true. In the LFS strain, there was no difference in gender response at

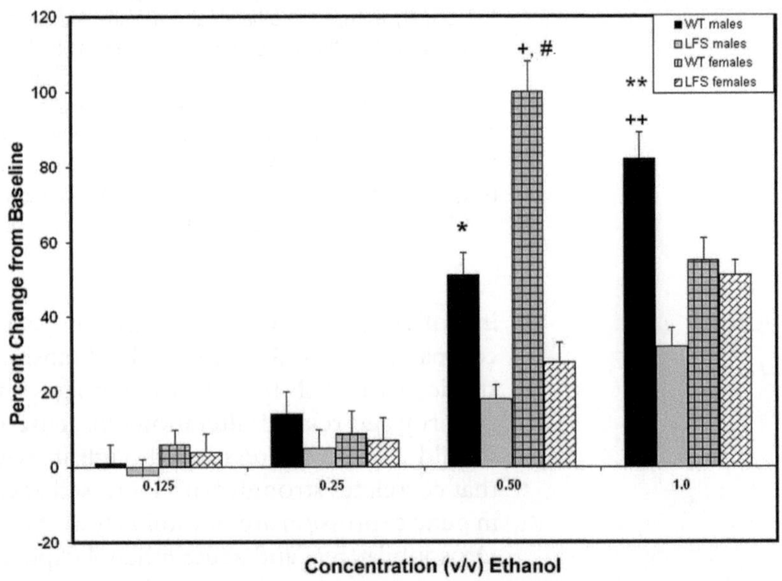

Fig. 6.3. Mean (± S.E.M.) percent change from baseline in WT and LFS fish separated by gender and acutely exposed to ethanol. Data were tested with three-way ANOVA which showed significant effects of gender [$F(1, 476) = 7.942$, $p = 0.005$], strain [$F(1, 476) = 64.627$, $p < 0.001$], and concentration [$F(3, 476) = 99.26$, $p < 0.001$]. Pair-wise comparisons of the 0.5 and 1.0% (v/v) ethanol concentrations were undertaken as, at these concentrations, fish of both strains and genders showed ethanol-related increases in nearest neighbor distance. In the WT strain, gender responses varied with respect to ethanol concentration as WT females were more sensitive than WT males at 0.5% (v/v) ethanol (+, $p < 0.001$) but less sensitive than WT males at 1.0% (v/v) ethanol (++, $p = 0.003$). In the LFS strain, ethanol-related increases were not gender dependent. Analysis of strain differences showed that at 0.5% (v/v) ethanol, change relative to baseline was greater in WT females (#, $p < 0.001$) and males (*, $p = 0.001$) compared to respective genders in the LFS strain. At the 1.0% (v/v) ethanol dose, change relative to baseline was significantly increased in WT compared to LFS males (**, $p < 0.001$).

0.5 or 1.0% (v/v) ethanol. In contrast to the previous study, WT males ($p < 0.001$) and females ($p = 0.001$) showed a greater percent change from baseline at 0.5% (v/v) ethanol than LFS male and female fish. A similar effect was shown at 1.0% (v/v) ethanol in male WT versus LFS fish ($p < 0.001$). These findings suggest either that separation by gender results in increased ethanol sensitivity in the WT compared to the LFS fish, or that, as is more likely, the WT and LFS populations respond similarly to acute ethanol exposure. Individual differences within the two groups are responsible for the crisscrossing effect between the WT and LFS fish in these studies (45, 61).

2.3. Startle Reaction

The startle reaction was assessed 2 h after ethanol treatment and showed that the WT and LFS fish demonstrated ethanol-induced decreases in the startle reaction (45). Data were converted to percent change from baseline for inter-strain testing and presentation here (**Fig. 6.4**) and effects of strain were noted [$F(7, 77) = 5.768$, $p = 0.005$] but could not be localized with post-hoc analysis. At 0.25% (v/v) ethanol, pair-wise comparisons showed a greater percent change in the LFS compared to the WT fish ($p < 0.001$). The presence of a decreased startle reaction in the WT and LFS strains correlates well with human and animal studies in which similar effects have been observed (53, 54). In fact, the startle behavior

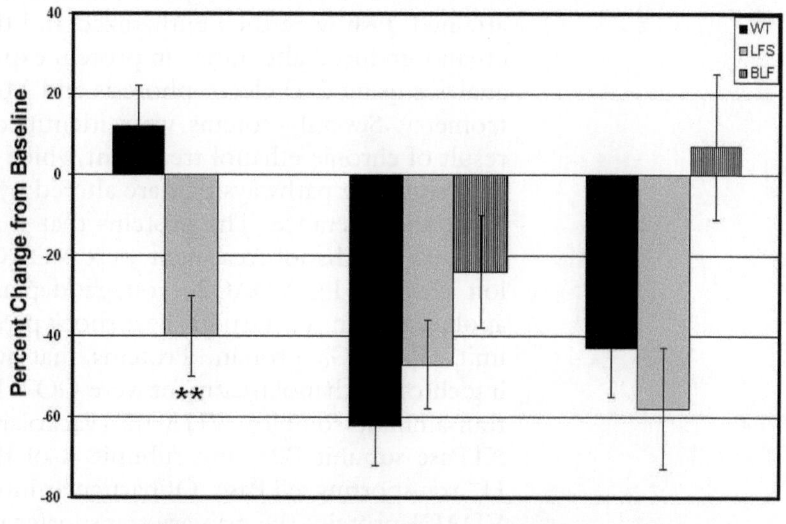

Fig. 6.4. Mean (±S.E.M.) percent change from baseline in results of the startle test in WT, LFS, and BLF fish exposed to 0.25, 0.50, and 1.0% (v/v) ethanol. BLF fish were not tested at 0.25% (v/v) ethanol. ANOVA with significant effect of concentration [$F(3, 152) = 6.641$, $p = 0.001$] and strain [$F(2, 152) = 6.552$, $p = 0.005$]. Pair-wise comparisons showed that at 0.25% ethanol, the startle reflex was more decreased in the LFS fish compared to the WT strain (**, $p < 0.001$).

of the zebrafish (56) may be a more appropriate model system in which to assess ethanol-induced alterations in the startle reaction, as it is often used in behavioral screens (62) and is likely simpler than a similar response in rats, in which 46 strain-dependent phenotypes of the startle reaction have been observed (63).

2.4. Chronic Study

The purpose of these studies was to determine the effects of chronic ethanol on swimming behavior. In one study (45), baseline measures for two groups/strain (8/group) of WT and LFS fish were taken prior to ethanol exposure. Fish were then transferred into a tank containing 0.5% (v/v) ethanol for a period of 2 weeks. The swimming behavior of the fish was tested after 1 and 2 weeks of ethanol exposure. Alcohol levels within the tanks were monitored and adjusted daily. WT fish had a significantly increased nearest neighbor distance following 1 and 2 weeks in alcohol but LFS fish did not show a significant change in nearest neighbor distance over the two full weeks of ethanol treatment (45). The conclusion reached was that during the first week of treatment, the LFS fish developed tolerance, or a declined response to alcohol disruption due to adaptive compensation in behavior and bodily functions (64–66). In our second chronic study, the LFS fish was used as a model because of its ability to develop tolerance (52). In this study, LFS fish were tested for baseline measures of swimming behavior, exposed to ethanol for 4 weeks, and tested again to ensure that tolerance had been attained. Fish were then euthanized and their brains assessed for ethanol-induced alterations in protein expression with proteomic analyses using 2-D electrophoresis and MALDI-TOF mass spectrometry. Several proteins were identified as up-regulated as a result of chronic ethanol treatment which may shed light on cellular signaling pathways that are altered by chronic ethanol treatment and tolerance. The proteins that were increased following 4 weeks of ethanol treatment were VDAC1 (voltage dependent ion channel 1), VDAC2 (voltage dependent ion channel 2), apolipoprotein A1, Hsp70 (heat shock protein 70), and the alpha unit of the G_0 protein. Proteins that were decreased following chronic ethanol treatment were GOT-1 (glutamic-oxaloacetic transaminase soluble), VHA-B2 (vacuolar type H^+ transporting ATPase subunit B2), and subunit A of the catylytic domain of H^+ transporting ATPase. Of particular interest is elevation of the VDAC1 protein, the primary transporter of the outer mitochondrial membrane. VDAC1 has been shown to be alcohol responsive (67) and increased in alcoholics (68).

In the current study, fish of the WT, LFS, and BLF strain were treated chronically with ethanol for the lengthy duration of 10 weeks to determine whether tolerance to ethanol in the LFS fish would persist and if the WT fish would develop tolerance to ethanol. The BLF fish were also included in this study, as their age-related response to acute ethanol exposure (**Fig. 6.2**) promoted them as an interesting model for chronic ethanol consumption. All fish were approximately 3 months of age at the initiation of the study. The Bonferroni correction for *p* values was used following one-way ANOVA to account for repeated testing of baseline against weeks of treatment. Observations in the WT strain (**Fig. 6.5**), similar to results previously observed after 2 weeks (45), showed increased nearest neighbor distance over baseline for the full 10 weeks.

The peak of disruption was at week 1, after which nearest neighbor distance decreased somewhat but remained significantly greater than baseline. In the LFS strain, complete tolerance to ethanol was achieved after 1 week as there was no difference relative to baseline from weeks 2–5. Development of tolerance here took slightly longer than in our previous study (45). Tolerance to alcohol, however, has a strong genetic component (69) and its timing is probably highly subject to individual variation. The development of complete tolerance to ethanol within the LFS strain in both studies, however, confirms the results of our previous studies (45, 52) and also a strain difference between the LFS and WT strain. The absence of tolerance between weeks 6 and 10 in the LFS strain is also notable, as it is suspected that by 6 weeks, constant submersion in 0.5% ethanol resulted in neuronal and

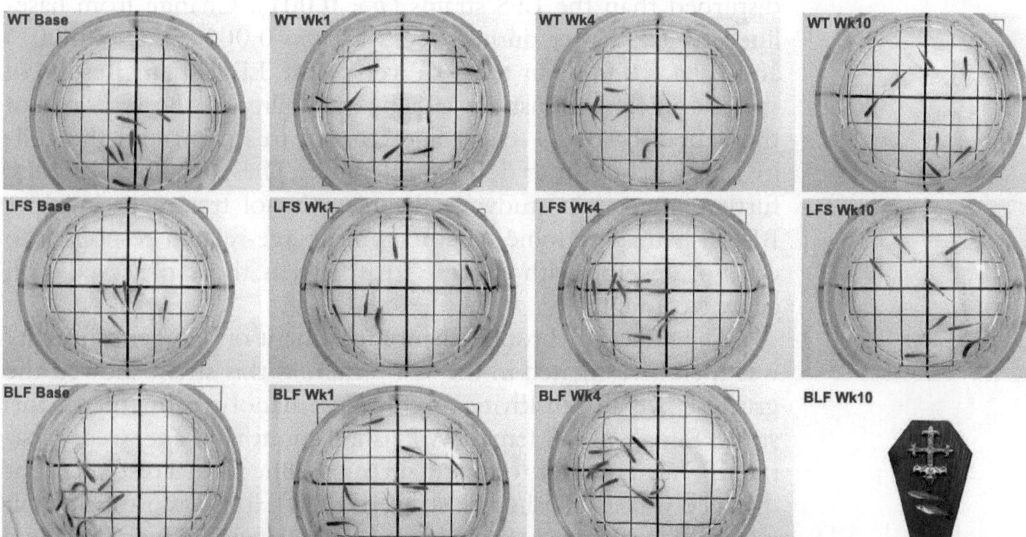

Fig. 6.5. Digital images of swimming behavior over the 10 weeks of chronic 0.5% (v/v) ethanol treatment in WT, LFS, and BLF strains of zebrafish. In WT fish, 0.5% chronic ethanol treatment disrupted swimming behavior and resulted in increased nearest neighbor distance from weeks 1 to 10. In the LFS strain, nearest neighbor distance was most disturbed at the onset and conclusion of treatment as shown here in weeks 1 and 10 where their nearest neighbor distance was increased over baseline. During weeks 2–5, the LFS fish developed tolerance to ethanol and a representative frame from week 4 showing normal swimming behaviors pictured here. In the BLF fish, tolerance developed during weeks 2 and 3 followed by increased disruption and death between weeks 5 and 6.

hepatic toxicity in the LFS strain, which prevented tolerance from being maintained. Chronic ethanol consumption in rodents and human beings results in neurotoxic and neurodegenerative effects (70–73).

In the BLF fish, complete tolerance to alcohol took longer to develop than in the LFS fish, occurring during treatment weeks 3 and 4. At week 5, nearest neighbor distance again increased above baseline and the BLF fish died between weeks 5 and 6. The BLF strain was the sole strain to display mortality. Fish death was gradual over the course of a week and was attributed to ethanol toxicity. This toxicity demonstrated as increases in nearest neighbor distance during week 5 relative to weeks 2 and 3 most probably involved the nervous system as behavior is largely a neuronal response. Ethanol toxicity in other ethanol-sensitive organs such as the liver, however, may have also invoked death. For comparison between strains, data were converted to percent change from baseline (**Fig. 6.6**) and tested with repeated measures ANOVA and revealed a significant effect of strain on change from baseline [$F(2, 193) = 206.863$, $p < 0.001$] and a strain × change from baseline interaction [$F(2, 193) = 3.257$, $p = 0.041$]. The WT group showed the largest change relative to baseline. For example, during weeks 1–5, WT change from baseline was significantly larger than the other two strains ($p < 0.02$) and, during weeks 7, 8, 9, and 10 WT zebrafish were considerably more disturbed than the LFS strains ($p < 0.001$). Change from baseline was also larger during weeks 1 ($p = 0.002$), 2 ($p < 0.001$) and 5 ($p < 0.001$) in the BLF versus the LFS strain. Results of these studies demonstrate genetic variability in the response of the zebrafish strains to chronic ethanol treatment and the ability of the zebrafish to develop tolerance. These results also ignite further interest in studying chronic ethanol treatment in older BLF fish to determine whether similar age-related responses to ethanol develop with chronic treatment as those observed with acute treatment.

The effect of chronic ethanol treatment on gender sensitivity was then investigated in the WT strain, as this strain showed the greatest overall sensitivity to chronic ethanol treatment. In this study, 8 male and 8 female WT fish, 4 months of age, were separated. One day prior to treatment, a baseline for nearest neighbor distance was established. Fish were then exposed to 0.5% (v/v) ethanol for the next 10 weeks and nearest neighbor distance was tested weekly. The Bonferroni correction was again applied to the p value to assess for ethanol-induced effects and confirmed that female and male WT fish were sensitive to ethanol effects. For comparison between the genders, data were converted to percent change from baseline. Repeated measures ANOVA showed a significant effect of gender on percent change from baseline [$F(1, 50) = 40.494$, $p < 0.001$] with females being more sensitive

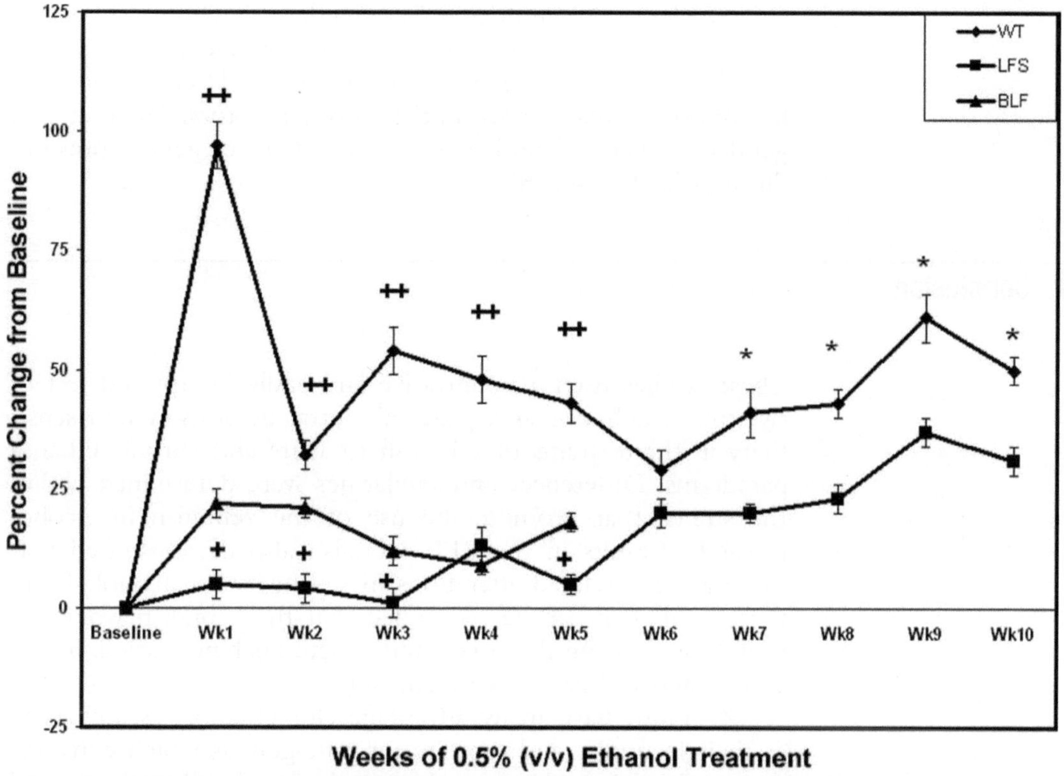

Fig. 6.6. Mean (±S.E.M.) percent change from baseline in the WT, LFS, and BLF zebrafish exposed to 0.5% (v/v) ethanol for 10 weeks. Repeated measures ANOVA showed a significant effect of strain on percent change from baseline [$F(2, 193) = 206.893$, $p < 0.001$]. Pair-wise comparisons show greater change from baseline in WT compared to BLF and LFS (++, $p < 0.001$) in weeks 1–5 of treatment. The degree of disruption was also larger in BLF compared to LFS at weeks 1, 2, 3, and 5 (+, $p < 0.001$). Ethanol treatment led to mortality in the BLF strains following 5 weeks of treatment. During weeks 7, 8, 9, and 10, pair-wise comparisons showed a greater change in the WT compared to the LFS strain (*, $p < 0.001$).

than the males. Pair-wise comparisons showed that female WT fish showed a greater percent change from baseline during 5 ($p < 0.001$), 6 ($p = 0.035$), 7 ($p < 0.001$), 8 ($p = 0.001$), and 10 ($p < 0.001$) weeks of treatment.

These data suggest that female zebrafish are more sensitive to chronic ethanol than male zebrafish. This is consistent with findings in humans as women develop ethanol-induced injuries to key organ systems at shorter durations and lower ethanol doses than men (10, 11). It is well recognized that pharmacokinetic factors, such as levels of blood alcohol and alcohol dehydrogenase, may contribute to the more pronounced ethanol effects that occur in human females compared to males (8, 9), but these have not been investigated in the zebrafish. It is known, however, that, when a time response curve was generated from groups of mixed gender fish, brain alcohol levels did not vary among the WT, LFS, and BLF strains (45). This result would probably not occur if there were large differences in brain alcohol levels between males and females as the number of fish/gender in each group was not controlled and multiple time points were assessed. It is assumed that, in addition to pharmacokinetics, gender differences in alcohol responses may be attributed to other factors. For example, gender has been identified as a strong factor of gene expression during withdrawal (15).

3. Conclusion

These studies used a noninvasive and easily interpreted test of swimming behavior as a gage of central nervous system sensitivity in three strains of zebrafish to acute and chronic ethanol paradigms. Differences and similarities were determined within the strains that promote the use of the zebrafish in alcohol research. Results in the BLF zebrafish also demonstrated that there are age-related alterations in sensitivity to alcohol. From these studies, it can be concluded that the zebrafish would be useful for studying the effects and genetics behind addiction, tolerance, and withdrawal from ethanol.

Zebrafish have many advantages for ethanol research such as their fecundity and capacity for mutagenesis which enhances the productivity of large-scale genetic screens. Transgenic and knockout fish are easily generated while orthologs and synteny between the zebrafish and human genomes may be more marked than between the human and mouse (27) genomes. Alcohol levels in zebrafish which absorb alcohol through their gills and skin remain at steady state for a longer period of time compared to the rapid escalations and depressions of blood alcohol levels that occur in rodent models. This is an advantage in morphologic and pharmacological studies (13, 40, 70, 71) where ethanol-related alterations at specific ethanol concentrations are assessed to determine the minimal ethanol concentrations required to develop pathologies. It is also an advantage in separating the effects of ethanol's specific actions from the effects of withdrawal from ethanol which occur in human binge drinking.

Acknowledgments

I would like to thank Dr. Richard Rabin, my collaborator, for his help throughout these studies and in preparation of this manuscript. I would also like to acknowledge the participation of Dr. Shereene Brown in the gender studies.

References

1. National Institute on Alcohol Abuse and Alcoholism (US), National Institutes of Health (US) and United States. Department of Health and Human Services. (2006) *National Epidemiologic Survey on Alcohol and Related Conditions*. Rockville, MD, US Department of Health and Human Services, National Institutes of Health, National Institute on Alcohol Abuse and Alcoholism.
2. Stratton, K. R., Howe, C. J. & Battaglia, F. C., Institute of Medicine (US). Division of Biobehavioral Sciences and Mental Disorders. Committee to Whom It May Concern: It May Concern: Study Fetal Alcohol Syndrome & National Institute on Alcohol Abuse and Alcoholism (US) (1996) *Fetal Alcohol Syndrome: Diagnosis, Epidemiology, Prevention, and Treatment. Summary*. Washington, DC, National Academy Press.
3. National Institute on Alcohol Abuse and Alcoholism (US) (2008) *Alcohol Research: A Lifespan Perspective*. Rockville, MD, US Department of Health and Human Services, National Institutes of Health, National Institute on Alcohol Abuse and Alcoholism.
4. Heath, A., Kk, B., Paf, M., Dinwiddie, S., Ws, S., Bierut, L., Statham, D., Dunne, M., Jb, W. & Martin, N. (1997) Genetic and environmental contributions to alcohol dependence risk in a national twin sample: consistency of findings in women and men. *Psychol Med* 27, 1381–1396.
5. Heath, A. C. & Martin, N. G. (1994) Genetic influences on alcohol consumption patterns and problem drinking: results from the Australian NH&MRC twin panel follow-up survey. *Ann N Y Acad Sci* 708, 72–85.
6. Kendler, K. S., Neale, M. C., Heath, A. C., Kessler, R. C. & Eaves, L. J. (1994) A twin-family study of alcoholism in women. *Am J Psychiatry* 151, 707–715.
7. Prescott, C. A. & Kendler, K. S. (1999) Genetic and environmental contributions to alcohol abuse and dependence in a population-based. *Am J Psychiatry* 156, 7p.
8. Goist, K. C., Jr. & Sutker, P. B. (1985) Acute alcohol intoxication and body composition in women and men. *Pharmacol Biochem Behav* 22, 811–814.
9. Jones, B. & Jones, M. (1976) Women and Alcohol. Intoxication, metabolism, and the menstrual cycle. In Greenblatt, M. & Ma, S. (Eds.) *Alcoholism Problems in Women and Children*. New York, NY, Grune & Sttratton.
10. Ashley, M. J., Olin, J. S., Le Riche, W. H., Kornaczewski, A., Schmidt, W. & Rankin, J. G. (1977) Morbidity in alcoholics. Evidence for accelerated development of physical disease in women. *Arch Int Med* 137, 883–887.
11. Blume, S. B. (1986) Women and alcohol. A review. *JAMA* 256, 1467–1470.
12. Colantoni, A., Idilman, R., de Maria, N., La Paglia, N., Belmonte, J., Wezeman, F., Emanuele, N., van Thiel, D. H., Kovacs, E. J. & Emanuele, M. A. (2003) Hepatic apoptosis and proliferation in male and female rats fed alcohol: role of cytokines. *Alcohol Clin Exp Res* 27, 1184–1189.
13. Dlugos, C. A. (2006b) Smooth endoplasmic reticulum dilation and degeneration in Purkinje neuron dendrites of aging ethanol-fed female rats. *Cerebellum* 5, 155–162.
14. Klintsova, A. Y., Cowell, R. M., Swain, R. A., Napper, R. M., Goodlett, C. R. & Greenough, W. T. (1998) Therapeutic effects of complex motor training on motor performance deficits induced by neonatal binge-like alcohol exposure in rats. I. Behavioral results. *Brain Res* 800, 48–61.
15. Hashimoto, J. G. & Wiren, K. M. (2008) Neurotoxic consequences of chronic alcohol withdrawal: expression profiling reveals importance of gender over withdrawal severity. *Neuropsychopharmacology* 33, 1084–1096.
16. White, A. M., Bae, J. G., Truesdale, M. C., Ahmad, S., Wilson, W. A. & Swartzwelder, H. S. (2002) Chronic-intermittent ethanol exposure during adolescence prevents normal developmental changes in sensitivity to ethanol-induced motor impairments. *Alcohol Clin Exp Res* 26, 960–968.
17. White, A. M., Ghia, A. J., Levin, E. D. & Swartzwelder, H. S. (2000) Binge pattern ethanol exposure in adolescent and adult rats: differential impact on subsequent responsiveness to ethanol. *Alcohol Clin Exp Res* 24, 1251–1256.
18. Breslow, R. A., Faden, V. B. & Smothers, B. (2003) Alcohol consumption by elderly Americans. *J Stud Alcohol* 64, 884–892.
19. Lovinger, D. M. & Crabbe, J. C. (2005) Laboratory models of alcoholism: treatment target identification and insight into mechanisms. *Nat Neurosci* 8, 1471–1480.
20. Tabakoff, B. & Hoffman, P. L. (2000) Animal models in alcohol research. *Alcohol Res Health J Natl Inst Alcohol Abuse Alcohol* 24, 77–84.
21. Lithgow, G. J. & Andersen, J. K. (2005) Models of oxidative stress in the biology of aging. *Drug Discov Today Dis Models* 2, 273–277.

22. Scholz, H., Franz, M. & Heberlein, U. (2005) The hangover gene defines a stress pathway required for ethanol tolerance development. *Nature* 436, 845–847.
23. Wolf, F. W. & Heberlein, U. (2003) Invertebrate models of drug abuse. *J Neurobiol* 54, 161–178.
24. Davies, A. G., Pierce-Shimomura, J. T., Kim, H., Vanhoven, M. K., Thiele, T. R., Bonci, A., Bargmann, C. I. & McIntire, S. L. (2003) A central role of the BK potassium channel in behavioral responses to ethanol in *C. elegans*. *Cell* 115, 655–666.
25. Corl, A. B., Rodan, A. R. & Heberlein, U. (2005) Insulin signaling in the nervous system regulates ethanol intoxication in *Drosophila melanogaster*. *Nat Neurosci* 8, 18–19.
26. Holley, S. A., Geisler, R. & Nusslein-Volhard, C. (2000) Control of her1 expression during zebrafish somitogenesis by a Delta-dependent oscillator and an independent wave-front activity. *Genes Dev* 14, 1678–1690.
27. Fishman, M. C. (2001) Genomics. Zebrafish – the canonical vertebrate. *Science* 294, 1290–1291.
28. Nüsslein-Volhard, C. & Dahm, R. (2002) *Zebrafish: A Practical Approach*. Oxford, Oxford University Press.
29. Rubinstein, A. L. (2003) Zebrafish: from disease modeling to drug discovery. *Curr Opin Drug Discov Dev* 6, 218–223.
30. Boehmler, W., Carr, T., Thisse, C., Thisse, B., Canfield, V. A. & Levenson, R. (2007) D4 dopamine receptor genes of zebrafish and effects of the antipsychotic clozapine on larval swimming behaviour. *Genes Brain Behav* 6, 155–166.
31. Giacomini, N. J., Rose, B., Kobayashi, K. & Guo, S. (2006) Antipsychotics produce locomotor impairment in larval zebrafish. *Neurotoxicol Teratol* 28, 245–250.
32. Parker, B. & Connaughton, V. P. (2007) Effects of nicotine on growth and development in larval zebrafish. *Zebrafish* 4, 59–68.
33. Langheinrich, U., Hennen, E., Stott, G. & Vacun, G. (2002) Zebrafish as a model organism for the identification and characterization of drugs and genes affecting p53 signaling. *Curr Biol* 12, 2023–2028.
34. Darland, T. & Dowling, J. E. (2001) Behavioral screening for cocaine sensitivity in mutagenized zebrafish. *Proc Natl Acad Sci USA* 98, 11691–11696.
35. Farber, S. A., Pack, M., Ho, S. Y., Johnson, I. D., Wagner, D. S., Dosch, R., Mullins, M. C., Hendrickson, H. S., Hendrickson, E. K. & Halpern, M. E. (2001) Genetic analysis of digestive physiology using fluorescent phospholipid reporters. *Science* 292, 1385–1388.
36. Bilotta, J., Barnett, J. A., Hancock, L. & Saszik, S. (2004) Ethanol exposure alters zebrafish development: a novel model of fetal alcohol syndrome. *Neurotoxicol Teratol* 26, 737–743.
37. Bilotta, J., Saszik, S., Givin, C. M., Hardesty, H. R. & Sutherland, S. E. (2002) Effects of embryonic exposure to ethanol on zebrafish visual function. *Neurotoxicol Teratol* 24, 759–766.
38. Blader, P. & Strahle, U. (1998) Ethanol impairs migration of the prechordal plate in the zebrafish embryo. *Dev Biol* 201, 185–201.
39. Carvan, M. J., 3rd, Loucks, E., Weber, D. N. & Williams, F. E. (2004) Ethanol effects on the developing zebrafish: neurobehavior and skeletal morphogenesis. *Neurotoxicol Teratol* 26, 757–768.
40. Dlugos, C. & Rabin, R. (2007) Ocular deficits associated with alcohol exposure during zebrafish development. *J Comp Neurol* 502, 497–506.
41. Gerlai, R., Lahav, M., Guo, S. & Rosenthal, A. (2000) Drinks like a fish: zebra fish *Danio rerio* as a behavior genetic model to study alcohol effects. *Pharmacol Biochem Behav* 67, 773–782.
42. Lockwood, B., Bjerke, S., Kobayashi, K. & Guo, S. (2004) Acute effects of alcohol on larval zebrafish: a genetic system for large-scale screening. *Pharmacol Biochem Behav* 77, 647–654.
43. Ryback, R. S. (1970) The use of fish, especially goldfish, in alcohol research. *Quart J Stud Alcohol* 31, 162–166.
44. Ryback, R., Percarpio, B. & Vitale, J. (1969) Equilibration and metabolism of ethanol in the goldfish. *Nature* 222, 1068–1070.
45. Dlugos, C. A. & Rabin, R. A. (2003) Ethanol effects on three strains of zebrafish: model system for genetic investigations. *Pharmacol Biochem Behav* 74, 471–480.
46. Scobie, S. R. & Bliss, D. K. (1974) Ethyl alcohol: relationships to memory for aversive learning in goldfish (*Carassius auratus*). *J Comp Physiol Psychol* 86, 867–874.
47. Rayes, A. E., Ryback, R. S. & Ingle, D. J. (1968) The effect of alcohol on aggression in *Betta splendens*. *Commun Behav Biol* 2, 141–146.
48. Peeke, H. V., Peeke, S. C., Avis, H. H. & Ellman, G. (1975) Alcohol, habituation and the patterning of aggressive responses in a cichlid fish. *Pharmacol Biochem Behav* 3, 1031–1036.

49. Gerlai, R., Lee, V. & Blaser, R. (2006) Effects of acute and chronic ethanol exposure on the behavior of adult zebrafish *Danio rerio*. *Pharmacol Biochem Behav* 85, 752–761.
50. Gerlai, R., Ahmad, F. & Prajapati, S. (2008) Differences in acute alcohol-induced behavioral responses among zebrafish populations. *Alcohol Clin Exp Res* 32, 1763–1773.
51. Ruhl, N. & McRobert, S. P. (2005) The effect of sex and shoal size on shoaling behaviour in *Danio rerio*. *J Fish Biol* 67, 1318–1326.
52. Damodaran, S., Dlugos, C. A., Wood, T. D. & Rabin, R. A. (2006) Effects of chronic ethanol administration on brain protein levels: a proteomic investigation using 2-D DIGE system. *Eur J Pharmacol* 547, 75–82.
53. Chester, J. A., Blose, A. M. & Froehlich, J. C. (2004) Acoustic startle reactivity during acute alcohol withdrawal in rats that differ in genetic predisposition toward alcohol drinking: effect of stimulus characteristics. *Alcohol Clin Exp Res* 28, 677–687.
54. Grillon, C., Sinha, R., Ameli, R. & O'malley, S. S. (2000) Effects of alcohol on baseline startle and prepulse inhibition in young men at risk for alcoholism and/or anxiety disorders. *J Stud Alcohol* 61, 46–54.
55. Westerfield, M. (2000) *The Zebrafish Book*. 4th ed. Eugene, Oregon, University of Oregon Press.
56. Eaton, R. & Hackett, J. (1984) The role of the Mauthner cell in fast-starts involving escape in teleost fishes. In Eaton, R. (Ed.) *Neural Mechanisms of Startle Behavior*. New York, NY, Plenum Press.
57. Tresnake, L. (1981) The long-finned zebra *Danio*. *Trop Fish Hobby* 29, 43–56.
58. Itzkowitz, M. & Iovine, M. K. (2007) Single gene mutations causing exaggerated fins also cause non-genetic changes in the display behavior of male zebrafish. *Behaviour* 144, 787–795.
59. Kim, S. N., Rhee, J. H., Song, Y. H., Park, D. Y., Hwang, M., Lee, S. L., Kim, J. E., Gim, B. S., Yoon, J. H., Kim, Y. J. & Kim-Ha, J. (2005) Age-dependent changes of gene expression in the Drosophila head. *Neurobiol Aging* 26, 1083–1091.
60. Rico, E. P., Rosemberg, D. B., Dias, R. D., Bogo, M. R. & Bonan, C. D. (2007) Ethanol alters acetylcholinesterase activity and gene expression in zebrafish brain. *Toxicol Lett* 174, 25–30.
61. Dlugos, C. A., Brown, S. & Rabin, R. A. (2010) Gender differences in ethanol-induced behavioral sensitivity in zebrafish. *Alcohol.*
62. Orger, M. B., Gahtan, E., Muto, A., Page-McCaw, P., Smear, M. C. & Baier, H. (2004) Behavioral screening assays in zebrafish. *Methods Cell Biol* 77, 53–68.
63. Glowa, J. & Hansen, C. (1993) Differences in response to an acoustic startle stimulus among forty-six rat strains. *Behav Genet* 24, 79–84.
64. American Psychiatric Association and American Psychiatric association. Task Force on DSM-IV. (1994) *Diagnostic and Statistical Manual of Mental Disorders: DSM-IV*. Washington, DC, American Psychiatric Association.
65. Tabakoff, B., Cornell, N. & Hoffman, P. L. (1986) Alcohol tolerance. *Ann Emerg Med* 15, 1005–1012.
66. Tabakoff, B. & Hoffman, P. L. (1988) Tolerance and the etiology of alcoholism: hypothesis and mechanism. *Alcohol Clin Exp Res* 12, 184–186.
67. Park, B., Jeong, S. K., Lee, W. S., Seong, J. K. & Paik, Y. K. (2004) A simple pattern classification method for alcohol-responsive proteins that are differentially expressed in mouse brain. *Proteomics* 4, 3369–3375.
68. Lewohl, J. M., van Dyk, D. D., Craft, G. E., Innes, D. J., Mayfield, R. D., Cobon, G., Harris, R. A. & Dodd, P. R. (2004) The application of proteomics to the human alcoholic brain. *Ann N Y Acad Sci* 1025, 14–26.
69. Le, A. D. & Kiianmaa, K. (1988) Characteristics of ethanol tolerance in alcohol drinking (AA) and alcohol avoiding (ANA) Rats. *Psychopharmacology* 94, 479–483.
70. Dlugos, C. A. (2006a) Ethanol-related smooth endoplasmic reticulum dilation in Purkinje dendrites of aging rats. [Erratum appears in Alcohol Clin Exp Res. 2006 Jun;30(6):1091]. *Alcohol Clin Exp Res* 30, 883–891.
71. Dlugos, C. A. (2008) Ethanol-related increases in degenerating bodies in the Purkinje neuron dendrites of aging rats. *Brain Res* 1221, 98–107.
72. Harper, C. & Kril, J. (1989) Patterns of neuronal loss in the cerebral cortex in chronic alcoholic patients. *J Neurol Sci* 92, 81–89.
73. Kril, J. J. & Halliday, G. M. (1999) Brain shrinkage in alcoholics: a decade on and what have we learned? *Prog Neurobiol* 58, 381–387.

Chapter 7

Conditioned Place Preference Models of Drug Dependence and Relapse to Drug Seeking: Studies with Nicotine and Ethanol

Caroline H. Brennan, Amit Parmar, Layla K.M. Kily, Arani Ananthathevan, Arti Doshi, and Salma Patel

Abstract

Addiction is a complex psychiatric disorder characterised by a spectrum of compulsive drug-seeking behaviours and a persistent tendency to relapse (return to drug taking) even after prolonged periods of abstinence. The most commonly used models for the study of drug reward and dependence involve drug self-administration paradigms in mice, rats or monkeys. However, assays using drug-induced conditioned place preference (CPP) have become increasingly popular due in part to the non-invasive and simple nature of the procedure. Using self-administration and conditioned place preference assays we and others have shown that zebrafish show reinforcement responses to common drugs of abuse including ethanol, nicotine, amphetamine, cocaine and opiates and are thus a suitable model for analysis of factors affecting 'reward'. Our work reviewed here further demonstrates that on prolonged exposure to nicotine or ethanol, zebrafish show persistent drug seeking in the face of adverse stimuli, and that drug seeking can be reinstated following extinction using stimuli that induce reinstatement in mammalian models and relapse in humans. Thus our work supports the use of zebrafish as a model system for the study of genetic/molecular mechanisms underlying vulnerability to drug dependence and addiction.

Key words: Dependence, conditioned place preference, nicotine, ethanol, zebrafish, reinstatement.

1. Introduction

1.1. Assessing Drug Reward and Dependence in Animal Models

Addiction is a complex psychiatric disorder characterised by a spectrum of compulsive drug-seeking behaviours and a persistent tendency to relapse (return to drug taking) even after prolonged periods of abstinence. The most commonly used models

A.V. Kalueff, J.M. Cachat (eds.), *Zebrafish Models in Neurobehavioral Research*, Neuromethods 52,
DOI 10.1007/978-1-60761-922-2_7, © Springer Science+Business Media, LLC 2011

for the study of drug-reward and dependence involve drug self-administration paradigms in mice, rats or monkeys. However, assays using drug-induced conditioned place preference (CPP) have become increasingly popular (1–4) due in part to the non-invasive and simple nature of the procedure.

Self-administration is an instrumental conditioning paradigm based on response learning (in that the drug is administered in a response-contingent way). In self-administration paradigms animals are trained to perform a task (commonly press a lever for a defined number of times) in order to obtain the drug. It may be argued that such self-administration paradigms more closely resemble the human condition; clearly they provide a face-relevant model. They also have the advantages of being able to assess increases in self-administration and the amount of work an animal is prepared to undertake in order to obtain the reward – thus more easily assessing the incentive value.

Conditioned place preference is based on Pavlovian stimulus learning: the drug is given in a response-independent manner; the animal is trained to associate, originally neutral but distinct, environmental cues with the presentation of the drug. Typically, the animal is allowed to explore the conditioning apparatus that consists of two arenas with distinct environmental cues and the basal preference (time spent in either chamber) is determined. Then, on alternate days the animal is confined to one or other of the chambers and given the drug or saline. Over repeated training sessions the animal learns to associate the environmental cue with experiencing the subjective effects of the drug. A change in place preference reflects the ability of the environmental cue to gain incentive value as a result of being repeatedly paired with the drug. Although the simple nature of conditioned place preference paradigms limits the variables that can be tested, it is this simplicity that has led to increased popularity of conditioned place preference assays for the study of reward and dependence. There is also good agreement between drugs and neural pathways that are inferred to be involved in reward and relapse based on self-administration assays and those inferred to be involved based on conditioned place preference assays (4–6).

Animal reinforcement models of drug seeking and taking have given great insight into the mechanisms underlying the rewarding effects of drugs of abuse. These studies have led to understanding of the primary targets of abused drugs and the key role played by the mesolimbic dopaminergic system in drug reward, as well as knowledge of adaptation that occurs in these systems on chronic exposure (7). However, addiction involves a loss of control of drug use such that drug taking becomes progressively habitual and then compulsive (see (8, 9) for reviews), and is characterised by a persistent tendency to relapse. Individuals differ in their vulnerability to progress to compulsive drug use as well as their vulnerability to relapse; thus there are individual differences in vulnerability to addiction. As discussed by Everitt et al. (8), studies of addiction must seek to understand these individual differences in propensity to compulsive drug taking and vulnerability to relapse and must therefore incorporate extended periods of drug taking in their behavioural models.

Drug taking in the face of adverse consequences can be seen as a measure of compulsive drug taking and is a key characteristic of addiction or 'dependence' as defined by DSM-IV, although many factors (see **Table 7.1**) are taken into account when diagnosing addiction. It is this compulsive aspect of drug addiction that is commonly used in animal models as an indicator of the establishment of dependence. (Note regarding terminology: Addiction involves both 'physical' dependence; adaptation as a result of normal physiological processes to oppose the presence of the drug, that contributes to physical withdrawal syndromes; as well as 'dependence' as defined in the DSM-IV that refers to compulsive drug taking. When used here 'dependence' refers to compulsive drug taking).

Table 7.1
Formal criteria for diagnosing substance-use disorders

According to the Diagnostic and Statistical Manual of Mental Disorders (DSM, issued by the American Psychiatric Association), the diagnosis of substance dependence requires at least three of seven criteria and the diagnosis of substance abuse requires one of four criteria. The criteria listed below are those described in the fourth edition of DSM (DSM-IV) published in 1994.
Criteria for substance dependence:
• The need for markedly increased amounts of the substance to achieve intoxication or desired effect, or diminished effect with continued use of the same amount (tolerance).
• Withdrawal syndrome or use of the substance to relieve or avoid withdrawal.
• One or more unsuccessful attempts to cut down or control.
• Use in larger amounts or over a longer period than intended.
• Important social, occupational or recreational activities are given up or reduced because of substance use.
• A large amount of time is spent in activities that are necessary to obtain, to use or to recover from the effects of the substance.
• Continued use despite knowledge of having persistent or recurrent physical or psychological problems that are caused or exacerbated by the substance.
Criteria for substance abuse:
• Recurrent use resulting in physically hazardous situations.
• Recurrent substance-related legal problems.
• Continued use despite persistent or recurrent social or interpersonal problems that are caused or exacerbated by the substance.
• Recurrent use resulting in failure to fulfil the main obligations at work, school or home.

Drug addiction is also characterised by a persistent tendency to relapse. Much of our understanding of the neurobiology of relapse has come from analysis of reinstatement models of relapse in mammalian laboratory animals. In reinstatement assays animals are trained to associate an activity with receipt of the drug and then undergo 'extinction training' during which the activity no longer elicits the drug reward and behaviour returns to basal levels. Subsequently, the effect of pharmacological and environmental stimuli to reinstate the non-reinforced activity (as a measure of reinstated drug seeking) is determined. The observations that factors that stimulate reinstatement in laboratory animals also induce relapse in humans, and compounds that attenuate reinstatement increase abstinence rates in human addicts demonstrate the criterion validity (see (10) for discussion) of reinstatement models and support their use for the study of relapse. Thus, as in humans, the three major stimuli that induce reinstatement in mammalian laboratory animals are re-exposure to the drug (drug priming), drug-associated cues and mild stress (2, 3, 5, 11–14).

1.2. Zebrafish as a Model for Reward and Dependence

A number of recent studies have investigated the potential of zebrafish as a model system for identifying factors influencing drug-associated reward (15–18). Zebrafish show conditioned place preference responses to cocaine (16), amphetamine (19), opiates (15) ethanol and nicotine (17) and the amphetamine-induced response is modified by pathways known to influence 'reward' in other systems (18). These results demonstrate the existence of a conserved drug-responsive 'reward' pathway in zebrafish.

In mammals the mesolimbic dopaminergic system has been shown to play a critical role in drug-induced reward and reinforcement. Thus, blockade of dopamine signalling within the nucleus accumbens using antagonists prevents drug-induced reward responses and direct stimulation of the ventral tegmental area can induce them (reviewed in e.g. (3)). The mammalian mesolimbic dopaminergic pathway consists of dopamine neurons that have their cell bodies in the midbrain ventral tegmental area and send projections to the ventral striatum (nucleus accumbens), pre-frontal cortex and amygdala (see **Fig. 7.1**). In anamniotes, including zebrafish, dopaminergic neurons are absent in the midbrain; however, dye-tracing experiments have identified a conserved ascending dopaminergic system in zebrafish that is essential for 'reward' responses. Here dopamine neurons of the posterior tuberculum of the dorsal hypothalamus project to the dorsal and ventral (limbic) striatum. These projections are proposed to represent the mesostriatal and mesolimbic systems, respectively (21). Based on similar tracing experiments, the dorsal-medial region of the telencephalon is considered to correspond to the mammalian amygdala (22, 23).

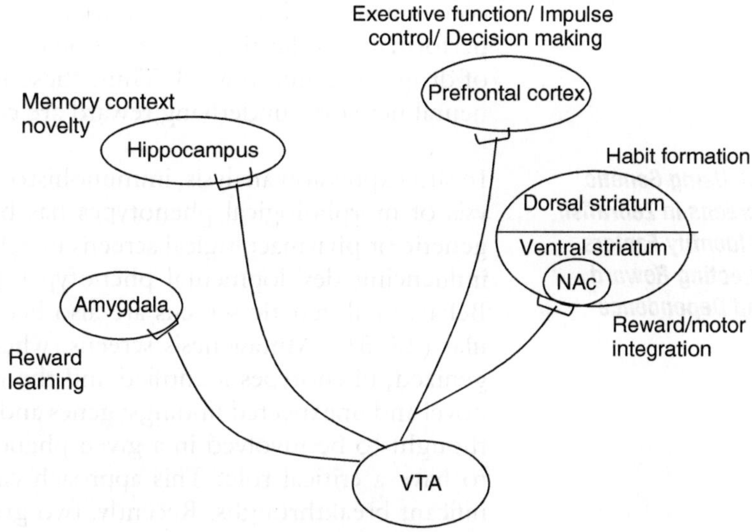

Fig. 7.1. Schematic view of dopaminergic projections involved in reward and addiction. VTA: ventral tegmental area, NAc: nucleus accumbens. Modified from Kelley (20). Co-ordinated signalling within the dopaminergic system shown here and brain glutamatergic systems is thought to integrate motivation, memory and learning so as to enhance the motivational value of memories thus reinforcing associated patterns of behaviour. Drugs of abuse, acting predominantly via increased dopaminergic transmission in the mesolimbic system exert their effects on these pathways and are apparently able to induce very long term, or even permanent, alterations in motivational networks, ultimately leading to changes in/loss of control of, behaviour (i.e. addiction).

Although details of the dopamine systems in zebrafish are not fully understood, evidence in support of the dopamine projection from the posterior tuberculum representing the mammalian mesolimbic projection comes from recent analysis of 'reward' responses in *too few* mutant fish. *Too few* homozygous mutant fish lack the fez1 transcription factor and lack dopamine and 5HT neurons in the hypothalamus (24, 25). Homozygous mutants are indistinguishable from their wild-type siblings in terms of size, morphology, anatomy, fertilisation, escape, feeding and prey-seeking responses but show a reduced reward response to opiates (15). Further evidence for conservation of neural networks involved in the regulation of reward comes from analysis of the acetylcholinesterase (AchE) mutant zebrafish (*AchE*). In mammals AchE terminates cholinergic synaptic transmission and AchE inhibitors block cocaine and morphine induced CPP suggesting a critical role of cholinergic systems in the regulation of 'reward' responses to drugs other than nicotine. *AchE* mutant fish have a loss of function mutation in the *AchE* gene (26). Homozygote fish die by 5 days post fertilisation but heterozygote fish are morphologically indistinguishable from wild-type siblings. These heterozygote fish have reduced AchE activity and show reduced reward responses to amphetamine (18) without involvement of concomitant defects in exploratory activity, learning, and visual performance indicating conservation of the cholinergic regulation of drug-associated reward. Thus, these findings suggest that key neural networks underlying reward are conserved in zebrafish.

1.3. Using Genetic Screens in Zebrafish to Identify Factors Affecting Reward and Dependence

In situ expression analysis, immunohistochemical staining or analysis of morphological phenotypes has been successfully used in genetic or pharmacological screens in zebrafish to identify factors influencing developmental phenotypes (e.g. reviewed in (27)). Behavioural genetic screens are also becoming increasingly popular (28–31). Mutagenesis screens, where the genome is mutagenized, phenotypes identified and the gene cloned, can lead to novel and unexpected findings: genes and pathways not previously thought to be involved in a given phenotype may be discovered to have a critical role. This approach can therefore lead to significant breakthroughs. Recently, two groups have demonstrated the feasibility of the use of behavioural genetic screens to identify factors affecting reward responses to drugs of abuse (16, 19). Darland and Dowling (16) used conditioned place preference in response to cocaine administration to screen for factors affecting cocaine sensitivity. Ninkovic et al. (18) used a similar approach to screen for factors affecting amphetamine reward: of the 1,128 mutagenised zebrafish genomes screened by Ninkovic et al. 26 were identified as potentially mutant in the reward response to amphetamine. Although potential effects on the visual system may influence conditioned place preference responses, these studies demonstrate the feasibility of applying behavioural genetic screens in zebrafish to the study of addiction if suitable behavioural assays that assess compulsive drug taking and relapse can be established.

We have examined zebrafish conditioned place preference responses to two of the most commonly abused drugs, nicotine and ethanol, with an aim of developing assays of addiction in zebrafish. We used persistent conditioned place preference despite adverse consequences as a criterion for dependence, and reinstatement or 'reactivation' of conditioned place preference following extinction as a model for relapse. Our data demonstrate that zebrafish are indeed a relevant model for the study of addiction and pave the way for the use of this system for the analysis of genetic factors influencing vulnerability to drug dependence and addiction.

2. Methods

2.1. Subjects and Maintenance of Zebrafish

Zebrafish were maintained according to established protocols (32). Experiments were performed on sex and age matched, 4–6 month old, 0.5–0.7 g Tuebingen or Queen Mary wild-type zebrafish (originally sourced from a local pet shop) bred in house. The fish were maintained at 28 ± 1°C on a 14 h light, 10 h dark cycle. All zebrafish were handled in accordance with home office licensing regulations.

2.2. Conditioned Place Preference Paradigm

A balanced conditioning paradigm modified from Darland and Dowling (16) was used to assess the reinforcing properties of ethanol or nicotine in zebrafish as described in Kily et al. (17). The testing apparatus was a 2 l rectangular tank (Aquatic Habitats, Apopka, FL, USA) that could be divided in half with a Perspex divider. Each end of the tank had distinct visual cues (1.5-cm diameter black spots uniformly distributed on all sides *versus* vertical 0.5-cm wide black and white stripes). Basal preference was determined for each fish each day over a 3-day period: individual fish were transferred to the conditioning apparatus and allowed to settle for 5 min before the time spent on a given side of the tank over the following 2 min being determined. After 3 days of baseline determination fish were subject to conditioning: individual fish were transferred to the testing tank, allowed to settle for 5 min before being restricted first to the preferred side for 20 min using a Perspex divider and then to the least preferred side for 20 min in the presence of either 30 μM (5 mg l^{-1}) nicotine, 175 mM [1.0% (v/v)] ethanol or fish-water. Drugs were added in a volume of 10 ml so as to give the desired final concentration. After treatment the fish were removed to fresh water in clean tanks and returned to the aquarium. To determine the reinforcing effects of ethanol or nicotine, the place preference of each fish was determined either the following day (for single exposure analysis) or after 1–4 weeks of daily conditioning. Any change in place preference was determined by subtracting the baseline time spent on the drug-treatment side from the final time spent on the drug-treatment side expressed as a percent of the testing period. Fish that showed a greater than 70% baseline preference for either side of the tank (i.e. spent more than 70% of the testing period on one side of the tank), approximately 10% of fish tested, were not used further. All fish tracking was performed manually with assessment of place preference performed by an observer blinded to the treatment conditions.

2.3. Mecamylamine Inhibition of Nicotine-Induced Place Preference

Following determination of basal preference the effect of pre-incubation in the nicotinic receptor antagonist mecamylamine on nicotine-induced place preference was determined. Individual fish were submerged in 30 ml of 100 μM mecamylamine in a 50 ml beaker for 5 min before being transferred to the conditioning tank. Fish were allowed to settle for 5 or 15 min in the conditioning tank before being restricted to their least preferred side and exposed to either 30 μM nicotine or saline for 20 min. The effect of pre-treatment with mecamylamine on changes in place preference following a single conditioning session or following repeated daily conditioning over a 3-week period was determined.

2.4. Conditioned Place Preference Despite an Adverse Stimulus

Removal from the tank to the air was used as punishment in an adverse stimulus test as described in Kily et al. (17). Following 4 weeks of conditioning, the effect of punishment compared with restriction on the number of returns made to the drug-treatment side over a 10-min period was determined. Single fish were placed in the conditioning apparatus, allowed a 5-min settling period and then each time the fish entered the drug-treatment side it was restricted to the non-drug-treatment side for 30 s using a Perspex divider. After 30 s the divider was removed and the fish allowed free access to the whole tank. The number of returns made over a 10-min period was determined. An hour later each fish was returned to the testing apparatus, allowed 5 min to settle and then each time the fish entered the drug-treatment side it was removed from the tank to the air for 3 s. On return to the tank, the fish was restricted to the non-drug-treatment side for 30 s to allow recovery. After this time the divider was removed and the fish allowed free access to the tank. Again the number of returns made over a 10-min period was determined. Tests were carried out on 18–20 fish for each treatment group with two parallel control groups.

2.5. Reactivation of Conditioned Place Preference Following Extinction

Following 4 weeks of daily conditioning fish were assessed for their conditioned place preference as described above. Fish were then subject to extinction training: individual fish were placed in the conditioning apparatus, allowed to settle for 5 min before the visual cues were placed round the tank. Fish were allowed free access to all areas of the tank (in the absence of a divider) for 20 min before being returned to their housing tanks. This procedure was repeated each day over a 2-week period. After 1 and 2 weeks the place preference of each fish was assessed (as described above). After 2 weeks preference had returned to, and remained, within 5% of basal and the ability of drug-priming to reactivate the conditioned place preference was assessed: individual fish were placed in a plain experimental tank (in the absence of a dividing panel) and allowed to settle for 5 min. Fish were then exposed to either 1% ethanol, 10 μM nicotine or fish water for 10 min. After this time fish were transferred to clean water in a fresh tank, allowed to settle for 3 min before the visual cues were placed round the tank. The time spent in the previously drug-conditioned side in 2-min increments over the next 15 min was determined. An increase in time spent in the drug-paired side compared to the extinction basal was taken as indicative of reactivation of conditioned place preference.

2.6. Statistical Analysis

CPP was analyzed using ANOVA followed by Tukey's post-hoc comparison and by paired or two sample *t*-test as appropriate. Conditioned place preference despite an adverse stimulus data were analyzed using two-way ANOVA with a repeat measure over condition (restricted *versus* punished) using Graphpad Prism 5, Instat (GraphPad, San Diego, CA, USA), followed by post-hoc two-sample or paired *t*-test, as appropriate, with Bonferroni adjustment.

3. Results and Discussion

Using self-administration (15) and conditioned place preference assays (16–18) we and others have shown that zebrafish show reinforcement responses to common drugs of abuse including ethanol, nicotine, amphetamine, cocaine and opiates, and are thus a suitable model for analysis of factors affecting 'reward'. Our work reviewed here further demonstrates that on prolonged exposure to nicotine or ethanol, zebrafish show persistent drug seeking in the face of adverse stimuli, and that drug seeking can be reinstated following extinction using stimuli that induce reinstatement in mammalian models and relapse in humans. Thus our work supports the use of zebrafish as a model system for the study of genetic/molecular mechanisms underlying vulnerability to drug dependence and addiction.

3.1. Prolonged Exposure to Ethanol or Nicotine Induces Persistent Conditioned Place Preference in Zebrafish

As described previously (17) zebrafish show conditioned place preference reinforcement responses to both ethanol and nicotine following either single or repeated drug exposure (**Fig. 7.2**). However, there are a number of criteria (see DSM-IV 1994, **Table 7.1**) that need to be met before conditioned place preference can be considered a model of dependence rather than reinforcement. These include the persistence of the response despite prolonged abstinence and conditioned place preference in the face of adverse consequences. We examined our model against these criteria following 1–4 weeks of daily conditioning using 1–3 weeks as a period of abstinence and 3 s removal from the tank as an adverse consequence. As the number of conditioning sessions increased both the magnitude (**Fig. 7.2**) and persistence (**Fig. 7.3**) of the conditioned place preference response increased. After 4 weeks of daily conditioning robust conditioned place preference that persisted over 3 weeks of abstinence was established (Kily et al., **Fig. 7.3**). Persistent drug seeking despite adverse consequences was seen after 4 weeks of daily conditioning (**Fig. 7.4**)

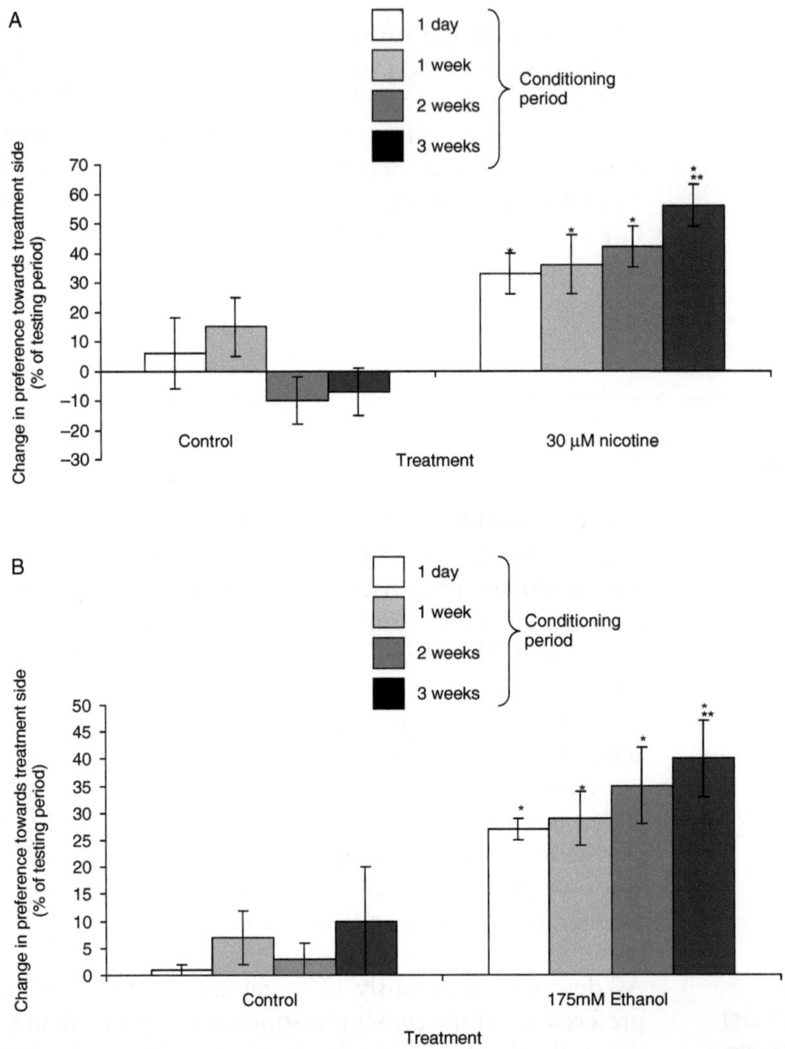

Fig. 7.2. Conditioned place preference induced by nicotine or ethanol exposure. **a** Nicotine. **b** Ethanol. For both nicotine-treated (**a**) and ethanol-treated (**b**) fish a single 20-min conditioning session or daily conditioning over a 1–3 week period induced significant change in preference for the treatment side (*$p < 0.05$, ANOVA). Conditioned place preference increased as the number of conditioning sessions increased such that the place preference seen after 3 weeks of conditioning was significantly greater than that seen after either a single exposure or after 1 week of conditioning (**$p < 0.05$).

but not after lesser treatment periods (not shown). This is consistent with the idea proposed by Everitt and Robbins (9) and others that, on prolonged exposure, there is a gradual loss of control of drug taking towards compulsion; not until fish had experienced 4 weeks of exposure did they exhibit compulsive drug taking (or dependence) defined by drug seeking in the face of adverse consequences. However, as in our adverse stimulus test the fish must learn not to enter the punishment area, it is also possible that the continued entrance into the punishment-paired area resulted

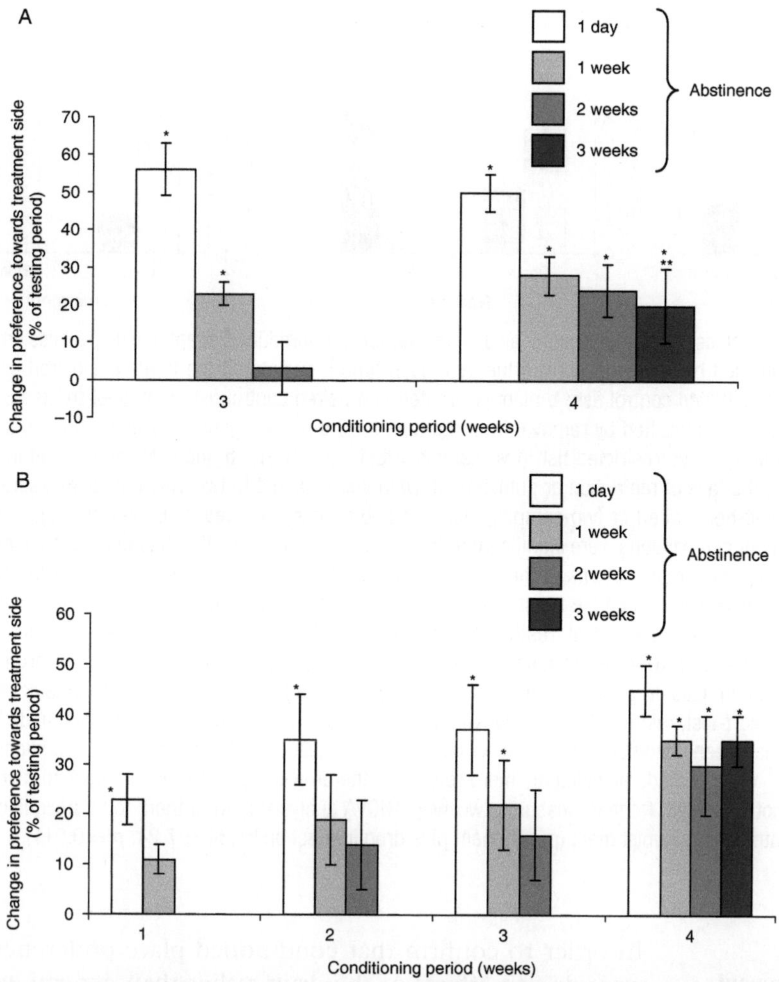

Fig. 7.3. Ethanol and nicotine induced conditioned place preference persists over prolonged periods of abstinence. **a** Nicotine. **b** Ethanol. **a** Fish that had been subject to 3 or 4 weeks of daily conditioning to 30 μM nicotine showed significant place preference for the drug-paired side 1 day or 1 week after last drug exposure. Fish that had been subject to 4 weeks of daily conditioning showed significant conditioned place preference that persisted over 3 weeks of abstinence (* $p < 0.05$). **b** Fish that had been subject to daily conditioning to 175 mM ethanol for 1 day or 1–4 weeks showed significant place preference for the drug-paired side 1 day after last drug treatment. Fish that had been subject to 3 or 4 weeks of conditioning showed persistent place preference 1 week after the last drug exposure. Fish that had been subject to 4 weeks of daily conditioning showed significant place preference that persisted over 3 weeks of abstinence (* $p < 0.05$).

from decreased ability to acquire new learnt behaviours. There is indeed evidence that repeated exposure to and withdrawal from drugs of abuse, including ethanol (33), leads to a decreased ability to acquire new learnt behaviours. This decreased ability to learn has been interpreted as due to effects on synaptic plasticity (33). Further work is required to determine whether decreased ability to learn contributed to the increased drug seeking despite adverse consequences seen here.

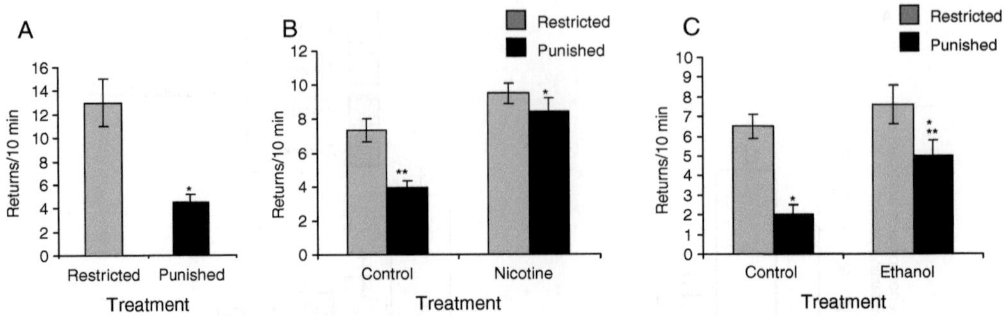

Fig. 7.4. Conditioned place preference despite an adverse stimulus. Reproduced/adapted with permission from Kily et al. (17). Fish were punished by 3-s removal from the tank each time they entered the treatment-paired side: **a** punished versus unpunished/restricted control fish; **b** nicotine-treated and paired control fish; **c** ethanol-treated and paired control fish. **a** Fish that were punished by removal from the tank for 3 s made significantly fewer returns to the treatment side compared to unpunished/restricted fish (two-sample *t*-test $*p < 0.01$). **b** and **c** Number of returns made to the drug-paired side in the face of restriction or punishment. Data were subject to two-way repeat-measures ANOVA analysis followed by post-hoc, paired or two-sample, *t*-test, as appropriate, followed by Bonferroni adjustment. Following Bonferroni adjustment comparisons were significant at the $p < 0.01$ level. **b** Fish that had been conditioned for 4 weeks with 30 μM nicotine made more returns to the drug-paired side than control fish when either restricted (two-sample *t*-test, $P=0.03$) or punished (two-sample *t*-test, $*p < 0.01$). 3-s removal from the tank caused a significant reduction in returns made by control fish (paired *t*-test, restricted compared with punished, $**p < 0.01$) but not nicotine-treated fish. Repeat-measures two-way ANOVA analysis showed there to be a significant interaction between drug treatment and punishment (punishment plus drug interaction $F1,34 = 8.74$, $p = 0.006$). **c** 3-s removal from the tank caused a significant reduction (paired *t*-test, restricted compared with punished, $*p < 0.01$) in number of returns made by both control fish and fish that had been conditioned for 4 weeks with 175 mM ethanol. Fish that had been conditioned for 4 weeks with 175 mmol l^{-1} ethanol made significantly more returns to the drug-paired side when punished (two-sample *t*-test $**p < 0.01$) but not restricted. Repeat measures two-way ANOVA analysis showed there to be a significant interaction between drug treatment and punishment (punishment plus drug interaction $F1,34 = 7.24$, $p = 0.011$).

3.2. Nicotinic Antagonists Prevent Nicotine-Induced Conditioned Place Preference

In order to confirm that conditioned place preference responses were due to effects of the drug rather than general environmental habituation, we examined the ability of pre-exposure to the nicotinic antagonist mecamylamine to prevent nicotine-induced conditioned place preference. As expected, nicotine-induced conditioned place preference was prevented by pre-exposure to the nicotine-receptor antagonist mecamylamine 15 min prior to conditioning (**Fig. 7.5a**). Interestingly, when conditioning was performed only 5 min after mecamylamine pre-treatment, fish exposed to mecamylamine alone showed a change in preference towards the conditioned side (**Fig. 7.5b**). There was no significant difference between the change in preference seen when mecamylamine pre-treatment was followed by conditioning to saline and when mecamylamine pre-treatment was followed by conditioning to 30 μM nicotine. Although detailed explanation of this result requires further investigation, it suggests that the neural networks regulating dopaminergic signalling and reinforcement in zebrafish are conserved with mammals. In mammalian systems nicotinic acetylcholine receptors are present on dopaminergic, GABAergic and glutamatergic neurons within the

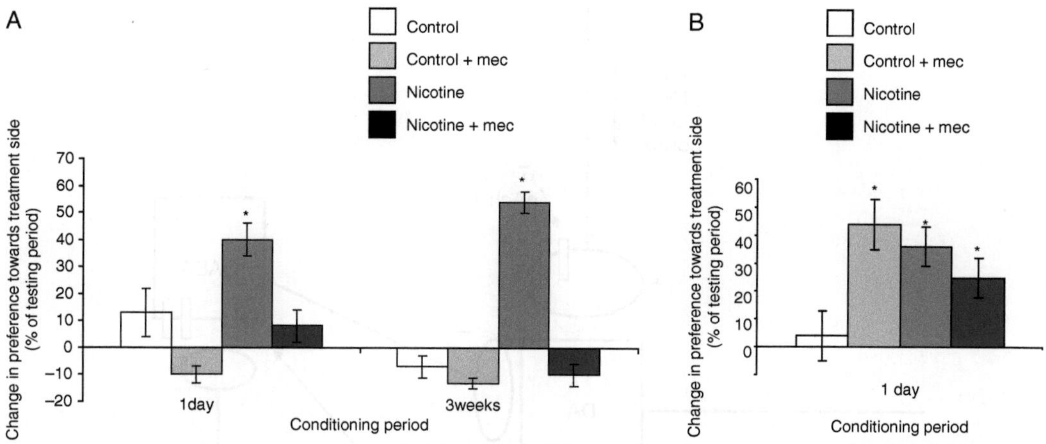

Fig. 7.5. Inhibition of nicotine-induced conditioned place preference by pre-exposure to mecamylamine. **a** 15-min exposure to mecamylamine 15 min prior to conditioning. **b** 15-min exposure to mecamylamine 5 min prior to conditioning. **a** A single 20-min exposure to 30 μM nicotine or daily conditioning over a 3-week period induced a significant ($p < 0.05$) change in place preference for the drug-paired side. 15-min pre-exposure to mecamylamine before the conditioning sessions prevented this nicotine-induced conditioning place preference. **b** A single 20-min exposure to 30 μM nicotine induced a significant ($p < 0.05$) change in place preference for the drug-paired side. 15-min pre-exposure to mecamylamine 5 min before the conditioning session reduced but did not prevent the nicotine-induced place preference. 15-min exposure to mecamylamine 5 min prior to conditioning led to a significant change in preference for the treatment-paired side (in this case saline).

ventral tegmental area. The major endogenous cholinergic input appears to contact GABAergic rather than dopaminergic neurons with nicotinic stimulation of GABAergic neurons inhibiting dopaminergic neuron firing (34, 35). Thus, inhibition of endogenous nicotinic tone would be predicted to lead to inhibition of these GABAergic neurons and an increase in dopamine release (see **Fig. 7.6**) that may be sufficient to cause reinforcement responses. Furthermore, as the kinetics of mecamylamine action at different nicotinic receptors differs such that in rats non-α7 receptors present on GABAergic interneurons appear to be more sensitive to inhibition by mecamylamine (36) than those present on glutamatergic interneurons, these data are consistent with a similar organisation of nicotinic receptor subtypes in the zebrafish ventral tegmental area: nicotinic receptors present on GABAergic neurons were more sensitive to inhibition by the dose of mecamylamine used here than hypothesized nicotinic receptors present on glutamatergic neurons. Thus these data suggest that cholinergic regulation of the neural networks regulating dopaminergic signalling and reinforcement in zebrafish are conserved with mammals.

3.3. Reactivation of Conditioned Place Preference in Zebrafish

As stated above, relapse to drug taking even after prolonged abstinence is a characteristic of addiction and is a major problem for its treatment. Our understanding of the neurobiology underlying relapse, as well as the identification of several therapeutics,

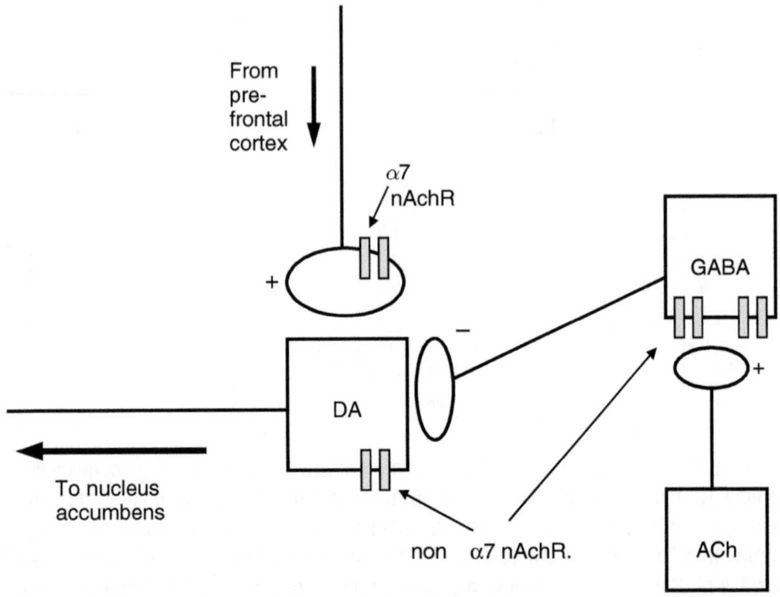

Fig. 7.6. Schematic illustrating the role of nicotinic receptors in the control of dopamine neuron activity in the ventral tegmental area. Based on Mansvelder and McGehee (35). Dopaminergic (DA) neurons receive input from excitatory glutamatergic (Glu) neurons from the prefrontal cortex as well as inhibitory input from GABAergic neurons. Endogenous acetylcholine (ACh) release from brainstem cholinergic neurons contributes to the GABAergic input to ventral tegmental area dopaminergic neurons. Non-α7 nicotinic acetylcholine receptors (nAChR) present on both dopaminergic and GABAergic neurons can excite dopaminergic and GABAergic neurons directly, while α7 receptors can enhance release from glutamatergic terminals. In the presence of low levels of nicotine the non-α7 receptors desensitize rapidly effectively inhibiting GABAergic inputs to the dopamine neurons. The α7 nAChRs are less sensitive to desensitization at low nicotine concentrations. Thus low doses of nicotine enhance glutamatergic inputs relative to GABAergic inputs leading to a net increase in excitation of the dopaminergic neurons.

has come in large part from analysis done in reinstatement models of relapse in laboratory animals. Reinstatement of drug seeking following extinction either by periods of abstinence or by active training is an established model for relapse (3, 10). Our data demonstrating that zebrafish show persistent conditioned place preference and conserved persistent changes in gene expression (17) following repeat exposure to ethanol or nicotine suggests that zebrafish may also show stimulus-induced reinstatement/reactivation following extinction. Thus, to determine whether zebrafish can be used as a model system to assess factors contributing to vulnerability to relapse we have conducted preliminary studies to determine whether drug-induced conditioned place preference could be reactivated in zebrafish using single non-contingent drug exposure. As shown in **Fig. 7.7** ethanol-induced conditioned place preference, and to a lesser extent nicotine-induced conditioned place preference, can indeed be reactivated in zebrafish by non-contingent drug exposure as in mammals.

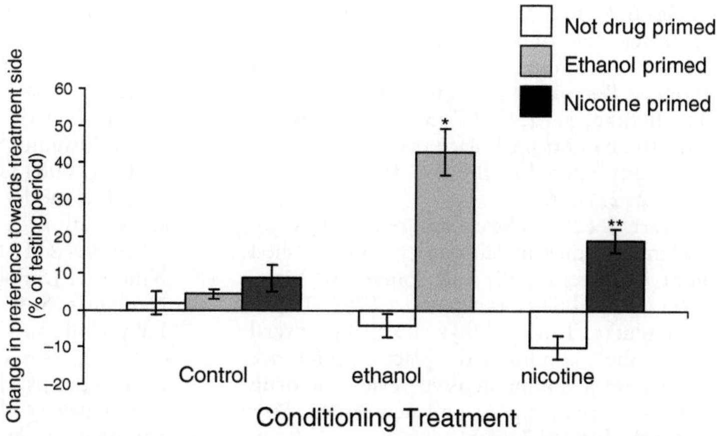

Fig. 7.7. Drug-primed reinstatement of conditioned place preference. Following 4 weeks of daily 20-min exposure to either 1% v/v ethanol or 30 μM nicotine fish showed a 50 or 35% change in preference for the ethanol or nicotine-paired side respectively (not shown). This preference was extinguished by daily conditioning in the absence of any drug until preference returned to basal and remained there for a one-week period (not shown). 10-min exposure to either 1% ethanol or 10 μM nicotine reinstated the ethanol-induced (* $p < 0.05$) or nicotine-induced (** $p < 0.1$) conditioned place preference.

4. Conclusion

In summary, our work demonstrates that zebrafish show robust reinforcement responses to both ethanol and nicotine, the two most commonly abused drugs in society today. We also demonstrate that on repeated exposure to either of these drugs zebrafish show the addiction-related behaviours of drug seeking despite adverse consequences (compulsive drug seeking) and stimulus-induced relapse following prolonged periods of withdrawal. Thus, our work establishes zebrafish as a behavioural model system for the analysis of neurobiological mechanisms underlying addiction. Our assays, coupled with the relative ease with which one can perform genetic analysis, including forward mutagenesis screening, and the existence of detailed microsatellite maps, make zebrafish an attractive model for studies addressing factors influencing vulnerability to addiction.

References

1. Mueller, D., Perdikaris, D. & Stewart, J. (2002) Persistence and drug-induced reinstatement of a morphine-induced conditioned place preference. *Behav Brain Res* 136, 389–397.

2. Mueller, D. & Stewart, J. (2000) Cocaine-induced conditioned place preference: reinstatement by priming injections of cocaine after extinction. *Behav Brain Res* 115, 39–47.

3. O'Brien, C. P. & Gardner, E. L. (2005) Critical assessment of how to study addiction and its treatment: human and non-human animal models. *Pharmacol Ther* 108, 18–58.
4. Tzschentke, T. M. (2007) Measuring reward with the conditioned place preference (CPP) paradigm: update of the last decade. *Addict Biol* 12, 227–462.
5. Stewart, J. (2003) Stress and relapse to drug seeking: studies in laboratory animals shed light on mechanisms and sources of long-term vulnerability. *Am J Addict* 12, 1–17.
6. Tzschentke, T. M. (1998) Measuring reward with the conditioned place preference paradigm: a comprehensive review of drug effects, recent progress and new issues. *Prog Neurobiol* 56, 613–672.
7. Nestler, E. J. (2004) Molecular mechanisms of drug addiction. *Neuropharmacology* 47(Suppl 1), 24–32.
8. Everitt, B. J., Belin, D., Economidou, D., Pelloux, Y., Dalley, J. W. & Robbins, T. W. (2008) Review. Neural mechanisms underlying the vulnerability to develop compulsive drug-seeking habits and addiction. *Philos Trans R Soc Lond B Biol Sci* 363, 3125–3135.
9. Everitt, B. J. & Robbins, T. W. (2005) Neural systems of reinforcement for drug addiction: from actions to habits to compulsion. *Nat Neurosci* 8, 1481–1489.
10. Epstein, D. H., Preston, K. L., Stewart, J. & Shaham, Y. (2006) Toward a model of drug relapse: an assessment of the validity of the reinstatement procedure. *Psychopharmacology (Berl)* 189, 1–16.
11. Sanchez, C. J. & Sorg, B. A. (2001) Conditioned fear stimuli reinstate cocaine-induced conditioned place preference. *Brain Res* 908, 86–92.
12. Stewart, J. (2000) Pathways to relapse: the neurobiology of drug- and stress-induced relapse to drug-taking. *J Psychiatry Neurosci* 25, 125–136.
13. Wang, J., Fang, Q., Liu, Z. & Lu, L. (2006) Region-specific effects of brain corticotropin-releasing factor receptor type 1 blockade on footshock-stress- or drug-priming-induced reinstatement of morphine conditioned place preference in rats. *Psychopharmacology (Berl)* 185, 19–28.
14. Wise, R. A. (2004) Dopamine, learning and motivation. *Nat Rev Neurosci* 5, 483–494.
15. Bretaud, S., Li, Q., Lockwood, B. L., Kobayashi, K., Lin, E. & Guo, S. (2007) A choice behavior for morphine reveals experience-dependent drug preference and underlying neural substrates in developing larval zebrafish. *Neuroscience* 146, 1109–1116.
16. Darland, T. & Dowling, J. E. (2001) Behavioral screening for cocaine sensitivity in mutagenized zebrafish. *Proc Natl Acad Sci USA* 98, 11691–11696.
17. Kily, L. J., Cowe, Y. C., Hussain, O., Patel, S., McElwaine, S., Cotter, F. E. & Brennan, C. H. (2008) Gene expression changes in a zebrafish model of drug dependency suggest conservation of neuro-adaptation pathways. *J Exp Biol* 211, 1623–1634.
18. Ninkovic, J., Folchert, A., Makhankov, Y. V., Neuhauss, S. C., Sillaber, I., Straehle, U. & Bally-Cuif, L. (2006) Genetic identification of AChE as a positive modulator of addiction to the psychostimulant D-amphetamine in zebrafish. *J Neurobiol* 66, 463–475.
19. Ninkovic, J. & Bally-Cuif, L. (2006) The zebrafish as a model system for assessing the reinforcing properties of drugs of abuse. *Methods* 39, 262–274.
20. Kelley, A. E. (2004) Memory and addiction: shared neural circuitry and molecular mechanisms. *Neuron* 44, 161–179.
21. Rink, E. & Wullimann, M. F. (2002) Connections of the ventral telencephalon and tyrosine hydroxylase distribution in the zebrafish brain (Danio rerio) lead to identification of an ascending dopaminergic system in a teleost. *Brain Res Bull* 57, 385–387.
22. Peitsaro, N., Kaslin, J., Anichtchik, O. V. & Panula, P. (2003) Modulation of the histaminergic system and behaviour by alpha-fluoromethylhistidine in zebrafish. *J Neurochem* 86, 432–441.
23. Portavella, M., Vargas, J. P., Torres, B. & Salas, C. (2002) The effects of telencephalic pallial lesions on spatial, temporal, and emotional learning in goldfish. *Brain Res Bull* 57, 397–399.
24. Levkowitz, G., Zeller, J., Sirotkin, H. I., French, D., Schilbach, S., Hashimoto, H., Hibi, M., Talbot, W. S. & Rosenthal, A. (2003) Zinc finger protein too few controls the development of monoaminergic neurons. *Nat Neurosci* 6, 28–33.
25. Rink, E. & Guo, S. (2004) The too few mutant selectively affects subgroups of monoaminergic neurons in the zebrafish forebrain. *Neuroscience* 127, 147–154.
26. Behra, M., Cousin, X., Bertrand, C., Vonesch, J. L., Biellmann, D., Chatonnet, A. & Strahle, U. (2002) Acetylcholinesterase is required for neuronal and muscular development in the zebrafish embryo. *Nat Neurosci* 5, 111–118.
27. Lieschke, G. J. & Currie, P. D. (2007) Animal models of human disease: zebrafish swim into view. *Nat Rev Genet* 8, 353–367.

28. Baraban, S. C., Dinday, M. T., Castro, P. A., Chege, S., Guyenet, S. & Taylor, M. R. (2007) A large-scale mutagenesis screen to identify seizure-resistant zebrafish. *Epilepsia* 48, 1151–1157.
29. DeBruyne, J., Hurd, M. W., Gutierrez, L., Kaneko, M., Tan, Y., Wells, D. E. & Cahill, G. M. (2004) Isolation and phenogenetics of a novel circadian rhythm mutant in zebrafish. *J Neurogenet* 18, 403–428.
30. Muto, A., Orger, M. B., Wehman, A. M., Smear, M. C., Kay, J. N., Page-McCaw, P. S., Gahtan, E., Xiao, T., Nevin, L. M., Gosse, N. J. et al. (2005) Forward genetic analysis of visual behavior in zebrafish. *PLoS Genet* 1, e66.
31. Renier, C., Faraco, J. H., Bourgin, P., Motley, T., Bonaventure, P., Rosa, F. & Mignot, E. (2007) Genomic and functional conservation of sedative-hypnotic targets in the zebrafish. *Pharmacogenet Genomics* 17, 237–253.
32. Westerfield, M. (1995) *The Zebrafish Book: Guide for the Laboratory Use of Zebrafish (Danio rerio)*. 3rd ed. Oregon, University of Oregon Press.
33. Stephens, D. N. & Duka, T. (2008) Review. Cognitive and emotional consequences of binge drinking: role of amygdala and prefrontal cortex. *Philos Trans R Soc Lond B Biol Sci* 363, 3169–3179.
34. Fiorillo, C. D. & Williams, J. T. (2000) Cholinergic inhibition of ventral midbrain dopamine neurons. *J Neurosci* 20, 7855–7860.
35. Mansvelder, H. D. & McGehee, D. S. (2002) Cellular and synaptic mechanisms of nicotine addiction. *J Neurobiol* 53, 606–617.
36. de Rover, M., Lodder, J. C., Kits, K. S., Schoffelmeer, A. N. & Brussaard, A. B. (2002) Cholinergic modulation of nucleus accumbens medium spiny neurons. *Eur J Neurosci* 16, 2279–2290.

Chapter 8

Zebrafish Biogenic Amine Transporters and Behavior in Novel Environments: Targets of Reuptake Inhibitors and Pesticide Action as Tools for Neurotoxicology Research

Georgianna G. Gould

Abstract

Central monoamine systems (e.g., dopamine, serotonin, norepinephrine) are associated with motivation, locomotion, social behavior, emotion, and mood. Biogenic amine transporters regulate neurotransmission by removing neurotransmitters from synapses and extracellular fluid. Despite evolutionary divergence, teleost fish and mammalian transporter proteins appear similar, particularly at active binding sites. However, it is not clear if the similarities extend to functional responses, reuptake-inhibiting drugs, or involvement in delayed neurotoxic responses to pesticide exposures. Under certain exposure conditions, alterations in expression and function of these transporters may be more sensitive biomarkers of pesticide exposure or neurodegenerative disease risk than acetylcholinesterase inhibition. Zebrafish (*Danio rerio*) behavioral assays targeting associative responses such as anxiety are useful as pharmacological and toxicological screens, or for studying modulation of behavior by central neurotransmitter systems. In novel environments, zebrafish go to tank bottoms and dark backgrounds, a stereotypical behavior (attributed to predator anxiety) forming the basis of the novel light/dark aquatic plus maze characterized in this chapter. Such behavioral paradigms are an essential component to establish zebrafish as pharmacological and toxicological research models. Herein adult zebrafish are exposed to reuptake inhibitors and representative organochloride, organophosphate, or pyrethroid pesticides at 1 μg day^{-1} for 21 days, tested for anxious response in the light/dark plus maze, then assayed for dopamine and serotonin transporter density by radioligand binding. Exposures to these compounds variably affect dopamine and serotonin transporter density and alter behavior in the maze as compared to controls.

Key words: Dopamine, serotonin, norepinephrine, monoamine systems, amine transporters, acetylcholinesterase inhibition, mammalian homology, SERT, DAT, transporter mechanisms, Parkinson's disease, pesticide, light-dark plus-maze, expression regulation, toxicology.

A.V. Kalueff, J.M. Cachat (eds.), *Zebrafish Models in Neurobehavioral Research*, Neuromethods 52,
DOI 10.1007/978-1-60761-922-2_8, © Springer Science+Business Media, LLC 2011

1. Occupational Pesticide Exposures and Elevated Risk for Parkinson's Disease

For agricultural and horticultural workers, formulators, or others routinely using pesticides, repeated occupational exposures pose long-term neurotoxicological health risks. Epidemiological studies indicate that individuals working with pesticides and their immediate families are at higher risk of developing neurodegenerative disorders than the population at large (1–4). There is a strong association between Parkinson's disease and pesticide exposure, but insufficient evidence for a causal link, due in part to the broad chemical spectrum of pesticides used (5–7). Aside from the paraquat-like neurotoxin MPTP (1-methyl-4-phenyl-1,2,3,6-tetrahydropyridine), mechanisms linking exposures to neurodegeneration are evasive (8). Other Parkinson's risk factors, such as farming, drinking well water, and rural living, are also associated with environmental pesticide exposure (6, 9). Risks of neurotoxicological effects from low-to-moderate level pesticide exposures increase with frequency, and depend upon chemical composition and the person's vulnerability at the time of exposure.

Typical occupational pesticide exposures are in the ppb–ppt range, occurring semiannually over weeks or days (10–14). In general, pesticide exposures collectively increase the overall risk of developing neurodegenerative disorders by 2–3% (5, 15). Further elevation of risk is linked to specific pesticides, high-dose poisoning events, and exposure to mixtures (8, 16–19). With pesticide mixtures, additive or synergistic effects may enhance the adverse properties of individual compounds (e.g., (20, 21)). In animal models, certain pesticides or combinations reproduce anatomical, neurochemical, behavioral, and/or pathological features of Parkinson's disease (22–24), allowing mechanistic studies of pathology.

Cognitive and emotional impairments may also result from pesticide-induced alterations of the central nervous system, the mechanisms of which can be uncovered through exposure and behavioral studies in animal models. High-dose exposure to organophosphate pesticides is associated with depression and anxiety disorders in farmworkers (25–27), and anhedonia is evident in rats following perinatal exposure to chlorpyrifos (28). Further, chronic low-level pesticide exposures are associated with impairments in concentration, learning, problem solving, and memory that last long after discontinuation of exposure, and can be reproduced in rodent behavioral parallels (29, 30). Hence, while mechanistic links between pesticide exposure, neurodegenerative, and psychiatric disorders are evasive, the association seen in workers is reproducible in animal models. Comprehensive exposure testing in representative systems is needed to establish criteria for acceptable levels of occupational and environmental exposure, and to estimate long-term health risks.

2. Potentiation and Exposures During Critical Stages of Brain Development

Elevated risk for Parkinson's disease from "take-home" environmental pesticide exposure during childhood or prenatal development has been documented in agricultural families (16, 31–35). Parkinson's arising late in life may stem from cumulative effects of earlier pesticide exposures. In rodent models, pesticide or MPTP pre-exposure can exacerbate dopaminergic degradation in subsequent pesticide exposures (e.g., (8)). Early exposure to some pesticide classes may persistently increase dopamine transporter (DAT) expression or function, providing a ready window for MPP+-like metabolites to enter dopamine (DA) neurons, rendering them more vulnerable to subsequent neurotoxin challenges (36–40). Pre-exposure to certain pesticides during critical stages of brain development can amplify these effects in animal models, and serotonin transporters (SERT) may be similarly upregulated (41–44). Potentiation by repeated pesticide exposures is consistent with a "Multiple Hit Hypothesis" of Parkinson's disease, in which age-related DA neuron degeneration is accelerated in a stepwise fashion by various factors (16, 24, 45, 46).

The underpinnings of pesticide potentiation mechanisms can be readily studied in zebrafish (*Danio rerio*) with their rapid development, but the sensitivity of their monoamine systems to different pesticide classes needs further characterization (47). Also, exposure techniques and dosing levels for hydrophobic pesticides (and other compounds) require refinement in zebrafish toxicity studies to reduce uptake variability and solvent effects (48, 49). Nevertheless, headway has been made: for example, zebrafish larvae, like humans and primates, are sensitive to Parkinson's-inducing MPTP, while rodents are less so (47, 49–52). As in rodents (41, 42, 53), zebrafish SERTs are sensitive to organophosphates like chlorpyrifos (54). Given that children of workers occupationally exposed to pesticides are themselves also exposed, it is important to model exposure during critical stages of development for comprehensive environmental risk assessment. Early developmental exposure of zebrafish to chlorpyrifos impairs swimming activity and spatial learning into adulthood (55, 56). Characterization of potentiation-inducing circumstances in repeated pesticide exposure scenarios may stimulate safer practice guidelines, and lead to better approaches to neuroprotection for pesticide workers and their families.

3. Sensitivity of Acetyl-cholinesterase versus Biogenic Amine Systems

Inhibition of acetylcholinesterase (AChE), the enzyme responsible for breakdown of the neurotransmitter acetylcholine, is often used as a biomarker of acute high-dose pesticide exposures. However, AChE inhibition may be too crude an indicator to monitor and identify low-to-mid level repeated pesticide exposures, while biogenic amine system receptors and transporters may be more sensitive indicators. Clinical evidence suggests resultant mood and cognition impairments from pesticide exposures are not due to AChE inhibition, but may instead be due to functional changes in biogenic amine systems (18, 29, 57). Chlorpyrifos exposure during critical stages of brain development, at doses below the threshold for AChE activity inhibition, persistently alters monoamine metabolism, serotonin receptor and transporter expression in rats (41, 42, 53). It also produces behavior consistent with animal models of depression, and impairs associative learning, despite psychostimulant reward, in conditioned place preference tasks and spatial learning (28, 30, 58). This suggests that mammalian biogenic amine systems and associated emotional or cognitive behaviors may be more prone to chronic low-level pesticide exposure effects than AChE activity. Zebrafish AChE activity is inhibited by chronic (144–250 day) organophosphate bath exposure (parathion) at $\geq 1\ \mu g\ l^{-1}$ concentrations, as is the AChE of fathead minnows, a standard EPA aquatic toxicity model (59, 60). Whether zebrafish biogenic amine systems are also more sensitive than their AChE activity to sub-acute, low-to-mid level pesticide exposures is of interest.

Central dopamine and serotonin systems are associated with motivation, locomotion, social behavior, emotion, and mood. Serotonin (SERT) and dopamine (DAT) transporters take up serotonin (5-HT) or dopamine (DA) from synapses and extracellular fluid in brain into neurons for reuse. Chronic administration of 5-HT or DA reuptake inhibitors reduces SERT or DAT binding (61, 62). In rodents, during all life stages, chronic organochloride exposures persistently increase DAT density in brain, and deplete DA levels (3, 63–65). DAT is upregulated for up to a month following discontinuation of low-dose (1.5 mg kg^{-1}), subchronic pyrethroid exposures (37, 63, 66). Chlorpyrifos exposures during critical stages of brain development increase SERT levels into adulthood (41, 42, 44, 53). If human and zebrafish DAT and SERT are also more sensitive than AChE activity to intermittent, developmental, or chronic low-dose pesticide exposures, their expanded use as biomarkers could improve exposure modeling and monitoring.

4. Zebrafish to the Rescue: Teleost Serotonin and Dopamine Systems

Zebrafish (*Danio rerio*) are an excellent model organism for toxicological studies of pesticide exposures and pharmacological screening because of their shared properties with mammalian central neurotransmitter systems, rapid development, ease of culture and dosing, and recognition as a refinement to the use of mammals to reduce animal suffering (67–73). Despite over 350 million years of divergence and two gene duplication events (74, 75), biochemical and molecular properties of critical DA and 5-HT system proteins are largely conserved among zebrafish and mammals, particularly at active binding sites (76, 77). Structurally, zebrafish monoamine systems are simpler, yet they exhibit parallels with mammals (49). Neural fibers connecting part of the zebrafish diencephalon (midbrain equivalent) and pallial regions of the telencephalon (paleostriatum) resemble rodent ascending limbic pathways, and function in learning memory and locomotion (78–80). Differences such as zebrafish dopamine neurons being expressed in diencephalon instead of mesencephalon, and an apparent lack of epinephrine (adrenaline) neurotransmission in zebrafish brain are to be expected (49). Yet in balance, if most functional binding properties and pathways are indeed conserved, zebrafish hold great promise for high throughput toxicity and pharmaceutical screening.

The fish serotonin (5-HT) system is associated with social interactions, mating, learning, and memory (81–84). There are multiple high-affinity binding sites for 5-HT in the teleost brain (85). A high-affinity binding site for the 5-HT reuptake inhibitor [^{3}H] paroxetine occurs in teleost lymphocytes (86). Two serotonin transporters (SERTa and SERTb) with 77% shared sequence identity have been cloned from zebrafish brain and appear to share functional properties with mammalian SERTs (87). SERTa, by phylogenetic sequence analysis is more homologous to human, mouse, rat, and chicken SERTs than SERTb, which shares 75% sequence similarity with other vertebrate SERTs (87). Zebrafish embryonic and adult SERTs are located presynaptically, SERTa is highly expressed in the raphae and diencephalon, while SERTb is limited to the medulla and retina (88). Functionally, the pharmacological profile of zebrafish and mammalian SERTs is largely conserved, except for the zebrafish SERTa expressing high affinity for desipramine and related compounds (89).

Dopamine in fish is associated with locomotion, foraging, and addictive behavior (90–92). A region anatomically similar to the DAT-rich striatum occurs in the zebrafish telencephalon (93). A zebrafish DAT was cloned from cDNA to facilitate ontogenetic studies of the embryonic DA system (94). The zebrafish DAT and human DAT appear homologous, with 76% amino acid identity (94). Following withdrawal from repeated cocaine administration, zebrafish DAT expression decreases and DA levels increase (95). In MPTP-challenged zebrafish, DAT was protected by the DA reuptake inhibitor nomifensine, but not by the norepinephrine reuptake inhibitor desipramine, hence pharmacologically it is more similar to human DAT than the nematode *Caenorhabditis elegans* DAT, which is blocked by both inhibitors (47). Zebrafish monoamine systems are responsive to pesticide exposures at occupationally relevant doses, particularly during embryonic development (47, 50, 51). Thus, there is a great extent of conservation between teleost and mammalian central biogenic amine system function, receptors, and transporters.

Biogenic amine systems are involved in both learning and substance dependency, and zebrafish are particularly amenable to associative learning tasks, and as their motivation and reward pathways closely parallel those in mammals, they are increasingly used in addiction studies (68, 96–98). Multiple versions of conditioned place preference paradigms for zebrafish have been developed to examine drug sensitivity and to study learning behavior and its genetic or pharmacological impairment (e.g. (99–101)). Exposure of embryonic zebrafish to the pesticide chlorpyrifos produced learning deficiencies in adult zebrafish (55, 56, 71). Conditioned place preference tests in zebrafish may also reveal pesticide-induced learning deficiencies that involve alterations in dopaminergic and serotonergic systems, consistent with clinical reports of impaired cognition following pesticide exposures (29, 30). Hence, given that most functional binding properties and pathways are either conserved with or closely parallel mammalian systems, zebrafish hold great promise for high throughput toxicity testing with both neurochemical and behavioral endpoints.

5. The Aquatic Light/Dark Plus Maze: A New Tool for Pharmacology and Toxicology Screening with Zebrafish

Anxiety affects emotion and cognition and is associated with stereotyped behavioral responses in humans and animal models. The elevated plus maze for rodents is a widely used example of an anxiety-based behavioral performance test sensitive to drugs, genetic manipulations, neurotoxins, and other factors affecting emotionality (102). Anxious responses can be altered by administration of benzodiazepines, serotonin or norepinephrine reuptake inhibitors (see e.g., (103)) and other compounds, including pesticides such as chlorpyrifos (28). One approach to modeling anxiety in fish focuses on associative learning wherein fish are confronted with predator odor, alarm-pheromones, fleeing conspecifics, or electric shock (49, 104–106); the perceptible cues employed are tied to very real environmental risks, so anxious behavior is reflexive. An alternative approach places zebrafish in unfamiliar environments and observes their response. Individual zebrafish introduced into novel tanks initially swim close to walls and tank bottoms, a thigmotaxic and putatively anxious predator avoidance response (107). Levin et al. (108) compared the vertical location of nicotine-treated zebrafish to control zebrafish in a novel dive tank for 5 min: upon introduction, untreated fish dove to tank bottoms, and swam higher in the tanks after several minutes, while nicotine-exposed fish swam high sooner in a dose-dependent manner. This assay holds great promise for rapid pharmacological and toxicological screening for anxiolytic properties of compounds in zebrafish.

A new and different novel environment-based test is described here for zebrafish that can be used alone or in conjunction with the dive tank of Levin et al. (108) to reveal several dimensions of defensive versus exploratory behavior. It is a light-dark plus-maze for zebrafish (and similar species) based upon the rodent elevated plus-maze anxiety test, with black arms analogous to closed arms, and white arms in place of open arms. The aquatic plus-maze is predicated on the innate preference of zebrafish for dark backgrounds when introduced into novel environments (109), which can be altered by exposure to ethanol and other anxiolytic compounds (68). In the study presented in this chapter and in validation studies with other drugs, adult zebrafish were used, but the maze at full size is also amenable for testing older juvenile zebrafish, and adults of other fish species such as goldfish (small) or fathead minnows. The concept that the maze can be scaled down for more efficient and effective testing of larval and young juvenile fish is currently being validated. In the following study, zebrafish are chronically exposed to the pesticides chlorpyrifos, dieldrin, the SERT uptake inhibitor sertraline (a demonstrated anxiolytic in rodent elevated plus maze), and the DAT uptake inhibitor GBR 12909 for 21 days via spiked diet. Zebrafish are subsequently tested, first in the dive tank, then in the aquatic light/dark plus maze. With some exceptions, results from the two tests generally agree for most exposures performed. However, the instances in which the two tests yield different results present a tantalizing basis for further dimensional discrimination and interpretation of innate anxious responses in novel environments.

6. Effects of Chronic Dietary Pesticide or Pharmaceutical Exposure on Adult Zebrafish SERT, DAT and Behavior in Novel Environments

6.1. Introduction

Chronic exposure to low-to-moderate levels of pesticides may elicit neurobiological changes in all vertebrate central biogenic amine systems. The aim of this study was to determine if chronic exposure to pesticides can produce neurobiological changes in zebrafish central biogenic amine systems. The working hypothesis was that zebrafish SERT and DAT, like rodent SERT and DAT, might be more prone to altered expression and/or function from chronic pesticide exposures from one or more commonly used classes (organophosphates, organochlorides, or pyrethroids), than AChE activity might be. Such effects, on serotonergic and/or dopaminergic systems may manifest in altered behavior in a novel environment-based anxiety test. Both the novel dive tank and aquatic light-dark plus maze tests were used to determine if such effects occurred with these pesticide exposures. If similarities between rodent and fish biogenic amine systems extend to their response to pesticides, then chronic exposures should produce similar neurobiological changes in zebrafish biogenic amine systems that may affect their behavior in novel environments.

6.2. Methods

Animals: Zebrafish 60–120 days old were obtained from Aquatic Eco-Systems (Apopka, FL). Zebrafish of mixed sex were housed in groups of 6 per 3-l tank in a benchtop aquatic habitat on a 14:10 light/dark cycle with lights on at 7:00 AM CST. Fish were fed Wardley Total Tropical flake food (Hartz Mountain, Secaucus, NJ) twice a day at approximately 8:00 AM and 17:00 PM CST. Water temperature was maintained at 27°C. Habitat water was filtered and deionized through a Nanopure water filtration system (Barnstead, Dubuque, IA), and supplemented with 0.2 g l^{-1} Instant Ocean Salt (Aquatic Eco-Systems) to reduce osmotic stress (51). Zebrafish acclimated to their new environment for 21 days prior to commencement of experiments. The Institutional Animal Care and Use Committee at the University of Texas Health Science Center, San Antonio, in accordance with the NIH Office of Laboratory Animal Welfare (110) guidelines, approved all animal procedures.

Pesticide Exposures: For chronic pesticide exposures, zebrafish were housed in groups of 6 (3 males and 3 females) in 3-l aquariums with under-gravel filtration systems, airstones, and pumps (Hawkeye International, Fenton, MO). Aquaria were filled with deionized water supplemented with 0.2 g l^{-1} Instant Ocean Salt (tank water). Every day 1 l of water from each tank was removed and replenished with fresh tank water to keep nitrogen waste levels low. Zebrafish received 1 μg day^{-1} of either sertraline (Pfizer, Groton, CT), GBR 12909 (Sigma, St. Louis, MO), chlorpyrifos, dieldrin, or permethrin (Chem Service, West Chester, PA), or a mixture of all three pesticides at 0.33 μg day^{-1} each for 21 days. Pesticides or pharmaceuticals were mixed into gel food (Aquatic-Ecosystems, Apopka, FL) at a concentration 1 μg/10 mg, because in a pilot study, single adult zebrafish readily consumed about 7–10 mg of food within 5 min.

The gel fish food was generally mixed as follows: To administer 1 μg day^{-1}, 1 mg of drug or pesticide was dissolved in 250 μl ethanol and 3,750 μl deionized water. This solution was added to 6 g of gel food mix to achieve 10 g of food with a dose of 1 μg/10 mg. The food was mixed with a spatula in 20-ml drug vials. Vials were stored at 4°C with the lids off for 48 h prior to use to allow ethanol to evaporate and food to solidify. Twice each day 30 mg of spiked food was weighed, crumbled, and crushed into small pieces in a plastic weigh boat, and fed to the zebrafish in the aquaria. Control zebrafish were fed the gel fish food mixed and weighed as for treatment groups except the food contained no pesticides or pharmaceuticals.

Quantitative Autoradiography of Zebrafish SERT and DAT: Following 21 days of pesticide exposure, fish were housed in clean tanks with fresh tank water and were fed Wardley Total Tropical flake food (Hartz Mountain, Secaucus, NJ) for 2 days to allow residual drug from their systems to wash out. Fish were then rapidly decapitated, their lower jaw severed at the mandible, and the dorsum of the head (containing the brain) was rinsed in ice-cold saline, patted dry with paper towel, and frozen on crushed dry ice. Heads containing brain were stored at –80°C until sectioning. Coronal sections (20 μm) from frozen fish brains (contained within skull) were cut in a cryostat (Reichert Jung-Leica, Deerfield, IL). Sections were mounted onto gelatin-coated, chilled microscope slides in a series of 16 slides each, with 6–7 brain sections from rostral to caudal on each slide, and dried under vacuum for 18 h at 4°C. Sections on slides were stored at –80°C until use. A zebrafish brain atlas (111) was used to identify pertinent anatomical structures in autoradiograms.

Serotonin Transporters: Serotonin transporter (SERT) density was determined in zebrafish brains as previously described (112). Brain sections were prewashed for 1 h in 4°C 50 mM Tris–HCl, 120 mM NaCl buffer, pH 7.4, incubated at 4°C in 1 nM [^{3}H] cyanoimipramine (80–85 Ci mmol^{-1}; American Radiolabeled Chemicals, St. Louis, MO) in the same buffer at 4°C pH 7.4, for 18–24 h. Nonspecific binding was determined using 5 μm sertraline (Pfizer, Groton) and was 15–20% of total binding.

Dopamine Transporters: Dopamine transporter (DAT) density was measured in the zebrafish brains as described in (113). Briefly, sections on slides were pre-incubated in 50 mM sodium phosphate buffer (9.5 mM $NaH_2PO_4{*}H_2O$, 40.5 mM Na_2PO_4), pH 7.4 buffer at 4°C for 20 min. Sections on slides were incubated for 2 h in the same buffer containing 10 nM [^{3}H] WIN 35428 (86 Ci mmol^{-1}; Perkin Elmer-NEN). Nonspecific binding was determined using 5 μM mazindol (a dopamine and norepinephrine antagonist, Sigma) and was 15–20% of total binding.

Quantitative Autoradiography Data Analysis: In both assays, sections were washed in buffer at 4°C, dipped in ice-cold deionized water and dried on a slide warmer. Tritium labeled sections (on slides) were exposed to Kodak MR film (Kodak, Rochester) for 6 weeks along with [^{3}H] standards (American Radiolabeled Chemicals) calibrated against [^{3}H] brain mash per Geary et al. (114). Films were developed using Kodak GBX developer. Optical densities of brain images were converted to fmol mg^{-1} protein and autoradiograms were captured as digital images and quantified using NIH image-based Scion Image 1.60c (http://scioncorp.com) on a Macintosh G4 running OS 9.2. ANOVA was used to analyze the effects of the chronic pesticide and drug treatments on the brain regions quantified using Statistica (StatSoft, Tulsa, OK).

Homogenate Binding to Zebrafish DAT: Brains from an additional exposure of 6 zebrafish each to 1 μg meal^{-1} dieldrin, GBR 12909, or control diet were assayed for DAT site density by homogenate binding, using the DAT and SERT specific ligand [^{125}I] RTI-55 (Perkin-Elmer, Boston, MA). Binding assay procedures were modified from Boja et al. (115). Samples were pooled from freshly collected brains of 3 zebrafish adults each. Brains were homogenized in 30 mM sodium phosphate buffer containing 0.32 M sucrose, pH 7.4, that was chilled to 4°C. The homogenate was centrifuged at 30,600 × *g* (16 K rpm for a SS34 rotor), resuspended, and recentrifuged once to wash. The final pellet was resuspended in 200 μl of buffer. Protein concentration was assayed with Bradford reagent (Sigma, St. Louis, MO, as in 116) and measured on a spectrophotometer (Beckman Instruments, Fullerton, CA). The incubation solution contained the radioligand [^{125}I] RTI-55 at a concentration of 50 pM in 30 mM sodium phosphate buffer containing 0.32 M sucrose. To specify DAT binding, 50 nM citalopram (Forrest Labs, Jersey City, NJ) was added to the assay to mask SERT binding. Each assay was performed in triplicate. Nonspecific binding was determined by addition of 10 μM mazindol to a subset of tubes. Incubation took place for 2 h at room temperature. The incubation was stopped with 4 ml of 30 mM sodium phosphate wash buffer, pH 7.4 at 4°C. Labeled homogenates were captured by filtration under vacuum using a Brandel tissue harvester (Gaithersburg, MD) onto glass fiber filters (Schleicher and Schuell#25, Keene, NH) presoaked in 5% polyethyleneimine. Filters were then washed with two additional 4 ml aliquots of wash buffer. [^{125}I] Radioactivity trapped by the filters was determined using a Packard Cobra II auto gamma counter (Packard Instrument Co., Downers Grove, IL) with a counting efficiency of 40%. Homogenate binding data was analyzed by one-way ANOVA with Statistica (StatSoft, Tulsa, OK).

Acetylcholinesterase Assays: Fish brain and muscle tissue samples were assayed for AChE activity as described in Wheelock et al. (117) and Ellman et al. (118), with modifications. Frozen samples were weighed and added to 2 ml of 50 mM Tris–HCl buffer pH 8, and homogenized on ice by hand with 20 strokes in a Potter Elrehijem homogenizer. Homogenates were centrifuged for 2 min at 5,000 × *g*. A 15 μl aliquot of supernatant was assayed for protein using Bradford reagent (Sigma, St. Louis, MO) with bovine serum albumin as a standard in 96 well microplates, and absorbance was measured at 595 nm on a BioAssay microplate reader (HTS 7000, Perkin Elmer, Boston, MA). Of the remaining supernatant, 100 μl was diluted 1:10, and 100 μl of the dilution was assayed for AChE activity, using as directed the Amplex Red Acetylcholine/Acetylcholinesterase Assay Kit (Molecular Probes-Invitrogen, Eugene, OR) in 96 well microplates (Costar, Corning, NY). After 30 min of incubation, fluorescence was measured in the microplate reader with excitation set at 540 nm and emission detection at 595 nm. Statistical comparisons were made by ANOVA using Statistica (StatSoft).

Novel Environment Behavioral Assays: Zebrafish behavior in novel environments was tested in one round of two novel environment assays performed back to back, taking approximately 15 min for each fish to complete. Zebrafish were tested 24 h after their last dietary exposure to pesticide or pharmaceutical. Behavior was recorded using a 7 megapixel digital camera (HP Photosmart R742, OfficeMax, USA), mounted on a tripod (Targus, Anaheim, CA) for the dive tank assays, or on a Kaiser RS1 copy stand (B&H Photo and Video, New York, NY) for the light/dark plus maze. Water filling both mazes was refreshed with habitat water after each fish's trial.

Dive Tank: Zebrafish were transferred from home tanks into a 1-l beaker containing 500 ml of tank water for 3–5 min (this step is for consistency, in other studies we exposed fish to drugs during this time). From the beaker, fish were transferred by net into a trapezoidal 4-l fish tank (Aquascene 1, TopFin, Phoenix, AZ) prefilled to a depth of 18 cm with tank water, which they had never been housed in before. The transparent tank bottom was placed on a black countertop, and a white board was placed vertically behind the tank and opposite the camera and observer. Lines dividing the tank into thirds were drawn on the outside to facilitate behavioral scoring. Fish were observed in the dive tank for 5 min, during which they were timed and digitally recorded for the amount of time spent in the top 2/3 versus bottom 1/3 of the dive tank.

Light/Dark Aquatic Plus Maze: The plus maze is a module of the offset cross maze available through Ezra Scientific (www.ezrascientific.com, San Antonio, TX). The whole offset cross maze is constructed of clear 0.32 cm (1/8″) Acrylex acrylic and is 28″ × 20″ high, subdivided into 4″ × 4″ modules by drop-in doors. For the light/dark plus maze paradigm, the offset cross maze is configured in a module with a 10 × 10 cm (4″ × 4″) square center section that serves as the starting position for the fish, and four arms consisting of two additional 10 × 10 cm sections opened, and all other sections closed off by sliding doors. The maze is 10 cm deep, and was filled to a uniform depth of 5 cm with home tank water. Two opposite arm bottoms and sides were lined with black polypropylene and the other two opposing arms were lined with white polypropylene sections cut from file folders (OfficeMax, USA). Poly sections were clipped to the sides with binder clips and placed on the bottom of the maze. The gray background of the copy stand (Kaiser RS1) showed through the middle 10 cm^2 section of the maze. A 60 W clip-on light was perched on the copy stand above the maze and behind the camera.

After completing the dive tank, each fish was netted and placed into the center of the light/dark plus maze to begin a 5 min trial. The number of entries into each arm and the amount of time the fish spent in the white versus center and dark arms of the maze was recorded and filmed. Fish behavior in the light/dark plus maze was scored as in the rodent elevated plus maze (102, 103).

6.3. Results

SERT and DAT Density Following Dietary Pesticide Exposures: Serotonin transporter density was significantly reduced in zebrafish treated for 21 days with 1 μg day^{-1} sertraline (SRT) or chlorpyrifos (CPF) in the optic tectum and periventricular hypothalamus (caudal zone), ($F_{(6,19)} = 2.4$ and 3.7, $p = 0.05$ and 0.013, Fisher's LSD $p < 0.05$). This brain region corresponded to cross section#168 in the zebrafish brain atlas (111). These results are shown in **Fig. 8.1a, b**.

Consistent with the findings of Rink and Wullimann (78, 79), DAT density was sparsely labeled by radioligand in several areas of the zebrafish forebrain. In the dorsal telencephalic area (cross section#92 in zebrafish brain atlas), there was a thin layer expressing

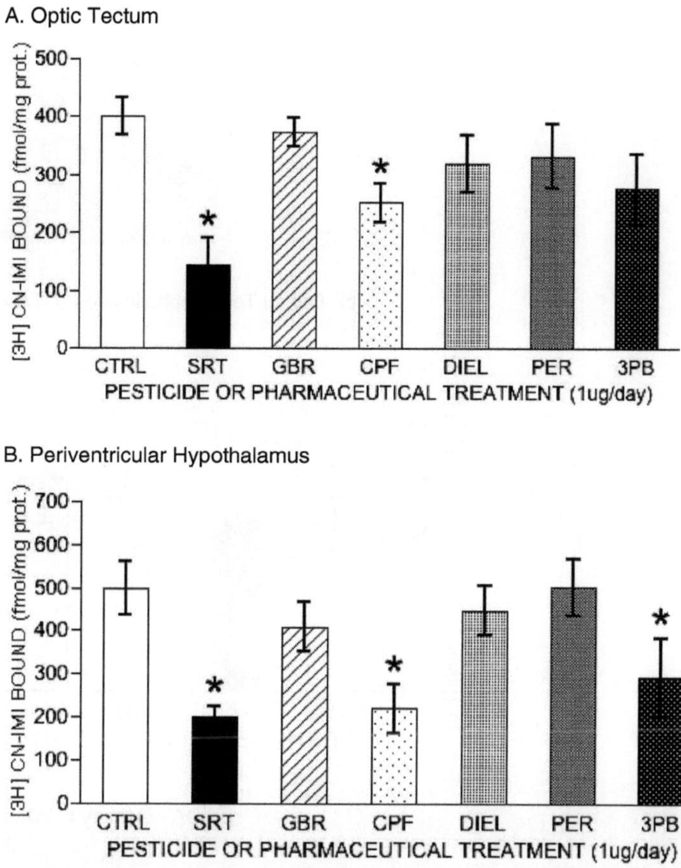

Fig. 8.1. SERT Density in the zebrafish optic tectum (a) and hypothalamus (b) following chronic dietary exposure to 1 μg day^{-1} pesticides or pharmaceuticals. SERT was labeled with 1 nM [^{3}H] cyanoimipramine (CN-IMI). *N* = 4, CTRL = control, SRT = sertraline, GBR = GBR 12909, CPF = chlorpyrifos, DIEL = dieldrin, PER = permethrin, 3 PB = 3 pesticide blend (1/3 dose each). *Bars* represent means and *lines* are standard error. SRT and CPF treatment significantly reduced SERT density in both brain regions, while the 3 PB reduced SERT density in the hypothalamus (**b**) but not in the optic tectum (**a**).

high density [^{3}H] WIN 35428 labeling in a series of small spots resembling cell body clusters (**Fig. 8.2a**). Densities of dark spots in the dorsum of each brain section were measured, the averages of which are shown in **Fig. 8.2b**. Following dietary exposure, GBR 12909 significantly reduced DAT density, while dieldrin significantly increased DAT density, as measured by [^{3}H] WIN 35428 binding in the dorsal telencephalon ($F_{(6,21)} = 2.6$, $p = 0.05$, Fisher's LSD $p < 0.05$), shown in **Fig. 8.2b**. When whole brain membrane homogenates were assayed for DAT density, labeled by [^{125}I] RTI-55, after chronic dietary exposures, again GBR 12909 reduced DAT density while dieldrin increased DAT density ($F_{(1,2)} = 80$, $p = 0.012$, Fisher's LSD $p < 0.03$).

Acetylcholinesterase (AChE) Activity: AChE activity in muscle was significantly inhibited for all pesticide and drug treatment

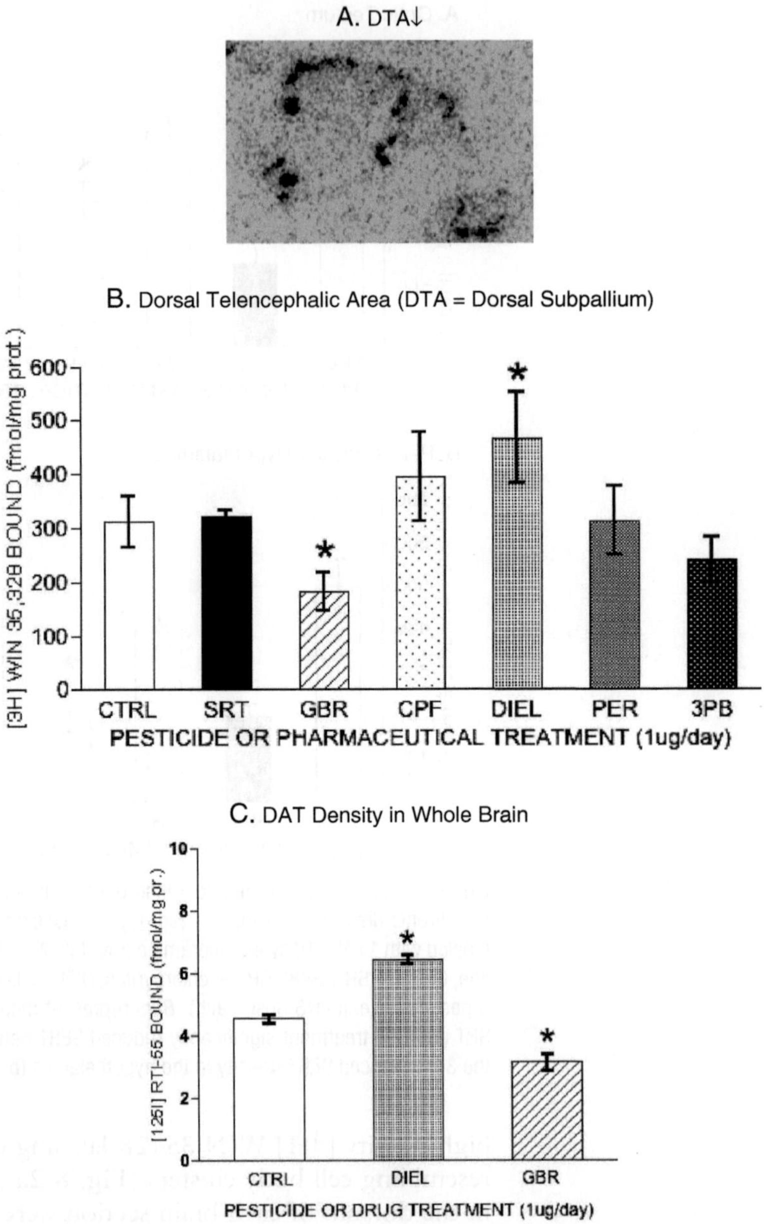

Fig. 8.2. **a** and **b** Measurement of DAT density in zebrafish dorsal telencephalon by autoradiography ($N = 4$), and c DAT density by homogenate binding in whole brain membranes ($N = 2$). Autoradiogram insert **a** shows typical [^{125}I] RTI-55 labeling in a control fish brain (DTA = dorsal telencephalic area, VTA = ventral telencephalic area), **b** shows mean and standard error for DAT density labeled by [^{3}H] WIN 35428 in the telencephalon (for treatment abbreviations, see **Fig. 8.1**), and **c** shows whole brain DAT density labeled with [^{125}I] RTI-55.

groups ($F_{(6,16)} = 4.17$, $p = 0.01$, Fisher's LSD $p < 0.03$) at the 1 μg day^{-1} dietary concentration, shown in **Fig. 8.3**. This indicates that with this chronic, dietary mode of administration, this

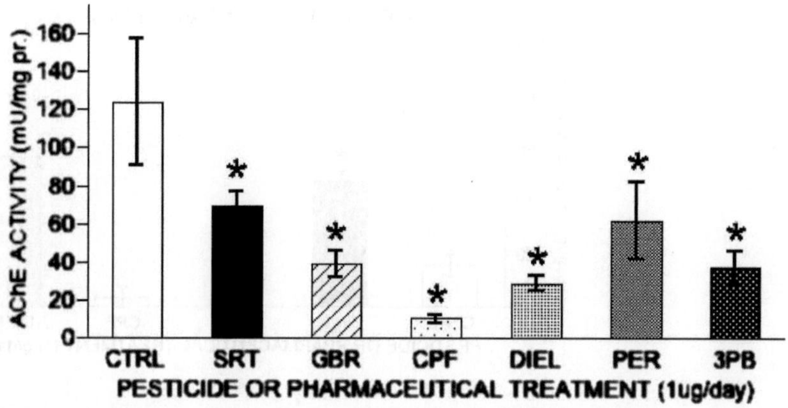

Fig. 8.3. Inhibition of AChE activity in zebrafish muscle tissue following 21-day dietary exposure to drugs or pesticides at 1 μg day^{-1}. For abbreviations, see **Fig. 8.1**. $N = 3–4$. Bars are mean milli-units of activity mg^{-1} protein, measured by fluorescence following excitation.

dose produced effects consistent with high dose pesticide exposures. AChE activity in zebrafish muscle and brain has a correlation coefficient of $r^2 = 0.85$, hence muscle tissue can be assayed for AChE activity so brain tissue can be conserved for assays of SERT and DAT density and function.

Zebrafish Behavior in Novel Environments: In the dive tank, zebrafish chronically exposed to 1 μg day^{-1} dieldrin in diet spent significantly more time in the top 2/3 of the dive tank than control fish ($F_{(4,24)} = 5.3$, $p = 0.003$, Fisher's LSD $p < 0.005$), see **Fig. 8.4**.

In the light/dark aquatic plus maze, as shown in **Fig. 8.5**, chronic dietary exposure to 1 μg day^{-1} of sertraline or dieldrin significantly increased the percent of total crosses into white arms ($F_{(4,24)} = 11.8, p = 0.00002$, A), and the amount of time spent in

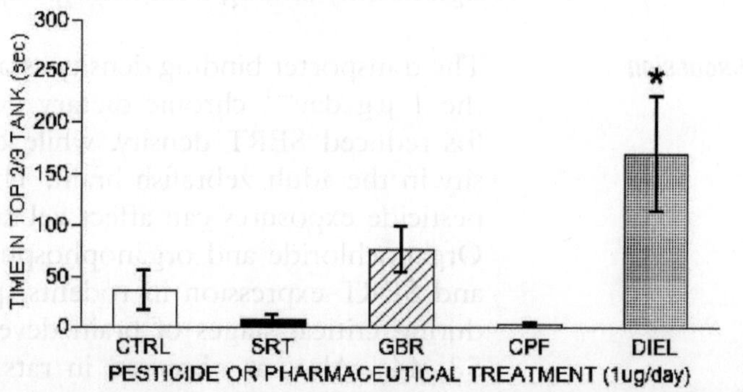

Fig. 8.4. Zebrafish behavior in the dive tank following 21-day dietary administration of pesticides or drugs at 1 μg day^{-1}. $N = 4–7$. *Bars* represent mean and *lines* show standard error. See list of abbreviations in **Fig. 8.1**. Dieldrin (DIEL) significantly increased the amount of time spent in the top 2/3 of the tank ($p < 0.05$) relative to the control (CTRL) group.

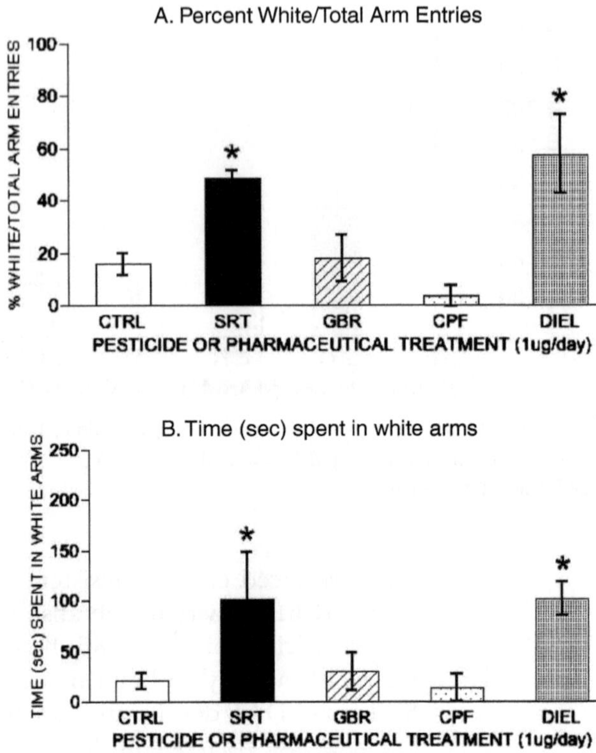

Fig. 8.5. Zebrafish behavior in the light/dark plus maze following 21-day dietary administration of pesticides or drugs at 1 μg day^{-1}. $N = 4$–7. See list of abbreviations in **Fig. 8.1**. **a** Shows % white entries/total arm entries, and **b** shows time spent in the white arms of the aquatic plus maze. Both sertraline (SRT) and dieldrin (DIEL) significantly increased the amount of time spent in white arms and the number of entries into white arms ($p < 0.05$).

white arms ($F_{(4,24)} = 2.9$, $p = 0.04$, B). Fishers LSD $p < 0.05$ for both parameters. The total number of line crosses did not differ significantly among treatment groups.

6.4. Discussion

The transporter binding density assay results revealed that among the 1 μg day^{-1} chronic dietary pesticide exposures, chlorpyrifos reduced SERT density, while dieldrin increased DAT density in the adult zebrafish brain. This demonstrates that chronic pesticide exposures can affect zebrafish biogenic amine systems. Organochloride and organophosphate exposures can alter DAT and SERT expression in rodents, particularly when they occur during critical stages of brain development (3, 41, 42, 63–65, 53, 44). Also, as observed in rats, sertraline and GBR 12909 reduced SERT and DAT density, respectively, in the zebrafish brain (61, 62). The reduction of SERT and DAT by sertraline and GBR 12909 are likely mediated through their long-term binding to and blockade of SERT and DAT, for which they have high affinity. In contrast, organochlorides, organophosphates, and pyrethroids generally exhibit low affinity for DAT and SERT, so their actions at those sites are likely indirect (36, 54, 65).

Reduction of SERT density by chlorpyrifos in the adult zebrafish brain could indicate degeneration of 5-HT neurons, compensatory downregulation in response to intracellular events, or possibly reduced 5-HT turnover or neurotransmission. If zebrafish larvae were exposed to chlorpyrifos during critical stages of brain development, it is possible that SERT might be upregulated in some brain regions, as it is in rats (42). This would be consistent with the notion that SERT upregulation could predispose serotonergic neurons to take up more pesticide metabolites, paralleling events that occur in DA neurons with MPP+ (24, 45, 46, 119). Upregulation of DAT in the zebrafish brain following dieldrin exposure is consistent with the hypothesis that organochlorides increase DAT expression, thereby potentiating DA neurons to greater damage from subsequent neurotoxin exposures, as in rodent models (43), and demonstrates further that this process can also occur in the mature brain.

In adult zebrafish, we did not observe permethrin to affect either SERT or DAT density. Rodent DAT was persistently upregulated after exposure to pyrethroids, which often were administered in juvenile or young adult animals (37, 63, 66). Hence pyrethroid-induced DAT upregulation may depend upon exposure occurring during critical stages of brain development, and may be further explored through larval or juvenile exposures in zebrafish. The three-pesticide blend at 0.33 μg day^{-1} for each pesticide likewise did not produce any significant synergistic or additive changes in SERT or DAT density in the adult zebrafish brain, perhaps due to dose, interactive properties, or timing of exposure.

It remains to be determined whether effects of pesticides in bath exposures are comparable to these dietary exposures, and what zebrafish dosing equivalents should be used to best model human exposures. Serum level measurements of pesticides or drugs by HPLC are not always feasible due to low blood volume, but zebrafish brain can be analyzed for pesticide content by GC-MS, as was performed in native Texas fish to demonstrate biomagnification of pharmaceuticals in wastewater streams (120). Alternatively, use of ELISA immunohistochemical assays to measure levels of specific pesticides or drugs in brain tissue following administration can provide valuable pharmacokinetic information. Upon establishment of appropriate dosing protocols, DAT and/or SERT will likely be affected in a measurable way within occupationally relevant dose equivalents, particularly if exposures occur during developmental windows of vulnerability. Then zebrafish DAT and SERT may be useful biomarkers, among other measures, for modeling effects of occupational or developmental pesticide exposures.

A major goal of this study was to observe the effects of exposure below the threshold of AChE inhibition, and we were not able to achieve this with a dietary dose of 1 μg day^{-1}. Autoradiographic analysis and further AChE activity measurements are underway to examine 0.1 μg day^{-1} dietary exposure for 21 days to the same pesticides and drugs examined in the current study (see **Fig. 8.1**). Sub-AChE inhibiting exposures produced alterations in SERT of rats only during critical stages of development (41, 42), hence we may not observe such effects with lower doses in adult zebrafish. In future studies, zebrafish larvae and juveniles will be exposed to these pesticides dissolved directly into habitat water, allowing examination of a wider range of dose levels, with fish reared to adulthood or senescence for assessment. Following the Braunbeck et al. (121) and Ton et al. (73) models for high throughput assays, such studies could be conducted to measure long-term effects of pesticide exposures occurring at a range of ages, doses, and mixtures and reveal mechanisms of potentiation responses through a vast array of endpoints in mature fish.

One important endpoint is measurement of various behavioral effects in adult fish following different exposure scenarios. Chlorpyrifos-induced AChE inhibition in salmon was significantly correlated with reduced swimming and feeding activity (122). In contrast, low levels of parathion-induced AChE inhibition increased activity and food consumption in zebrafish, with no effects on any other standard toxicological parameters, leading the authors to conclude that while AChE inhibition is a good biomarker for pesticide exposure, it is not a good marker of effects from such exposures (60). AChE activity is positively correlated with addiction, as evidenced by amphetamine studies in AChE-mutant zebrafish and alcohol-induced AChE activity upregulation (123, 124). Through zebrafish behavioral assays such as associative learning or novel environment anxiety tests, the relative role that AChE inhibition may play in the cognitive and emotional impairments experienced by workers exposed to pesticides could be better characterized (25, 26, 28–30).

Pesticide or pharmaceutical exposures may also alter innate anxiety responses. Generalized anxiety disorders are associated with pesticide exposures in farmworkers, particularly to organophosphates (26, 124). In rats, acute chlorpyrifos exposure had anxiogenic effects on behavior in the elevated plus maze, evidenced by fewer open-arm entries and less time spent in open arms (126). In contrast, imaging studies on patients taking selective serotonin reuptake inhibitors indicated that threat-cue processing by the amygdala is impaired (127). When monoamine oxygenase (MAO) was inhibited by deprenyl treatment in larval zebrafish serotonin levels increased, swimming activity decreased, and larvae swam closer to the surface (52). Such behavioral responses might be mediated through extracellular serotonin levels in brain.

In the elevated plus maze, benzodiazepines and the norepinephrine reuptake inhibitor desipramine tend to increase time spent in open arms, and open-arm entries (an anxiolytic response), while paradoxically, serotonin reuptake inhibitors tend to produce anxiogenic responses in rats (103, 128–130). The novel light/dark plus maze for zebrafish is based upon the rodent elevated plus maze. It measures innate response of zebrafish upon introduction into a novel environment, which is a balance between defensive or anxious behavior (fish would remain in dark arms and cross few lines (109)), and exploratory behavior (fish will visit all arms and cross many lines, with white and dark arm visits equally likely (107)), and the effect of drug or pesticide treatment on that response. The novel aquatic light/dark plus maze is not a learning task, but like the elevated plus maze, is an observation of innate behavior in potentially risky new environments.

Results from the dive tank and light/dark plus maze indicate that long-term pesticide exposures can affect zebrafish behaviors that are tied to activities in limbic circuits that involve monoamine systems. In the dive tank, zebrafish exposed to dieldrin spent significantly more time at the top of the tank. One concern in using fish vertical location in the dive tank, as per Levin et al. (108), is that some drug or pesticide treatments might affect swim bladder function through cholinergic and monoaminergic neurons that innervate it, thereby confounding detection of anxiolytic effects (131). However, in the light/dark plus maze, the increased number of white-arm entries, and time spent in white arms by dieldrin-treated fish also demonstrate that it has anxiolytic properties for fish in novel environments. Since the total number of line crosses did not differ significantly among treatment groups, we can assume mobility was similar among them and therefore sedative effects of the pesticide or drug treatments were negligible.

Chlorpyrifos did not increase risky behavior, but neither did it produce anxiogenic effects in zebrafish, which would have been evidenced by significantly reduced exploratory behavior in either novel environment. Anxiogenic responses may be difficult to detect with these protocols, since mean time spent at the top of the dive tank, white-arm entries, and time spent in white arms tend to be low in control fish. It is possible that longer observation times, perhaps 10 min for each test, might improve the chances of detecting anxiogenic effects. Sertraline did not impair swimming activity, as fluoxetine did for larval zebrafish (132). Sertraline, which blocks SERT function directly, and dieldrin, which upregulates DAT through an indirect mechanism, both had anxiolytic effects in the light/dark plus maze: both compounds increased entries into and time spent in white arms, potentially risky behavior for zebrafish in unfamiliar environments (109). Patients taking the serotonin reuptake inhibitor citalopram showed reduced amygdala response to subconscious threat cues (127), consistent with sertraline increasing white-arm entries and time for zebrafish in the light/dark plus maze.

In summary, while there is a strong correlation, the link between pesticide exposure and increased risk of developing a neurodegenerative disorder remains unclear. Use of zebrafish larvae and adults in exposure studies to examine this link has advantages over rodent models, including ease of access, observation, dosing, and the potential to examine effects of pesticides on biogenic amine system development and behavior. Biogenic amine system components, particularly the DAT and SERT, are conserved in fish and are affected by pesticides, especially organochlorides and organophosphates. Mechanistic studies of the chronic neurotoxicological effects of pesticide exposure are needed to understand their relationship to the development of Parkinson's disease and other neurodegenerative disorders. Addition of biogenic amine system assays in zebrafish, plus assessment of anxiety and learning behavior could represent a significant advancement for environmental health risk assessment modeling of chronic low-level pesticide exposures as well as exposures occurring during critical stages of brain development. Effects of exposures to other pesticide classes and mixtures may further affect DA, 5-HT, and other biogenic amine systems, and could be similarly studied through high throughput zebrafish binding assays and behavioral tests. Zebrafish monoamine transporters may be excellent biomarkers of neurotoxicological effects from pesticide exposures. Combined measurement in zebrafish of monoamine transporter expression, function, and AChE activity, together with behavioral measures such as anxiety response or associative learning tasks will be a powerful approach to assessing the role of timing, dose, and pesticide class in potentiative responses and extrapyramidal effects of pesticide exposures in the central nervous system.

7. Conclusions

Zebrafish hold vast potential as model organisms for the study of Parkinson's disease etiology and treatment. Emphasis has been placed on genetic approaches, and mutanogenesis used as a tool to produce disease phenotypes consistent with identified gene mutations in subpopulations of Parkinson's patients. For example zebrafish mutants producing deficient Parkin protein have fewer ascending neurons in the posterior tuberculum of the diencephalon (133). PINK1 kinase enzyme mutant zebrafish (produced through morpholino oligonucleotide knockdown) exhibited an oxidative stress phenotype and some dopaminergic neuron loss that was rescued by antioxidant drugs or human PINK1 mRNA (134). Zebrafish in which the DJ-1 protein, which is mutated in a rare recessive form of early onset Parkinson's, has been knocked down are more susceptible to oxidative stress and are useful for further studies of cell-death pathways (135, 136). Other proteins affected by gene mutations in rare heritable forms of Parkinson's disease have been cloned and their expression characterized in zebrafish embryos, such as ubiquitin C-terminal hydroxylase (137). Use of these mutant forms and genetic manipulations will advance the understanding of processes and cellular mechanisms occurring in rare forms of the disease, and may be more broadly applicable to larger populations of affected individuals. Such studies could lead to improved treatment or prevention of progressive dopaminergic neurodegeneration.

Another approach has been to use neurotoxins such as MPTP or 6-OHDA to reproduce Parkinson's disease like states in zebrafish (138). Exposure of zebrafish embryos or adults to MPTP decreased locomotion and dopamine neuron numbers, but similar exposures to rotenone and paraquat did not (51). MPTP and 6-OHDA intramuscular injections (20–25 mg kg^{-1}) reduced brain dopamine levels and locomotor activity but did not trigger cell death pathways in adult zebrafish (50). However, protection (blockade of DAT or MAO) or DAT removal indicates MPTP-induced neurodegenerative pathways are conserved among zebrafish and mammals (47), and are consistent with potentiation through DAT upregulation. MPP^{+}-induced DA neurodegeneration can be visualized via vesicular monoamine transporters labeled with fluorescent proteins in larvae (139). These models can be used to understand disease mechanisms and to begin developing interventions, such as neuroprotection against MPTP-induced dopaminergic neuron loss by adenosine receptor antagonists like caffeine (140).

In zebrafish models of environmental pesticide or neurotoxin exposure contributions to neurodegenerative diseases, focal behavioral endpoints have been locomotor activity that can be readily quantified en masse. For example, locomotor impairment is often measured by monitoring swimming speed or frequency following neurotoxin or pesticide treatments. Chlorpyrifos larval and MPTP adult exposures decreased or slowed swimming activity (e.g., (51, 56)). MPTP and 6-OHDA injected systemically induced zebrafish to be less mobile, to change direction more, and to swim in the center of plates as opposed to the edges, behavior that is readily video-tracked and computer analyzed for high throughput (49). Bath exposures of dechorionated embryos to μM concentrations MPTP or MPP+ reduced swim speed, dopamine levels, and numbers of diencephalic dopaminergic neurons, these effects could be blocked by deprenyl (141). However, most pesticide or neurotoxin contact effects have been explored primarily through the use of embryos and larvae, and occasionally adults in acute-exposure studies. Few studies have examined the long-term effects of neurotoxin or pesticide exposures in zebrafish, or have focused on mechanisms associated with environmental, in contrast to genetic origins of neurodegenerative pathways. Given the complex nature of Parkinson's disease in which both genetics and environment interplay to produce a range of severity and onset, use of zebrafish to model genetic mutations, environmental exposures, and their interactions in studies encompassing the complete lifespan of the fish should accelerate characterization of the diverse pathological mechanisms of the disease and aid in development of improved interventions.

High throughput pesticide exposure studies in zebrafish, with combined neurochemical and behavioral endpoints might be the key to discovering the etiology of and developing treatments for Parkinson's disease or related neurodegenerative disorders of environmental origin. The nervous system can be very sensitive to certain types of environmental contaminants, especially during vulnerable stages of the life cycle, such as brain development (65, 142). Chronic low-dose exposures to pesticides, by modulating expression of the dopamine transporter, or other monoamine transporters, can render the nervous system more susceptible to further injury by a later exposure (8, 36, 37, 119). Because these types of adverse effects may manifest only after prolonged and serial pesticide or neurotoxin contact, acute risk assessments may underestimate the threat posed by long-term, low-dose, or serial exposures to different pesticide classes. In zebrafish, due to their abbreviated life cycle, it is feasible to conduct juvenile exposures and observe long-term effects in adults or even in geriatric fish. This advantage has been underemphasized and underutilized in applications of the zebrafish to studies of neurodegenerative disorders.

Acknowledgments

This study was funded by a NIOSH ERC Pilot Project research training grant awarded by the Southwest Center for Occupational and Environmental Health at the University of Texas School of Public Health in Houston (Grant No. T42CCT610417 from

NIOSH/CDC to SWCOEH). I would like to thank Dr. Alan Frazer for sponsoring this research in his laboratory at UTHSCSA, Ezra Scientific LLC for technical assistance in developing and building the zebrafish offset cross maze, and Jim Sackerman, Ngoc Nhung Nguyen, Adam Long, Kelly Lawless, and Dr. Bob Benno at William Paterson University, New Jersey, for their technical assistance and intellectual input in validating the dive tank and light/dark plus maze tests.

References

1. Baldereschi, M., Di Carlo, A., Vanni, P., Ghetti, A., Carbonin, P., Amaducci, L. & Inzitari, D. (2003) Lifestyle-related risk factors for Parkinson's disease: a population-based study. *Acta Neurol Scand* 108, 239–244.
2. Hubble, J. P., Cao, T., Hassanein, R. E., Neuberger, J. S. & Koller, W. C. (1993) Risk factors for Parkinson's disease. *Neurology* 43, 1693–1697.
3. Kanthasamy, A. G., Kitazawa, M., Kanthasamy, A. & Anantharam, V. (2005) Dieldrin-induced neurotoxicity: relevance to Parkinson's disease pathogenesis. *Neurotoxicology* 26, 701–719.
4. Petrovitch, H., Ross, G. W., Abbott, R. D., Sanderson, W. T., Sharp, D. S., Tanner, C. M., Masaki, K. H., Blanchette, P. L., Popper, J. S., Foley, D., Launer, L. & White, L. R. (2002) Plantation work and risk of Parkinson disease in a population-based longitudinal study. *Arch Neurol* 59, 1787–1792.
5. Ascherio, A., Chen, H., Weisskopf, M. G., O'Reilly, E., McCullough, M. L., Calle, E. E., Schwarzschild, M. A. & Thun, M. J. (2006) Pesticide exposure and risk for Parkinson's disease. *Ann Neurol* 60, 197–203.
6. Brown, T. P., Rumsby, P. C., Capleton, A. C., Rushton, L. & Levy, L. S. (2006) Pesticides and Parkinson's disease – is there a link? *Environ Health Perspect* 114(2), 156–164, Feb.
7. Hatcher, J. M., Pennell, K. D. & Miller, G. W. (2008) Parkinson's disease and pesticides: a toxicological perspective. *Trends Pharmacol Sci* 29, 322–329.
8. Shepherd, K. R., Lee, E. S., Schmued, L., Jiao, Y., Ali, S. F., Oriaku, E. T., Lamango, N. S., Soliman, K. F. & Charlton, C. G. (2006) The potentiating effects of 1-methyl-4-phenyl-1,2,3,6-tetrahydropyridine (MPTP) on paraquat-induced neurochemical and behavioral changes in mice. *Pharmacol Biochem Behav* 83, 349–359.
9. Gorell, J. M., Johnson, C. C., Rybicki, B. A., Peterson, E. L. & Richardson, R. J. (1998) The risk of Parkinson's disease with exposure to pesticides, farming, well water, and rural living. *Neurology* 50, 1346–1350.
10. Alexander, B. H., Burns, C. J., Bartels, M. J., Acquavella, J. F., Mandel, J. S., Gustin, C. & Baker, B. (2006) Chlorpyrifos exposure in farm families: results from the farm family exposure study. *J Exp Sci Environ Epidemiol* 16, 447–456.
11. Bouvier, G., Blanchard, O., Momas, I. & Seta, N. (2006a) Environmental and biological monitoring of exposure to organophosphorus pesticides: application to occupationally and non-occupationally exposed adult populations. *J Exp Sci Environ Epidemiol* 16(5), 417–426.
12. Bouvier, G., Blanchard, O., Momas, I. & Seta, N. (2006b) Pesticide exposure of non-occupationally exposed subjects compared to some occupational exposure: a French pilot study. *Sci Total Environ* 366, 74–91.
13. Fenske, R. A., Lu, C., Curl, C. L., Shirai, J. H. & Kissel, J. C. (2005) Biologic monitoring to characterize organophosphorus pesticide exposure among children and workers: an analysis of recent studies in Washington State. *Environ Health Perspect* 113, 1651–1657.
14. Jirachaiyabhas, V., Visuthismajarn, P., Hore, P. & Robson, M. G. (2004) Organophosphate pesticide exposures of traditional and integrated pest management farmers from working air conditions: a case study in Thailand. *Int J Occup Environ Health* 10, 289–295.
15. Coghlan, A. (2005) Exposure to pesticides can cause Parkinson's. *New Scientist* 2501, 14.
16. Costello, S., Cockburn, M., Bronstein, J., Zhang, X. & Ritz, B. (2009) Parkinson's disease and residential exposure to maneb and paraquat from agricultural applications in the

central valley of California. *Am J Epidemiol* 169(8), 919–926, Apr 15.
17. Kamel, F., Tanner, C., Umbach, D., Hoppin, J., Alavanja, M., Blair, A., Comyns, K., Goldman, S., Korell, M., Langston, J., Ross, G. & Sandler, D. (2006) Pesticide exposure and self-reported Parkinson's disease in the agricultural health study. *Am J Epidemiol* 165, 364–374.
18. Kamel, F. & Hoppin, J. A. (2004) Association of pesticide exposure with neurologic dysfunction and disease. *Environ Health Perspect* 112, 950–958.
19. Seidler, A., Hellenbrand, W., Robra, B. P., Vieregge, P., Nischan, P., Joerg, J., Oertel, W. H., Ulm, G. & Schneider, E. (1996) Possible environmental, occupational, and other etiologic factors for Parkinson's disease: a case-control study in Germany. *Neurology* 46, 1275–1284.
20. Abou-Donia, M. B., Wilmarth, K. R., Jensen, K. F., Oehme, F. W. & Kurt, T. L. (1996) Neurotoxicity resulting from coexposure to pyridostigmine bromide, deet, and permethrin: implications of Gulf War chemical exposures. *J Toxicol Environ Health* 48, 35–56.
21. Chen, C. & Lu, C. (2002) An analysis of the combined effects of organic toxicants. *Sci Total Environ* 289, 123–132.
22. Betarbet, R., Sherer, T. B., MacKenzie, G., Garcia-Osuna, M., Panov, A. V. & Greenamyre, J. I. (2000) Chronic systemic pesticide exposure reproduces features of Parkinson's disease. *Nat Neurosci* 3, 1301–1306.
23. Betarbet, R., Sherer, T. B. & Greenamyre, J. T. (2002) Animal models of Parkinson's disease. *Bioessays* 24, 308–318.
24. Cory-Slechta, D. A., Thiruchelvam, M., Richfield, E. K., Barlow, B. K. & Brooks, A. I. (2005) Developmental pesticide exposures and the Parkinson's disease phenotype. *Birth Defects Res A Clin Mol Teratol* 73, 136–139.
25. Beseler, C. L., Stallones, L., Hoppin, J. A., Alavanja, M. C., Blair, A., Keefe, T. & Kamel, F. (2008) Depression and pesticide exposures among private pesticide applicators enrolled in the Agricultural Health Study. *Environ Health Perspect* 116, 1713–1719.
26. Salvi, R. M., Lara, D. R., Ghisolfi, E. S., Portela, L. V., Dias, R. D. & Souza, D. O. (2003) Neuropsychiatric evaluation in subjects chronically exposed to organophosphate pesticides. *Toxicol Sci* 72, 267–271.
27. Stallones, L. & Beseler, C. (2002) Pesticide poisoning and depressive symptoms among farm residents. *Ann Epid* 12, 389–394.
28. Aldridge, J. E., Levin, E. D., Seidler, F. J. & Slotkin, T. A. (2005a) Developmental exposure of rats to chlorpyrifos leads to behavioral alterations in adulthood, involving serotonergic mechanisms and resembling animal models of depression. *Environ Health Perspect* 113, 527–531.
29. Alavanja, M. C., Hoppin, J. A. & Kamel, F. (2004) Health effects of chronic pesticide exposure: cancer and neurotoxicity. *Annu Rev Public Health* 25, 155–197.
30. Sánchez-Santed, F., Cañadas, F., Flores, P., López-Grancha, M. & Cardona, D. (2004) Long-term functional neurotoxicity of paraoxon and chlorpyrifos: behavioural and pharmacological evidence. *Neurotoxicol Teratol* 26, 305–317.
31. Bradman, A., Eskenazi, B., Barr, D. B., Bravo, R., Castorina, R., Chevrier, J., Kogut, K., Harnly, M. E. & McKone, T. E. (2005) Organophosphate urinary metabolite levels during pregnancy and after delivery in women living in an agricultural community. *Environ Health Perspect* 113, 1802–1807.
32. Crisostomo, L. & Molina, V. V. (2002) Pregnancy outcomes among farming households of Nueva Ecija with conventional pesticide use versus integrated pest management. *Int J Occup Environ Health* 8, 232–242.
33. Curl, C. L., Fenske, R. A., Kissel, J. C., Shirai, J. H., Moate, T. F., Griffith, W., Coronado, G. & Thompson, B. (2002) Evaluation of take-home organophosphorus pesticide exposure among agricultural workers and their children. *Environ Health Perspect* 110, A787–A792.
34. Fenske, R. A. (1997) Pesticide exposure assessment of workers and their families. *Occup Med* 12, 221–237.
35. Shipp, E. M., Cooper, S. P., Burau, K. D. & Bolin, J. N. (2005) Pesticide safety training and access to field sanitation among migrant farmworker mothers from Starr County, Texas. *J Agric Saf Health* 11, 51–60.
36. Elwan, M. A., Richardson, J. R., Guillot, T. S., Caudle, W. M. & Miller, G. W. (2006) Pyrethroid pesticide-induced alterations in dopamine transporter function. *Toxicol Appl Pharmacol* 211, 188–197.
37. Gillette, J. S. & Bloomquist, J. R. (2003) Differential up-regulation of striatal dopamine transporter and alpha-synuclein by the pyrethroid insecticide permethrin. *Toxicol Appl Pharmacol* 192, 287–293.
38. Miller, G. W., Gainetdinov, R. R., Levey, A. I. & Caron, M. G. (1999) Dopamine transporters and neuronal injury. *Trends Pharmacol Sci* 20, 424–429.

39. Ossowska, K., Wardas, J., Smialowska, M., Kuter, K., Lenda, T., Wieronska, J. M., Zieba, B., Nowak, P., Dabrowska, J., Bortel, A., Kwiecinski, A. & Wolfarth, S. (2005) A slowly developing dysfunction of dopaminergic nigrostriatal neurons induced by long-term paraquat administration in rats: an animal model of preclinical stages of Parkinson's disease? *Eur J Neurosci* 22, 1294–1304.
40. Uhl, G. R. (1998) Hypothesis: the role of dopaminergic transporters in selective vulnerability of cells in Parkinson's disease. *Ann Neurol* 43, 555–560.
41. Aldridge, J. E., Seidler, F. J., Meyer, A., Thillai, I. & Slotkin, T. A. (2003) Serotonergic systems targeted by developmental exposure to chlorpyrifos: effects during different critical periods. *Environ Health Perspect* 111, 1736–1743.
42. Aldridge, J. E., Seidler, F. J. & Slotkin, T. A. (2004) Developmental exposure to chlorpyrifos elicits sex-selective alterations of serotonergic synaptic function in adulthood: critical periods and regional selectivity for effects on the serotonin transporter, receptor subtypes, and cell signaling. *Environ Health Perspect* 112, 148–155.
43. Richardson, J. R., Caudle, W. M., Wang, M., Dean, E. D., Pennell, K. D. & Miller, G. W. (2006) Developmental exposure to the pesticide dieldrin alters the dopamine system and increases neurotoxicity in an animal model of Parkinson's disease. *FASEB J* 20, 1695–1697.
44. Slotkin, T. A. & Seidler, F. J. (2005) The alterations in CNS serotonergic mechanisms caused by neonatal chlorpyrifos exposure are permanent. *Brain Res Dev Brain Res* 158, 115–119.
45. Carvey, P. M., Punati, A. & Newman, M. B. (2006) Progressive dopamine neuron loss in Parkinson's disease: the multiple hit hypothesis. *Cell Transplant* 15, 239–250.
46. Sulzer, D. (2007) Multiple hit hypotheses for dopamine neuron loss in Parkinson's disease. *Trends Neurosci* 30, 244–250.
47. McKinley, E. T., Baranowski, T. C., Blavo, D. O., Cato, C., Doan, T. N. & Rubinstein, A. L. (2005) Neuroprotection of MPTP-induced toxicity in zebrafish dopaminergic neurons. *Brain Res Mol Brain Res* 141, 128–137.
48. Doshna, C., Benbow, J., DePasquale, M., Okerberg, C., Turnquist, S., Stedman, D., Chapin, B., Sivaraman, L., Waldron, G., Navetta, K., Brady, J., Banker, M., Casimiro-Garcia, A., Hill, A., Jones, M., Ball, J. & Aleo, M. (2009) Multi-phase analysis of uptake and toxicity in zebrafish: relationship to compound physical-chemical properties. The Toxicologist, Poster Abstract#377, Society of Toxicology 2009 Annual Meeting, Baltimore MD.
49. Panula, P., Sallinen, V., Sundvik, M., Kolehmainen, J., Torkko, V., Tiittula, A., Moshnyakov, M. & Podlasz, P. (2006) Modulatory neurotransmitter systems and behavior: towards zebrafish models of neurodegenerative diseases. *Zebrafish* 3, 235–247.
50. Anichtchik, O. V., Kaslin, J., Pietsaro, N., Scheinin, M. & Panula, P. (2004) Neurochemical and behavioral changes in zebrafish *Danio rerio* after systematic administration of 6-hydroxydopamine and 1-methyl-4-phenyl-1,2,3,6-tetrahydropyridine. *J Neurochem* 88, 443–453.
51. Bretaud, S., Lee, S. & Guo, S. (2004) Sensitivity of zebrafish to environmental toxins implicated in Parkinson's disease. *Neurotox Teratol* 26, 857–864.
52. Sallinen, V., Sundvik, M., Reenilä, I., Peitsaro, N., Khrustalyov, D., Anichtchik, O., Toleikyte, G., Kaslin, J. & Panula, P. (2009a) Hyperserotonergic phenotype after monoamine oxidase inhibition in larval zebrafish. *J Neurochem* 109, 403–415.
53. Aldridge, J. E., Meyer, A., Seidler, F. J. & Slotkin, T. A. (2005b) Alterations in central nervous system serotonergic and dopaminergic synaptic activity in adulthood after prenatal or neonatal chlorpyrifos exposure. *Environ Health Perspect* 113, 1027–1031.
54. Gould, G. G., Brooks, B. W. & Frazer, A. (2007) [^{3}H] citalopram binding to serotonin transporter sites in minnow brains. *Basic Clin Pharmacol Toxicol* 101, 203–210.
55. Levin, E. D., Chrysanthis, E., Yacisin, K. & Linney, E. (2003) Chlorpyrifos exposure of developing zebrafish: effects on survival and long-term effects on response latency and spatial discrimination. *Neurotoxicol Teratol* 25, 51–57.
56. Levin, E. D., Swain, H. A., Donerly, S. & Linney, E. (2004) Developmental chlorpyrifos effects on hatchling zebrafish swimming behavior. *Neurotoxicol Teratol* 26, 719–723.
57. Ciesielski, S., Loomis, D. P., Mims, S. R. & Auer, A. (1994) Pesticide exposures, cholinesterase depression, and symptoms among North Carolina migrant farmworkers. *Am J Publ Health* 84, 446–451.
58. Cañadas, F., Cardona, D., Davila, E. & Sanchez-Santed, F. (2005) Long-term neurotoxicity of chlorpyrifos: spatial learning impairment on repeated acquisition in a water maze. *Toxicol Sci* 85, 944–951.

59. Jarvinen, A. W., Nordling, B. R. & Henry, M. E. (1983) Chronic toxicity of Dursban (chlorpyrifos) to the fathead minnow and the resultant acetylcholinesterase inhibition. *Ecotoxicol Environ Safety* 7, 423–434.
60. Roex, E. W., Keijzers, R. & van Gestel, C. A. (2003) Acetylcholinesterase inhibition and increased food consumption rate in the zebrafish, *Danio rerio*, after chronic exposure to parathion. *Aquat Toxicol* 64, 451–460.
61. Benmansour, S., Cecchi, M., Morilak, D., Gerhardt, G. A., Javors, M. A., Gould, G. G. & Frazer, A. (1999) Effects of chronic antidepressant treatments on serotonin transporter function, density, and mRNA. *J Neurosci* 19, 10494–10501.
62. Kunko, P. M., Loeloff, R. J. & Izenwasser, S. (1997) Chronic administration of the selective dopamine uptake inhibitor GBR 12,909, but not cocaine, produces marked decreases in dopamine transporter density. *Naunyn Schmiedebergs Arch Pharmacol* 356, 562–569.
63. Bloomquist, J. R., Barlow, R. L., Gillette, J. S., Li, W. & Kirby, M. L. (2002) Selective effects of insecticides on nigrostriatal dopaminergic nerve pathways. *Neurotoxicology* 23, 537–544.
64. Kitazawa, M., Anantharam, V. & Kanthasamy, A. G. (2003) Dieldrin induces apoptosis by promoting caspase-3-dependent proteolytic cleavage of protein kinase Cgamma in dopaminergic cells: relevance to oxidative stress and dopaminergic degradation. *Neuroscience* 119, 945–964.
65. Purkerson-Parker, S., McDaniel, K. L. & Moser, V. C. (2001) Dopamine transporter binding in the rat striatum is increased by gestational, perinatal, and adolescent exposure to heptachlor. *Toxicol Sci* 64, 216–223.
66. Karen, D. J., Li, W., Harp, P. R., Gillette, J. S. & Bloomquis, J. R. (2001) Striatal dopaminergic pathways as a target for the insecticides permethrin and chlorpyrifos. *Neurotoxicology* 22, 811–817.
67. Braunbeck, T., Boettcher, M., Hollert, H., Kosmehl, T., Lammer, E., Leist, E., Rudolf, M. & Seitz, N. (2005) Towards an alternative for the acute fish LC(50) test in chemical assessment: the fish embryo toxicity test goes multi-species – an update. *ALTEX* 22, 87–102.
68. Guo, S. (2004) Linking genes to brain, behavior and neurological diseases: what can we learn from zebrafish? *Genes Brain Behav* 3, 63–74.
69. Hinton, D. E., Kullman, S. W., Hardman, R. C., Volz, D. C., Chen, P. J., Carney, M. & Bencic, D. C. (2005) Resolving mechanisms of toxicity while pursuing ecotoxicological relevance? *Mar Pollut Bull* 51, 635–648.
70. Lardelli, M. (2005) Zebrafish – do we need another vertebrate model? *Anzccart News* 13, 1–3.
71. Linney, E., Upchurch, L. & Donerly, S. (2004) Zebrafish as a neurotoxicological model. *Neurotoxicol Teratol* 26, 709–718.
72. Repetto, G., del Peso, A. & Repetto, M. (2000) Alternative ecotoxicological methods for the evaluation, control and monitoring of environmental pollutants. *Ecotoxicol Environ Restor* 3, 47–51.
73. Ton, C., Lin, Y. & Willett, C. (2006) Zebrafish as a model for developmental neurotoxicity testing. *Birth Defects Res A Clin Mol Teratol* 76, 553–567.
74. Le Crom, S., Kapsimali, M., Barôme, P. O. & Vernier, P. (2003) Dopamine receptors for every species: gene duplications and functional diversification in Craniates. *J Struct Funct Genomics* 3, 161–176.
75. Ryu, S., Holzschuh, J., Mahler, J. & Driever, W. (2006) Genetic analysis of dopaminergic system development in zebrafish. *J Neural Transm Suppl* 70, 61–66.
76. Caveney, S., Cladman, W., Verellen, L. & Donly, C. (2006) Ancestry of neuronal monoamine transporters in the metazoa. *J Exp Biol* 209, 4858–4868.
77. Huggett, D. B., Cook, J. C., Ericson, J. E. & Williams, R. T. (2003) Theoretical model for prioritizing potential impacts of human pharmaceuticals to fish. *J Hum Ecol Risk Assess* 9, 1789–1799.
78. Rink, E. & Wullimann, M. F. (2002) Connections of the ventral telencephalon and tyrosine hydroxylase distribution in the zebrafish brain (*Danio rerio*) lead to identification of an ascending dopaminergic system in a teleost. *Brain Res Bull* 57, 385–387.
79. Rink, E. & Wullimann, M. F. (2004) Connections of the ventral telencephalon (subpallium) in the zebrafish (*Danio rerio*). *Brain Res* 1011, 206–220.
80. Salas, C., Broglio, C., Durán, E., Gómez, A., Ocaña, F. M., Jiménez-Moya, F. & Rodríguez, F. (2006) Neuropsychology of learning and memory in teleost fish. *Zebrafish* 3, 157–171.
81. Koutoku, T., Zhang, R., Tachibana, T., Oshima, Y. & Furuse, M. (2003) Effect of acute L-tryptophan exposure on the brain serotonergic system and behavior in the male medaka. *Zool Sci* 20, 121–124.
82. Lepage, O., Larson, E. T., Mayer, I. & Winberg, S. (2005) Serotonin, but not mela-

tonin, plays a role in shaping dominant-subordinate relationships and aggression in rainbow trout. *Horm Behav* 8, 233–242.
83. Perreault, H. A., Semsar, K. & Godwin, J. (2003) Fluoxetine treatment decreases territorial aggression in a coral reef fish. *Physiol Behav* 79, 719–724.
84. Winberg, S., Overli, O. & Lepage, O. (2001) Suppression of aggression in rainbow trout (*Oncorhynchus mykiss*) by dietary L-tryptophan. *J Exp Biol* 204, 3867–3876.
85. Winberg, S. & Nilsson, G. (1996) Multiple high-affinity binding sites for [3H] serotonin in the brain of a teleost fish, the Arctic charr (*Salvelinus alpinus*). *J Exp Biol* 199, 2429–2435.
86. Ferriere, F., Khan, N. A., Meyniel, J. P. & Deschaux, P. (1999) Characterisation of serotonin transport mechanisms in rainbow trout peripheral blood lymphocytes: role in PHA-induced lymphoproliferation. *Dev Comp Immunol* 23, 37–50.
87. Wang, Y., Takai, R., Yoshioka, H. & Shirabe, K. (2006) Characterization and expression of serotonin transporter genes in zebrafish. *Tohoku J Exp Med* 208, 267–274.
88. Norton, W. H., Folchert, A. & Bally-Cuif, L. (2008) Comparative analysis of serotonin receptor (HTR1A/HTR1B families) and transporter (slc6a4a/b) gene expression in the zebrafish brain. *J Comp Neurol* 511, 521–542.
89. Severinsen, K., Sinning, S., Müller, H. K. & Wiborg, O. (2008) Characterisation of the zebrafish serotonin transporter functionally links TM10 to the ligand binding site. *J Neurochem* 105, 1794–1805.
90. Boehmler, W., Obrecht-Pflumio, S., Canfield, V., Thisse, C., Thisse, B. & Levenson, R. (2004) Evolution and expression of D2 and D3 dopamine receptor genes in zebrafish. *Dev Dyn* 230, 481–493.
91. Clements, S. & Schreck, C. B. (2004) Evidence that GABA mediates dopaminergic and serotonergic pathways associated with locomotor activity in juvenile chinook salmon (*Oncorhynchus tshawytscha*). *Behav Neurosci* 118, 191–198.
92. Johansson, V., Winberg, S. & Bjornsson, B. T. (2005) Growth hormone-induced stimulation of swimming and feeding behaviour of rainbow trout is abolished by the D1 dopamine antagonist SCH23390. *Gen Comp Endocrinol* 141, 58–65.
93. Rink, E. & Wullimann, M. F. (2001) The teleostean (zebrafish) dopaminergic system ascending to the subpallium (striatum) is located in the basal diencephalon (posterior tuberculum). *Brain Res* 889, 316–330.
94. Holzschuh, J., Ryu, S., Aberger, F. & Driever, W. (2001) Dopamine transporter expression distinguishes dopaminergic neurons from other catecholaminergic neurons in the developing zebrafish embryo. *Mech Dev* 101, 237–243.
95. López Patiño, M. A., Yu, L., Yamamoto, B. K. & Zhdanova, I. V. (2008) Gender differences in zebrafish responses to cocaine withdrawal. *Physiol Behav* 95, 36–47.
96. Bretaud, S., Li, Q., Lockwood, B. L., Kobayashi, K., Lin, E. & Guo, S. (2007a) A choice behavior for morphine reveals experience-dependent drug preference and underlying neural substrates in developing larval zebrafish. *Neuroscience* 146, 1109–1116.
97. Kily, L. J., Cowe, Y. C., Hussain, O., Patel, S., McElwaine, S., Cotter, F. E. & Brennan, C. H. (2008) Gene expression changes in a zebrafish model of drug dependency suggest conservation of neuro-adaptation pathways. *J Exp Biol* 211, 1623–1634.
98. Ninkovic, J. & Bally-Cuif, L. (2006) The zebrafish as a model system for assessing the reinforcing properties of drugs of abuse. *Methods* 39, 262–274.
99. Colwill, R. M., Raymond, M. P., Ferreira, L. & Escudero, H. (2005) Visual discrimination learning in zebrafish (*Danio rerio*). *Behav Processes* 70, 19–31.
100. Darland, T. & Dowling, J. E. (2001) Behavioral screening for cocaine sensitivity in mutagenized zebrafish. *Proc Natl Acad Sci USA* 98, 11691–11696.
101. Lau, B., Bretaud, S., Huang, Y., Lin, E. & Guo, S. (2006) Dissociation of food and opiate preference by a genetic mutation in zebrafish. *Genes Brain Behav* 5, 497–505.
102. Carobrez, A. P. & Bertoglio, L. J. (2005) Ethological and temporal analyses of anxiety-like behavior: the elevated plus-maze model 20 years on. *Neurosci Biobehav Rev* 29, 1193–1205.
103. Bondi, C. O., Rodriguez, G., Gould, G. G., Frazer, A. & Morilak, D. A. (2008) Chronic unpredictable stress induces a cognitive deficit and anxiety-like behavior in rats that is prevented by chronic antidepressant drug treatment. *Neuropsychopharmacology* 33, 320–331.
104. Bass, S. L. & Gerlai, R. (2008) Zebrafish (*Danio rerio*) responds differentially to stimulus fish: the effects of sympatric and allopatric predators and harmless fish. *Behav Brain Res* 186, 107–117.
105. Ferrari, M. C., Messier, F. & Chivers, D. P. (2008) Can prey exhibit threat-sensitive generalization of predator recognition?

Extending the predator recognition continuum hypothesis. *Proc Biol Sci* 2008(275), 1811–1816.
106. Speedie, N. & Gerlai, R. (2008) Alarm substance induced behavioral responses in zebrafish (*Danio rerio*). *Behav Brain Res* 188, 168–177.
107. Peitsaro, N., Kaslin, J., Anichtchik, O. V. & Panula, P. (2003) Modulation of the histaminergic system and behaviour by alpha-fluoromethylhistidine in zebrafish. *J Neurochem* 86, 432–441.
108. Levin, E. D., Bencan, Z. & Cerutti, D. T. (2007) Anxiolytic effects of nicotine in zebrafish. *Physiol Behav* 90, 54–58.
109. Serra, E. L., Medalha, C. C. & Mattioli, R. (1999) Natural preference of zebrafish (*Danio rerio*) for a dark environment. *Braz J Med Biol Res* 32, 1551–1553.
110. Office of Laboratory Animal Welfare (2002) *Guide for the Care and Use of Laboratory Animals.* Bethesda, MD, Public Health Service Policy on Humane Care and Use of Laboratory Animals. National Institutes of Health.
111. Wullimann, M. F., Rupp, B. & Reichert, H. (1996) *The Neuroanatomy of the Zebrafish Brain: A Topological Atlas.* Boston, MA, Birkhäuser.
112. Kovachich, G. B., Aronson, C. E., Brunswick, D. J. & Frazer, A. (1988) Quantitative autoradiography of serotonin uptake sites in rat brain using [^{3}H] cyanoimipramine. *Brain Res* 454, 78–88.
113. Galici, R., Galli, A., Jones, D. J., Sanchez, T. A., Saunders, C., Frazer, A., Gould, G. G., Lin, R. Z. & France, C. P. (2003) Selective decreases in amphetamine self-administration and regulation of dopamine transporter function in diabetic rats. *Neuroendocrinology* 77, 132–140.
114. Geary, W. A., Tioga, A. W. & Wooten, G. F. (1985) Quantitative film autoradiography for tritium: methodological considerations. *Brain Res* 337, 99–108.
115. Boja, J. W., Mitchell, W. M., Patel, A., Kopajtic, T. A., Carroll, F. I., Lewin, A. H., Abraham, P. & Kuhar, M. J. (1992) High-affinity binding of [^{125}I] RTI-55 to dopamine and serotonin transporters in rat brain. *Synapse* 12, 27–36.
116. Bradford, M. M. (1976) A rapid and sensitive method for the quantitation of microgram quantities of protein utilizing the principle of protein-dye binding. *Anal Biochem* 72, 248–254.
117. Wheelock, C. E., Eder, K. J., Werner, I., Huang, H., Jones, P. D., Brammell, B. F., Elskus, A. A. & Hammock, B. D. (2005) Individual variability in esterase activity and CYP1A levels in Chinook salmon (*Oncorhynchus tshawytscha*) exposed to esfenvalerate and chlorpyrifos. *Aquat Toxicol* 74, 172–192.
118. Ellman, G. L., Courtney, K. D., Andres, V., Jr & Feather-Stone, R. M. (1961) A new and rapid colorimetric determination of acetylcholinesterase activity. *Biochem Pharmacol* 7, 88–95.
119. Cui, M., Aras, R., Christian, W. V., Rappold, P. M., Hatwar, M., Panza, J., Jackson-Lewis, V., Javitch, J. A., Ballatori, N., Przedborski, S. & Tieu, K. (2009) The organic cation transporter-3 is a pivotal modulator of neurodegeneration in the nigrostriatal dopaminergic pathway. *Proc Natl Acad Sci* 106, 8043–8048.
120. Brooks, B. W., Chambliss, C. K., Stanley, J. K., Ramirez, A., Banks, K. E., Johnson, R. D. & Lewis, R. J. (2005) Determination of select antidepressants in fish from an effluent-dominated stream. *Environ Toxicol Chem* 24, 464–469.
121. Braunbeck, T., Boettcher, M., Hollert, H., Kosmehl, T., Lammer, E., Leist, E., Rudolf, M. & Seitz, N. (2005) Towards an alternative for the acute fish LC(50) test in chemical assessment: the fish embryo toxicity test goes multi-species – an update. *ALTEX* 22, 87–102.
122. Sandahl, J. F., Baldwin, D. H., Jenkins, J. J. & Scholz, N. L. (2005) Comparative thresholds for acetylcholinesterase inhibition and behavioral impairment in coho salmon exposed to chlorpyrifos. *Environ Toxicol Chem* 24, 136–145.
123. Ninkovic, J., Folchert, A., Makhankov, Y. V., Neuhauss, S. C., Sillaber, I., Straehle, U. & Bally-Cuif, L. (2006) Genetic identification of AChE as a positive modulator of addiction to the psychostimulant D-amphetamine in zebrafish. *J Neurobiol* 66, 463–475.
124. Rico, E. P., Rosemberg, D. B., Dias, R. D., Bogo, M. R. & Bonan, C. D. (2007) Ethanol alters acetylcholinesterase activity and gene expression in zebrafish brain. *Toxicol Lett* 174, 25–30.
125. Beseler, C. & Stallones, L. (2003) Safety practices, neurological symptoms, and pesticide poisoning. *J Occup Environ Med* 45, 1079–1086.
126. Sánchez-Amate, M. C., Flores, P. & Sánchez-Santed, F. (2001) Effects of chlorpyrifos in the plus-maze model of anxiety. *Behav Pharmacol* 12, 285–292.
127. Harmer, C. J., Mackay, C. E., Reid, C. B., Cowen, P. J. & Goodwin, G. M. (2006) Antidepressant drug treatment modifies the

neural processing of nonconscious threat cues. *Biol Psychiatry* 59, 816–820.
128. Drapier, D., Bentué-Ferrer, D., Laviolle, B., Millet, B., Allain, H., Bourin, M. & Reymann, J. M. (2007) Effects of acute fluoxetine, paroxetine and desipramine on rats tested on the elevated plus-maze. *Behav Brain Res* 176, 202–209.
129. Kurt, M., Arik, A. C. & Celik, S. (2000) The effects of sertraline and fluoxetine on anxiety in the elevated plus-maze test in mice. *J Basic Clin Physiol Pharmacol* 11, 173–180.
130. Lapiz-Bluhm, M. D., Bondi, C. O., Doyen, J., Rodriguez, G. A., Bédard-Arana, T. & Morilak, D. A. (2008) Behavioural assays to model cognitive and affective dimensions of depression and anxiety in rats. *J Neuroendocrinol* 20, 1115–1137.
131. Finney, J. L., Robertson, G. N., McGee, C. A., Smith, F. M. & Croll, R. P. (2006) Structure and autonomic innervation of the swim bladder in the zebrafish (*Danio rerio*). *J Comp Neurol* 495, 587–606.
132. Airhart, M. J., Lee, D. H., Wilson, T. D., Miller, B. E., Miller, M. N. & Skalko, R. G. (2007) Movement disorders and neurochemical changes in zebrafish larvae after bath exposure to fluoxetine (PROZAC). *Neurotoxicol Teratol* 29, 652–664.
133. Flinn, L., Mortiboys, H., Volkmann, K., Köster, R. W., Ingham, P. W. & Bandmann, O. (2009) Complex I deficiency and dopaminergic neuronal cell loss in parkin-deficient zebrafish (*Danio rerio*). *Brain* 132, 1613–1623.
134. Anichtchik, O., Diekmann, H., Fleming, A., Roach, A., Goldsmith, P. & Rubinsztein, D. C. (2008) Loss of PINKl function affects development and results in neurodegeneration in zebrafish. *J Neurosci* 28, 8199–8207.
135. Bai, Q., Mullett, S. J., Garver, J. A., Hinkle, D. A. & Burton, E. A. (2006) Zebrafish DJ-1 is evolutionarily conserved and expressed in dopaminergic neurons. *Brain Res* 1113, 33–44.
136. Bretaud, S., Allen, C., Ingham, P. W. & Bandmann, O. (2007b) p53-dependent neuronal cell death in a DJ-1-deficient zebrafish model of Parkinson's disease. *J Neurochem* 100, 1626–1635.
137. Son, O. L., Kim, H. T., Ji, M. H., Yoo, K. W., Rhee, M. & Kim, C. H. (2003) Cloning and expression analysis of a Parkinson's disease gene, uch-L1, and its promoter in zebrafish. *Biochem Biophys Res Commun* 2003(312), 601–607.
138. Rubinstein, A. L. (2003) Zebrafish: from disease modeling to drug discovery. *Curr Opin Drug Discov Devel* 6, 218–223.
139. Wen, L., Wei, W., Gu, W., Huang, P., Ren, X., Zhang, Z., Zhu, Z., Lin, S. & Zhang, B. (2008) Visualization of monoaminergic neurons and neurotoxicity of MPTP in live transgenic zebrafish. *Dev Biol* 314(1), 84–92.
140. Boehmler, W., Petko, J., Woll, M., Frey, C., Thisse, B., Thisse, C., Canfield, V. A. & Levenson, R. (2009) Identification of zebrafish A2 adenosine receptors and expression in developing embryos. *Gene Expr Patterns* 9, 144–151.
141. Sallinen, V., Torkko, V., Sundvik, M., Reenilä, I., Khrustalyov, D., Kaslin, J. & Panula, P. (2009b) MPTP and MPP+ target specific aminergic cell populations in larval zebrafish. *J Neurochem* 108, 719–731.
142. Qiao, D., Seidler, F. J., Padilla, S. & Slotkin, T. A. (2002) Developmental neurotoxicity of chlorpyrifos: what is the vulnerable period? *Environ Health Perspect* 110, 1097–1103.

Chapter 9

Learned Recognition by Zebrafish and Other Cyprinids

Brian D. Wisenden

Abstract

Antipredator behavior is triggered by a combination of internal proximate mechanisms (anatomical receptors and the physiological processes that regulate their function) and external environmental cues that signal the context and timing of when behavior is likely to be effective. Responses to some external environmental triggers, such as the presence of conspecific chemical alarm cue, are governed strictly by genetic templates. Other external environmental triggers are learned through a special type of associative learning called releaser-induced recognition learning. Zebrafish are one of several model systems upon which this body of literature has been developed. Minnows (including zebrafish) associate danger with any novel stimulus (visual, chemical, or auditory) that is correlated with the presence of chemical alarm cue released from damaged epithelial tissue of conspecifics. Alarm cue is released only in the context of predation and serves as a reliable external environmental trigger for associating novel stimuli with predation risk. Minnows use learned recognition to learn about predator identity and about the chemical alarm cues of ecologically similar heterospecifics. Learning also occurs when alarm cues are released indirectly through the digestive tract of the predator. Behavioral and chemical responses to disturbance can also facilitate learned recognition. Learned recognition is an ideal system with which to study the molecular mechanisms that underlie the cognitive processes of learning and memory. Collectively, this suggests that zebrafish are a very promising model organism for future study.

Key words: Antipredator behavior, predator response, alarm cue, environmental cues, conditioned stimulus, unconditioned stimulus, releaser-induced recognition learning, learned recognition, associative learning, memory.

1. Introduction

The ultimate function of animal behavior is to contribute to reproductive success (**Fig. 9.1**). Animals make behavioral decisions about how to find food, survive encounters with predators and disease, secure a territory and mates and ultimately,

A.V. Kalueff, J.M. Cachat (eds.), *Zebrafish Models in Neurobehavioral Research*, Neuromethods 52,
DOI 10.1007/978-1-60761-922-2_9, © Springer Science+Business Media, LLC 2011

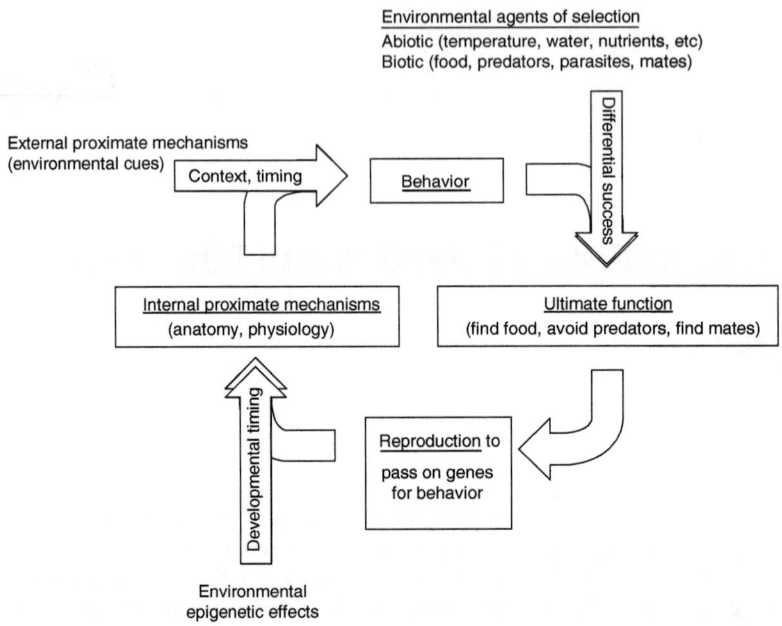

Fig. 9.1. Evolution of behavior is a consequence of interactions between genes and environment. See text for details.

to reproduce. Genes passed on through successful reproduction code for the anatomical structures, physiological interactions, and neural wiring that make behavior possible. In addition to genetic influences, rearing environment can also affect embryonic development and expression of behavioral phenotype. Taken together, these internal proximate mechanisms give the actor the capacity and inclination to behave. However, behavior must be expressed in the correct context and timing if it is to contribute to reproductive success.

External environmental cues such as evidence of food, presence of a predator or a reproductive rival trigger expression of context-appropriate behavior. In many cases, recognition and adaptive responses are regulated through hard-wired genetic templates. For example, detection of amino acids reliably indicates the presence of potential food (1). Sex pheromones released by one individual stimulate receptors and activate endocrine responses in other individuals. In these examples, stimulants trigger behavioral responses in a mechanical, stimulus-response manner. In the context of foraging and reproduction salient stimuli are limited to a narrow range of predictable candidates because all vertebrates consume similar macromolecules and use the same basic endocrine systems to regulate reproductive function.

Antipredator behavior differs from foraging and reproduction in two important ways. First, it is much more difficult to predict predation risk because risk varies tremendously over ecological space and time and also throughout ontogenetic life stages within the lifetime of an individual. Therefore, unlike foraging and reproduction, the size, shape, odor, and sound of a predator could be almost anything, especially for small fish such as minnows. Second, a mistake in antipredator behavior has different consequences for ultimate reproductive success than a mistake in foraging or reproductive behavior. An individual can recover from a missed meal or a lost mating opportunity and try again later. Failure to detect and avoid a predator is much more serious. Antipredator behavior therefore presents a unique ecological problem by presenting a wide array of potential behavioral cues and dire consequences for getting it wrong, even once.

In fishes, antipredator behavioral responses occur through a special type of associative learning, known as releaser-induced recognition learning (2, 3) that is able to accommodate almost limitless variation (**Fig. 9.2**). This form of learning is similar to

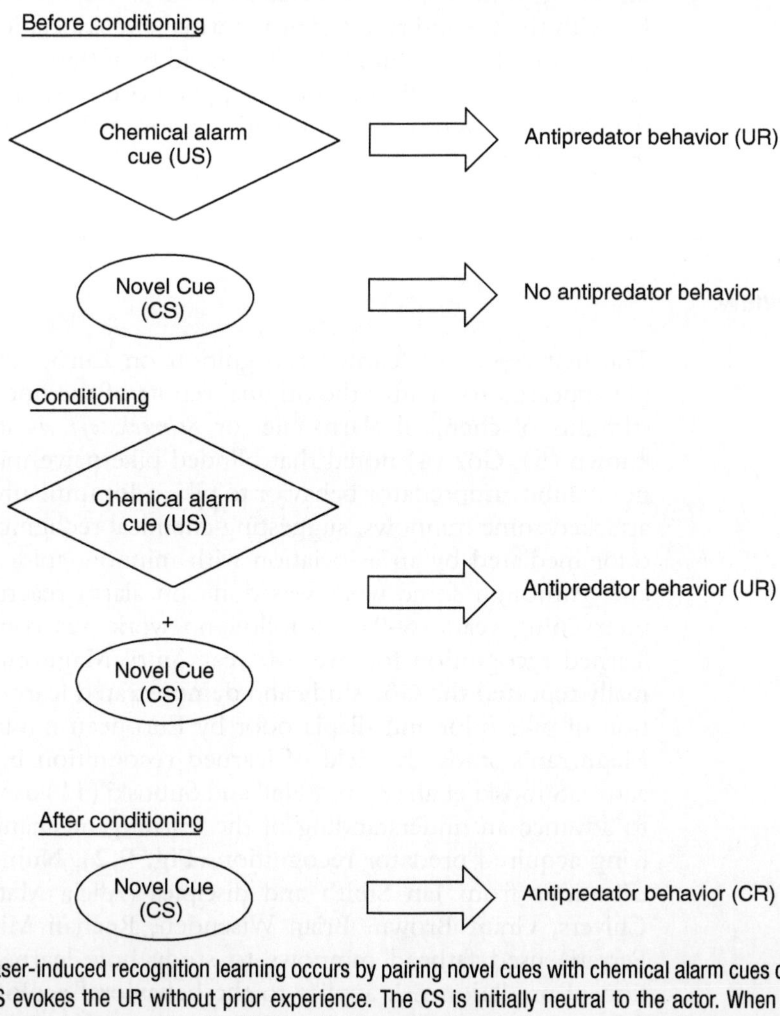

Fig. 9.2. Releaser-induced recognition learning occurs by pairing novel cues with chemical alarm cues derived from skin extract. The US evokes the UR without prior experience. The CS is initially neutral to the actor. When the US is paired with the CS the actor associates the CS with the UR. Thereafter the CS alone evokes the CR. Only one pairing event is necessary to achieve this effect.

classical conditioning in that an aversive stimulus (unconditioned stimulus, US) produces an unconditional behavioral response (UR) without prior experience. The US is paired in space and time with a neutral novel stimulus (conditioned stimulus, CS). Thereafter, the neutral novel stimulus (CS) elicits the same response (now called a conditioned response, or CR). In many fishes, including zebrafish, the US is chemical alarm cue released by damaged epidermal tissue. When predators grasp prey their teeth and other weaponry damage the skin. Damaged skin releases chemicals that are released in no other context and therefore serve as a reliable external environmental cue for the context and timing of antipredator behavior (**Fig. 9.1**). The conditioned stimulus can be in the visual, olfactory, or auditory sensory modalities, as discussed in the examples below. The remarkable property of releaser-induced recognition learning is that, unlike classical conditioning, only a single association event is required to pair the US with the CS and result in near permanent association between the CS and the CR (**Fig. 9.2**). The rapidity of this response is the result of steep selection gradients applied over evolutionary time that has consistently and efficiently removed slow learners from the gene pool.

2. A Brief Historical Review

The first report of learned recognition on European minnows (4) appeared soon after the original report of the unconditioned stimulus of chemical alarm cue, or *Schreckstoff*, as it was then known (5). Göz (4) noted that blinded pike-naive minnows did not exhibit antipredator behavior to pike odor until after pike had attacked some minnows, suggesting chemical recognition of pike odor mediated by an association with minnow epidermal alarm cue. Although some work was done on alarm reactions in the intervening years (6–9), no follow-up work was conducted on learned recognition for over 40 years until Magurran (10) formally repeated the Göz study and demonstrated learned recognition of pike odor and tilapia odor by European minnows. After Magurran's study the field of learned recognition became very active. Suboski et al. (3) and Hall and Suboski (11) used zebrafish to advance an understanding of the learning mechanism underlying acquired predator recognition (**Fig. 9.2**). Numerous contributions from Jan Smith and disciples (Alicia Mathis, Doug Chivers, Grant Brown, Brian Wisenden, Reehan Mirza, Maud Ferrari) used fathead minnows to study how learned recognition of predation risk applies to the behavioral ecology of small fishes.

Zebrafish are a convenient study organism for the study of alarm reactions (6–9) and learned recognition (e.g. (3, 11, 12)).

Hall and Suboski (11) used chemical alarm cue to condition zebrafish to form aversive associations with novel stimuli that do not occur in nature (the non-biological chemical cue morpholine, the visual stimulus of a flashing red light). These experiments not only established releaser-induced recognition learning as a special case of associative learning, they revealed the extreme plasticity of this learning mechanism.

3. Learned Recognition

Fathead minnows are the study organism of choice by behavioral ecologists because these minnows can be studied in the lab and in their ecological context in the field. Both fathead minnows and zebrafish are small-bodied cyprinids and obligate schooling fishes that thus far seem to be identical from the standpoint of learned recognition. Results from fathead minnows apply broadly to all cyprinids, including zebrafish.

Fathead minnows quickly form associations with a range of visual (13) chemosensory (14, 15), and auditory (16) stimuli. The concentration of the novel cue used to condition the fish does not seem to be important for the formation of an association. However, once conditioned to recognize it as dangerous, response intensity to predator odor is proportional to cue concentration (17).

Learned recognition of predator identity has been demonstrated in field populations (18–20). For example, every fish tested in a population of approximately 20,000 adult pike-naive fathead minnows showed a behavioral response to pike odor 14 days after 10 pike had been placed in their pond (19; **Fig. 9.3**). Subsequent study in a different pond with 78,000 minnows and newly stocked with 39 juvenile pike showed that fathead minnows acquire visual recognition of pike in 6–8 days and olfactory recognition of pike odor in 2–4 days (18). Minnows can learn to distinguish between two water samples collected 15 m apart based on subtle differences in water chemistry, suggesting the potential use of recognition learning to detect and avoid risky habitat (21). The presentation of US (chemical alarm cue) and CS (pike odor) do not have to occur simultaneously for associations to form because the US keeps the fish in a state receptive to novel CS for some undetermined time. Zebrafish formed associations with pike odor when pike odor was presented 5 min after the presentation of skin extract (12). However, the formation of associations is inhibited when the CS (e.g., predator odor) is continuously present so that the subject becomes habituated to the chemical cue. In this context, subsequent presentation of chemical alarm cues does not result in formation of an association between predator odor and predation risk (22).

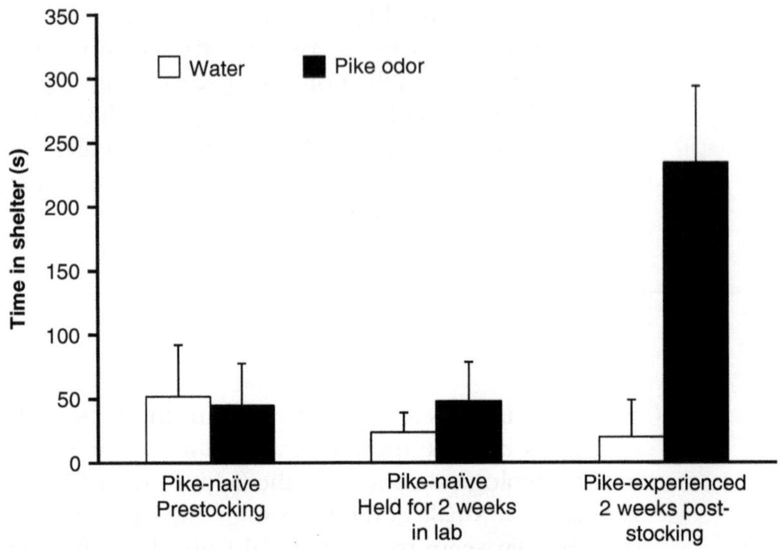

Fig. 9.3. Learned recognition of predators in a natural population of fathead minnows. Minnows sampled from a pond before a pike-free pond was stocked with 10 pike showed no response pike odor. Minnows removed from the pond before stocking and tested in the lab 14 days later showed no response to pike odor. Minnows that remained in the pond for 14 days after 10 pike were stocked showed a strong antipredator response to pike odor. Data are means ±1 SE (after 19).

When minnows do learn to associate risk with the odor of a novel predator, they generalize and extend recognition of that novel odor to other predator species (23). The intensity of the reaction to other predator species is proportional to the phylogenetic distance between the reference species and these other species.

The speed, permanence, and generality with which these associations form could present an ecological liability. A population of predators causes the release of chemical alarm cues several times per day per predator. How do prey species prevent associating danger with irrelevant stimuli? There is some evidence of constraints on the types of stimuli that fish associate with risk. Pike are more readily associated with danger than tilapia (10) or goldfish (13). Moving objects are associated with danger while stationary objects are ignored (24; **Fig. 9.4**).

4. Broader and More Subtle Manifestations of Learned Recognition

The US can be in the form of a dietary alarm cue. A dietary cue is epidermal chemical alarm cue, or its metabolites, that remain recognizable after they pass through the digestive tract of a predator. Dietary cues otherwise perform a similar function as alarm cues

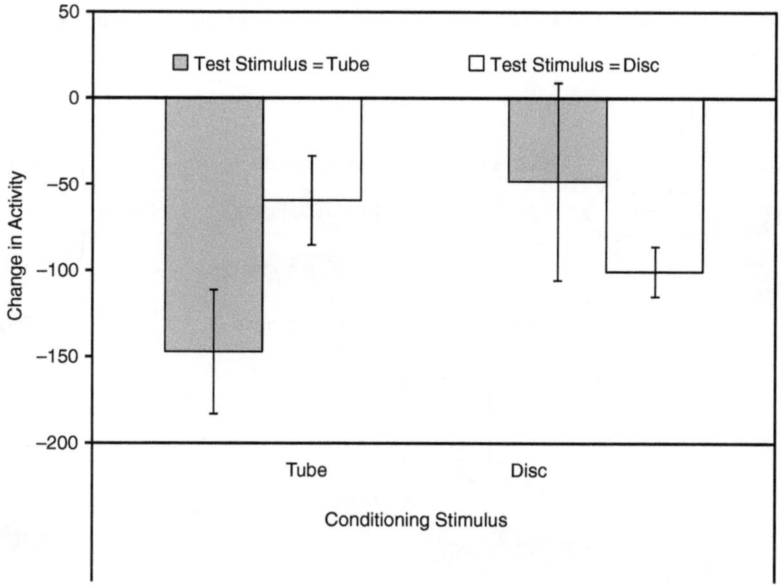

Fig. 9.4. Antipredator behavior includes reduction in activity because predators often locate prey by detecting motion. Fish were conditioned with chemical alarm cues and visual presentation of two objects: a disc and a horizontal tube. During the presentation one of the objects was repeatedly raised and lowered while the other object remained stationary. When these fish were retested with one only one of the two objects, if they had been conditioned with a moving tube (*shaded bars*) they responded to the tube but not the disc. If they had been conditioned with the moving disc, they subsequently responded to the disc but not the tube. Means ±1 SE are shown (after Wisenden and Harter (24)). These data were collected from fathead minnows, which are ecologically, phylogenetically, and behaviorally similar to zebrafish. Results from minnows are fully extendable to zebrafish.

in serving as a releaser of antipredator behavior and facilitating predator recognition (15, 25).

Prey fishes, such as minnows, learn to respond to alarm cue of other species. Prey fishes that are similar in size and ecology, have overlapping home ranges, and are vulnerable to similar predators form a "prey guild." Chemical alarm cue released by any species within a prey guild indicate danger to all species in the prey guild. Recognition of heterospecific chemical alarm cues are usually learned unless the two species are very closely related (23). Acquired recognition of heterospecific alarm cue can occur via three mechanisms. First, if the odor of the predator is already recognized as an indicator of predation risk, then an attack on a heterospecific will present an occasion where predator odor (US) is paired with the novel heterospecific chemical alarm cue (CS). Second, a known predator that releases dietary cues of novel heterospecific prey allows prey to associate predator odor (US) with the heterospecific dietary cue (CS) (26). Third, a novel predator that has a mixed diet of conspecifics and heterospecifics allows prey to associate risk with both the predator odor and the heterospecific alarm cue simultaneously (27).

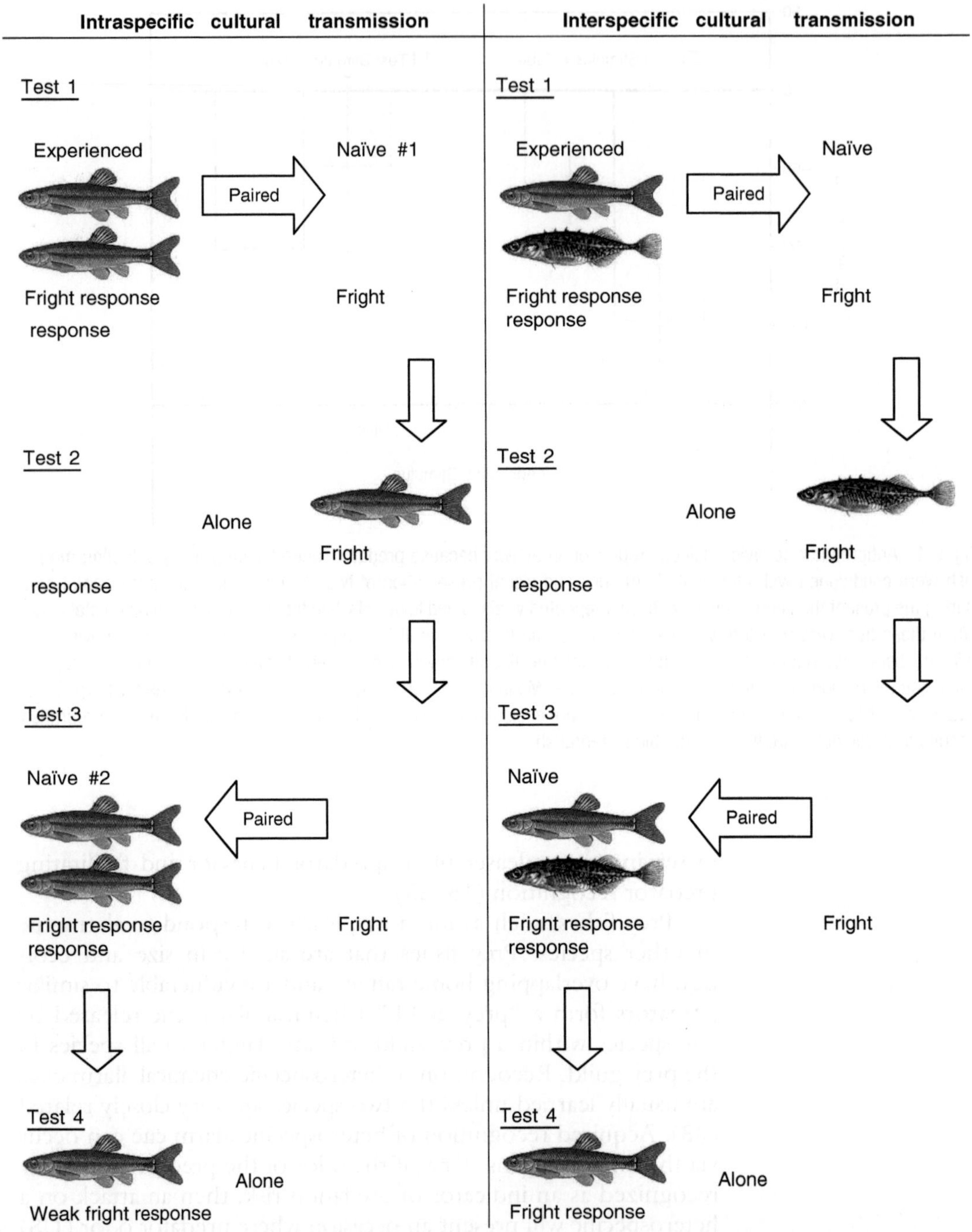

Fig. 9.5. Cultural transmission allows fish to learn predator recognition from the behavioral responses of conspecifics (*left-hand column*) and heterospecific brook stickleback (*right-hand column*). This figure shows the experimental design and results of four sequential tests that were conducted exposing test fish to the odor of northern pike. In the first test, the behavioral response of a pike-experienced individual serves as a model for a pike-naive individual to associate risk with pike odor. In test 2 the former naive observer responds to pike odor with alarm behavior. In test 3 the socially conditioned fish is paired with a fresh pike-naive individual. Test 4 shows that the formerly naive fish in test 1 is able to acquire and later transmit information about the association between pike odor and predation risk. After Mathis et al. (28).

Injury from predator attack is not the only way that prey can learn about predation risk. Responding to overt behaviors and to chemical cues released by disturbed animals gives prey early warning and more time to evade predation risk. In a shoal (a group of fish), when any one individual detects a predator, its behavioral response alerts the entire group to the presence of predation risk. Behavior of disturbed fish is a visual cue of disturbance. As such, alarm behavior itself can serve as a visual US that allows shoalmates to associate risk with any novel stimuli that are correlated in space and time with an alarm reaction. Moreover, social facilitation of learned recognition occurs between conspecifics and between heterospecifics in a mixed-species shoal (28; **Fig. 9.5**). For example, brook stickleback and fathead minnows form a well-studied prey guild. When one individual that is conditioned to recognize pike odor as dangerous (the "tutor" in a dyad) is paired with a second individual that is pike-naive (the "pupil" in the dyad), the behavioral response of the tutor to introduced pike odor allows the pupil to associate pike odor with danger (**Fig. 9.5**). As for all forms of associative learning discussed thus far, only one reinforcement event is necessary for learning to occur. The pupil can then be used as the tutor in a subsequent trial with a new naive individual.

Disturbed fish also release chemical cues that alert nearby conspecifics and heterospecifics of heightened risk of predation (29). Several studies have implicated urinary ammonia as the active ingredient in disturbance cue (30, 31). Learned recognition of indicators of predation risk, including predator identity, occurs by pairing disturbance cues (US) with novel stimuli (32).

5. Information Gathering

Because predator identity is very important information prey tolerate some risk to acquire it. For example, fleeing from an area where alarm cues are present reduces exposure to predation risk but also reduces access to information about the nature of that risk, including predator identity. As a result of this trade-off, zebrafish in long tanks without any water movement occupy shelters adjacent to the source of alarm cue rather than fleeing to distant shelters beyond the dispersion of chemical information (33). In the presence of flow, zebrafish hide at the downstream of the fluvarium where they can monitor chemical information from a safe distance. Fathead minnows and glowlight tetras perform predator inspection behavior whereby one or more prey approach a predator to gain information. Inspectors indeed acquire and use information to learn to recognize about predator identity during the act of inspection (34).

6. Future Directions for Zebrafish Research on Learned Recognition

The myriad ways in which minnow species acquire recognition of novel stimuli is now well established in the literature. The ease and simplicity of behavioral assays to measure learned recognition lay the foundation for an exciting new era of research on the molecular genetics of regulating mechanisms for the development and function of cognitive processes.

A recent line of research that is coming to the fore is the role of pre-hatching experience in facilitating learned recognition of predators and indicators of predation risk (e.g., frog embryos (35)). In other words, the chemical signature of predators and perhaps other indicators of predation risk influence development and impact the behavioral phenotype of the hatchling. This is one research area where the zebrafish model may be able to make a substantial contribution.

A conspicuous gap in this literature, noted for years, is the lack of a consensus on the chemical nature of chemical alarm cue in zebrafish and other species in the minnow family. That literature is not reviewed here (see (36) for a recent review). However, once the chemistry is known, we will then be able to explore the olfactory receptors and neural wiring responsible for information processing and memory storage of learned recognition.

References

1. Wisenden, B. D. & Stacey, N. E. (2005) Fish semiochemicals and the network concept. In McGregor, P. K. (Ed.) *Animal Communication Networks*. Cambridge, Cambridge University Press, pp. 540–567.
2. Suboski, M. D. (1990) Releaser-induced recognition learning. *Psychol Rev* 97, 271–284.
3. Suboski, M. D., Bain, S., Carty, A. E., McQuoid, L. M., Seelen, M. I. & Seifert, H. (1990) Alarm reaction in acquisition and social transmission of simulated-predator recognition by zebra danio fish (*Brachydanio rerio*). *J Comp Psychol* 104, 101–112.
4. Göz, H. (1941) Über den Art-und Individualgeruch bei Fischen. *Z Vergl Physiol* 29, 1–45.
5. von Frisch, K. (1938) Zur psychologie des Fische-Schwarmes. *Naturwissenschaften* 26, 601–606.
6. Dill, L. M. (1974) The escape response of the *zebra* danio (*Brachydanio rerio*). I. The stimulus for escape. *Anim Behav* 22, 711–722.
7. Gandolfi, G., Classon, L. J. & Rossi, A. C. (1968) The fright reaction of zebra fish. *Atti Soc Ital Sci Nat* 107, 74–88.
8. Pfeiffer, W. (1977) The distribution of fright reaction and alarm substance cells in fishes. *Copeia* 1977, 653–665.
9. Waldman, B. (1982) Quantitative and developmental analysis of the alarm reaction in the zebra danio, *Brachydanio rerio*. *Copeia* 1982, 1–9.
10. Magurran, A. E. (1989) Acquired recognition of predator odour in the European minnow (*Phoxinus phoxinus*). *Ethology* 82, 216–223.
11. Hall, D. & Suboski, M. D. (1995) Visual and olfactory stimuli in learned release of alarm reactions by zebra danio fish (*Brachydanio rerio*). *Neurobiol Learn Mem* 63, 229–240.
12. Korpi, N. L. & Wisenden, B. D. (2001) Learned recognition of novel predator odour by zebra danios, *Danio rerio*, following time-shifted presentation of alarm cue and predator odour. *Environ Biol Fishes* 61, 205–211.

13. Chivers, D. P. & Smith, R. J. F. (1994a) Fathead minnows (*Pimephales promelas*) acquire recognition when alarm substance is associated with the sight of unfamiliar fish. *Anim Behav* 48, 597–605.
14. Chivers, D. P. & Smith, R. J. F. (1994b) The role of experience and chemical alarm signaling in predator recognition by fathead minnows, *Pimephales promelas* Rafinesque. *J Fish Biol* 44, 273–285.
15. Mathis, A. & Smith, R. J. F. (1993) Fathead minnows, *Pimephales promelas*, learn to recognize northern pike, *Esox lucius*, as predators on the basis of chemical stimuli from minnows in the pike's diet. *Anim Behav* 46, 645–656.
16. Wisenden, B. D., Pogatshnik, J., Gibson, D., Bonacci, L., Schumacher, A. & Willett, A. (2008) Sound the alarm: learned association of predation risk with novel auditory stimuli by fathead minnows (*Pimephales promelas*) and glowlight tetras (*Hemigrammus erythrozonus*) after single simultaneous pairings with conspecific chemical alarm cues. *Environ Biol Fish* 81, 141–147.
17. Ferrari, M. C. O. & Chivers, D. P. (2006a) The role of learning in the acquisition of threat-sensitive responses to predator odors. *Behav Ecol Sociobiol* 60, 522–527.
18. Brown, G. E., Chivers, D. P. & Smith, R. J. F. (1997) Differential learning rates of chemical versus visual cues of a northern pike by fathead minnows in a natural habitat. *Environ Biol Fish* 49, 89–96.
19. Chivers, D. P. & Smith, R. J. F. (1995a) Free-living fathead minnows rapidly learn to recognize pike as predators. *J Fish Biol* 46, 949–954.
20. Leduc, A. O. H. C., Roh, E., Breau, C. & Brown, G. E. (2007) Learned recognition of a novel odour by wild juvenile Atlantic salmon (*Salmo salar*) under fully natural conditions. *Anim Behav* 73, 471–477.
21. Chivers, D. P. & Smith, R. J. F. (1995b) Fathead minnows (*Pimephales promelas*) learn to recognize chemical stimuli from high-risk habitats by the presence of alarm substance. *Behav Ecol* 6, 155–158.
22. Ferrari, M. C. O. & Chivers, D. P. (2006b) The role of latent inhibition in acquired predator recognition in fathead minnows. *Can J Zool* 84, 505–509.
23. Ferrari, M. C. O., Gonzalo, A., Messier, F. & Chivers, D. P. (2007) Generalization of learned predator recognition: an experimental test and framework for future studies. *Proc R Soc Lond B* 274, 1853–1859.
24. Wisenden, B. D. & Harter, K. R. (2001) Motion, not shape, facilitates association of predation risk with novel objects by fathead minnows (*Pimephales promelas*). *Ethology* 107, 357–364.
25. Chivers, D. P., Mirza, R. S. & Johnston, J. G. (2002) Learned recognition of heterospecific alarm cues enhances survival during encounters with predators. *Behaviour* 139, 929–938.
26. Mirza, R. S. & Chivers, D. P. (2003) Fathead minnows learn to recognize heterospecific alarm cues they detect in the diet of a known predator. *Behaviour* 140, 1359–1369.
27. Mirza, R. S. & Chivers, D. P. (2001) Learned recognition of heterospecific alarm signals: the importance of a mixed predator diet. *Ethology* 107, 1007–1018.
28. Mathis, A., Chivers, D. P. & Smith, R. J. F. (1996) Cultural transmission of predator recognition in fishes: intraspecific and interspecific learning. *Anim Behav* 51, 185–201.
29. Wisenden, B. D., Chivers, D. P. & Smith, R. J. F. (1995) Early warning in the predation sequence: a disturbance pheromone in Iowa darters (*Etheostoma exile*). *J Chem Ecol* 21, 1469–1480.
30. Hazlatt, B. A. (1990) Source and nature of disturbance-chemical system in crayfish. *J Chem Ecol* 16, 2263–2275.
31. Kiesecker, J. M., Chivers, D. P., Marco, A., Quilchanos, C., Anderson, M. T. & Blaustein, A. R. (1999) Identification of a disturbance signal in larval red-legged frogs *Rana aurora*. *Anim Behav* 57, 1295–1300.
32. Mirza, R. S. & Chivers, D. P. (2002) Behavioural responses to conspecific disturbance chemicals enhance survival of juvenile brook charr, *Salvelinus fontinalis*, during encounters with predators. *Behaviour* 139, 1099–1109.
33. Wisenden, B. D., Binstock, C. L., Knoll, K. E. & Linke, A. D. (2010) Risk-sensitive information gathering by zebrafish following release of chemical alarm cues. *Anim Behav* 79, 1101–1107.
34. Brown, G. E. & Godin, J.-G. J. (1999) Who dares, learns: chemical inspection behaviour and acquired predator recognition in a characin fish. *Anim Behav* 57, 475–481.
35. Mathis, A., Ferrari, M. C. O., Windel, N., Messier, F. & Chivers, D. P. (2008) Learning by embryos and the ghost of predation future. *Proc R Soc B* 275, 2603–2607.
36. Ferrari, M. C. O., Wisenden, B. D. & Chivers, D. P. (2010) Chemical ecology of predator–prey interactions in aquatic ecosystems: a review and prospectus. *Can J Zool* 88, 698–724.

Chapter 10

Inhibitory Avoidance and Color Discrimination Learning in Zebrafish

Luciana Cofiel and Rosana Mattioli

Abstract

The unique properties of zebrafish, such as its high reproductive capability, small size, simple genome, and relatively low maintenance cost, make it a cheap and effective genetic model for scientists to study. As a result, scientists have acquired an enormous quantity of zebrafish genetic information and developed numerous genetic tools for the zebrafish. With the large amount of zebrafish information available, one area that is lacking is behavioral characterization and, therefore, effective and reliable behavioral experiment models are lacking as well. After years of research on the zebrafish, our laboratory has discovered successful behavioral models for zebrafish and goldfish in the areas of learning and memory. This chapter discusses two of these models. The first is the Inhibitory Avoidance Experimental Model, and the second is a Color Discrimination Model. In the inhibitory avoidance paradigm, the animals had to learn to avoid an aversive stimulus present on the aquarium preferred compartment. Immediately after training, one group received saline and the other one did not receive the injection. On the test day, the time to cross to the preferred compartment was determined. The latency to enter the black compartment increased significantly on the second trial in relation to BL. No difference between the animal's latencies on T2, on the test of non-injected animals, or saline-treated animals was recorded, indicating that the animals did not forget the adverse experience from the previous day. The second experiment is a color discrimination model for zebrafish. On each of the 5 consecutive days of experiment (D1, D2, D3, D4, D5), animals had to associate the feeder indicated by the green light with food offering. The latency to enter the feeding area indicated by the green light decreased throughout the trials, with significant difference between D5 and D1 indicating that the animals were able to learn the task. Our results indicate that these tests represent suitable experimental models for the study of learning and memory in zebrafish.

Key words: Inhibitory avoidance paradigm, aversive stimulus, color discrimination paradigm, behavioral assay, neuroanatomy, conditioned place preference, color preference conditioning, learning, memory.

A.V. Kalueff, J.M. Cachat (eds.), *Zebrafish Models in Neurobehavioral Research*, Neuromethods 52,
DOI 10.1007/978-1-60761-922-2_10, © Springer Science+Business Media, LLC 2011

1. Introduction

Classical theories about cognition in fish stated that these animals' telencephalon would consist mainly of a subpallium ("paleostriatum") and a very reduced and primitive pallium ("paleocortex"), both entirely dominated by olfactory inputs and relatively simple neural circuits. Therefore, the behavior of the fish was considered essentially "reflex" or "instinctive" (1). However, studies using different methodologies and different experimental approaches showed that the forebrain of teleost fish is involved in emotional, social, and reproductive behavior, as well as in learning and memory, just as has occurred in mammals (see (2) for review). A large amount of evidence indicates that learning and memory systems, involving the optic tectum, the cerebellum, and the hippocampal and the amygdalar pallium, are strikingly similar among teleost fish and land vertebrates (1).

Moreover, circuitry and the pharmacology involved in behavior processes in fish often appear comparable to those operating to control similar behaviors in higher vertebrates (3), for example the histaminergic system (3). This observation makes teleosts valid models for an approach to many behaviors, with the additional hope that their simpler brain organization (in particular regarding forebrain anatomy) will reveal the neuronal bases underlying these functions (4). Based on this, interest in the use of fish in behavioral studies has been increasing in the past few years. In fact, teleosts such as the goldfish have been largely used in a variety of studies approaching cognitive processes for their relatively large and stereotactically well-characterized brain (1).

Opposite to features of the other teleost, zebrafish turned it into an interesting animal model to carry out genetic studies. It has an adult size of only about 3–5 cm, which allows maintenance of a large number of animals in the laboratory and it reproduces robustly. These two features essential to carry out this type of study are lacking in vertebrate model organisms such as mice (5). The embryo and larva of zebrafish are transparent and develop very rapidly. In 5 days, swimming and self-feeding larvae can be observed. This whole process unravels in a Petri dish (5).

Because of these special characteristics, a large amount of genetic information has been accumulated and numerous genetic tools have been developed for zebrafish (6). These tools have allowed the investigation of a number of human diseases using the zebrafish (7), including human degenerative diseases such as Alzheimer's disease (8) and Parkinson disease (9). Since zebrafish present adequate features for studying developmental processes, organ function, and human diseases (10, 11), but with a simpler central nervous system than the mammals, there is a need for the characterization of its behavior, as well as the development of reliable behavioral test methods. Our laboratory has developed efficient and practical experimental models to study learning and memory in zebrafish. In this chapter, we describe current experimental models to evaluate learning in zebrafish that proved to be efficient.

2. Methods

2.1. Fish and Maintenance

Fish used in both experiments were bought at a local pet shop and placed in 30-l aquariums (30 animals per aquarium), at 18–22°C with constant filtering and aeration, natural cycle of light (with approximately 13 h light/11 h dark), and fed five times a week with flake food (Wardly Corporation, NJ, USA). At least a 1-week acclimation period was allowed from the purchasing of the fish until the beginning of the experiment. Since it is difficult to distinguish the fish individually, before the beginning of the experiments, fish were individually placed in 25 cm long, 11.5 cm wide, and 15 cm high aquarium. Animal husbandry and all behavioral experiments were conducted in accordance with guidelines for the use of laboratory animals set by the Brazilian Neuroscience and Behavior Society (SBNeC), based on the US National Institute of Health guidelines for the care and use of laboratory animals.

In the "Inhibitory avoidance" procedure, 48 zebrafish from both sexes were used in a blind ratio unknown to the experimenters. In the "Color discrimination procedure," 16 fish were used.

3. Experiment 1: Inhibitory Avoidance Model for Zebrafish

3.1. Experimental Aquarium

A rectangular (30 cm long, 15 cm high, and 15 cm wide) aquarium was used. It was divided by a sliding door into two chambers, one black and one white. On the black side of the aquarium was a pulley system from where a weight of 45 g could be dropped (**Fig. 10.1**).

3.2. Procedure

Animals had two sessions of 10 min to explore both black and white sides of the experimental aquarium on two consecutive days before the beginning of the experiment. On the following

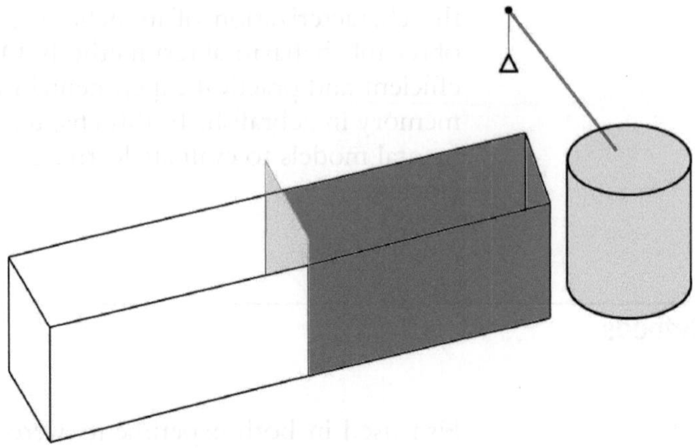

Fig. 10.1. Inhibitory avoidance aquarium.

day, animals were placed individually in the white chamber of the aquarium with the sliding door closed. After 30 s, the door was opened and the animals had access to the black compartment. Once the animal's entire body was inside the black compartment, the 45-g weight was dropped in front of the fish. This evoked an escape response and fish returned to the white compartment. Some fish presented a freezing response after this stimulus. If no escape response was evoked, fish were gently moved back to the white compartment using a fish net. This procedure was repeated two more times. Time to enter the black compartment was observed in the three trials (BL: baseline, T1: trial 1, and T2: trial 2). In order to verify if the animals recorded the presence of the aversive stimulus in the black compartment on the training day, a test (T) was preformed. On the following day, the animal was placed back in the white compartment and the time fish took to enter the black compartment was recorded again. All procedures were videotaped.

Two experimental groups were used in this experiment: one group of non-injected animals (NI) and one group that received intraperitoneal saline injection (SAL) immediately after the training trials. Volume of the injection was 1 ml kg^{-1} of body weight.

The saline-treated group was used to verify possible effects of the injection procedure, since this experimental model was developed for the study of histaminergic drugs on learning, and these drugs were also injected intraperitoneally. For the injection procedure, animals were captured using a fish net and removed from water. Drug was injected using a needle connected to a 10 μL syringe (Hamilton, model 7105KH, USA) through a polythene tube. The complex needle-tube-syringe was filled with distilled water, and the injection flow could be observed by the movement of an air bubble between the drug being administered and the distilled water. This procedure was carried out as quickly as possible, so that the time the fish remained out of water was kept to a minimum.

On the first training trial, it was observed that about one third of the animals took more than 10 min to cross from the white to the black compartment. These animals were excluded from the experiment.

3.3. Statistical Analysis

Statistical analysis was carried out using the GB-STAT School Pack software, version 1997. Data are reported as means ± SEM and differences were considered significant if the probability of error was <5% ($p = 0.05$).

Since the data obtained in this experiment was not homogeneously distributed, the non-parametric Friedman test followed by the Student Newman-Keuls multiple comparison test was used to verify differences between the training trials (BL, T1, and T2).

To investigate possible differences between test latencies of both experimental groups and the latency observed in T2, the non-parametric Kruskall-Wallis test was used.

3.4. Results

Acquisition of the task can be inferred from a gradual raise in the latencies to enter the black compartment during the training day. This was observed in these animals with significant difference between T2 and BL (Friedman $p = 0.0373$, DF = 2, $\chi^2 = 6.5769$; Student Newman-Keuls, $p < 0.05$) (**Fig. 10.2**).

The comparison between the latencies in T2, and in T of noninjected and saline-injected animals indicated no significant difference (Kruskall-Wallis, $p = 0.1993$, DF = 2, $\chi^2 = 3.2254$; **Fig. 10.3**).

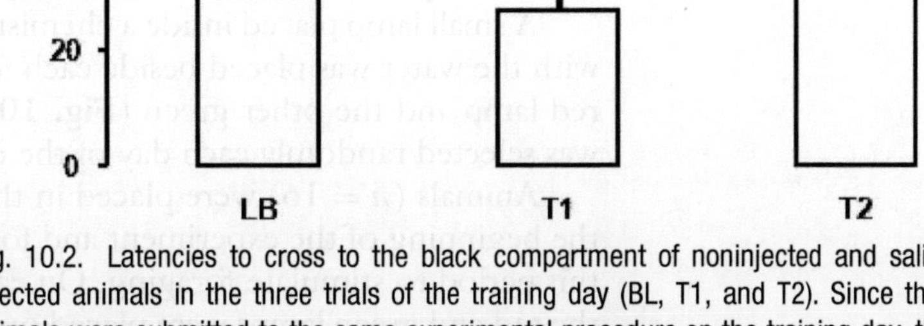

Fig. 10.2. Latencies to cross to the black compartment of noninjected and saline-injected animals in the three trials of the training day (BL, T1, and T2). Since these animals were submitted to the same experimental procedure on the training day, their data were analyzed as a group ($n = 48$). Data are reported as means ± SEM. * $p < 0.05$ compared to BL (Friedman test followed by the Student Newman-Keuls test).

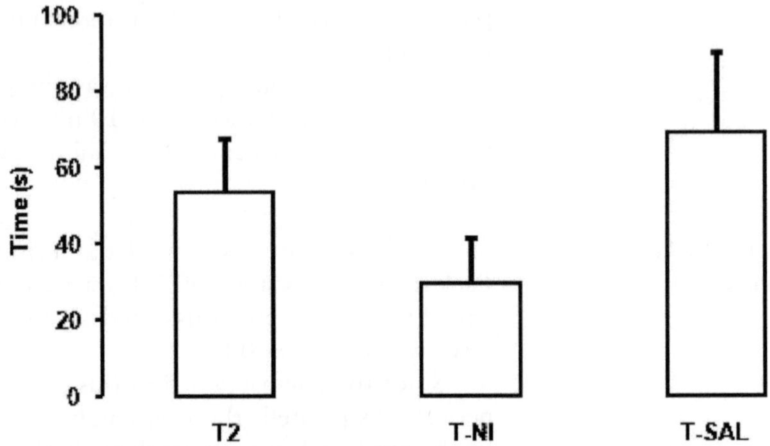

Fig. 10.3. Latencies to cross to the black compartment on the last training trial (T2, $n = 48$ – Since on the training day animals from the saline-injected and the noninjected group were submitted to the same experimental procedure, data from training trials were analyzed conjointly) and on the test trial of noninjected (T-NI; $n = 21$) and saline-injected animals (T-SAL; $n = 27$). Data are reported as means ± SEM (Kruskall-Wallis test).

4. Experiment 2: Color Preference Conditioning Model for Zebrafish

4.1. Experimental Aquarium

Two days before the beginning of the experiment, animals were individually placed in a 25 cm long, 11.5 cm wide, and 15 cm high aquarium, in which water was constantly aired. A transparent plastic barrier was fixed 6 cm from both ends of the aquarium, limiting two opposite feeding areas. A plastic transparent cylinder was placed inside each area, where one pellet of floating food was offered during the experiments. Unconsumed food was removed from the water after the end of the experiment, since it could increase organic waste and decrease water quality.

A small lamp placed inside a chemistry tube to prevent contact with the water was placed beside each feeder, being on one side a red lamp and the other green (**Fig. 10.4**). The color of the side was selected randomly each day of the experiment.

Animals ($n = 16$) were placed in the aquarium 2 days before the beginning of the experiment and food was not offered during this period to stimulate foraging. On each day of the experiment, the red and green lamps were placed on opposite sides and turned on. Food was offered 30 s later, always on the feeder indicated by the green light.

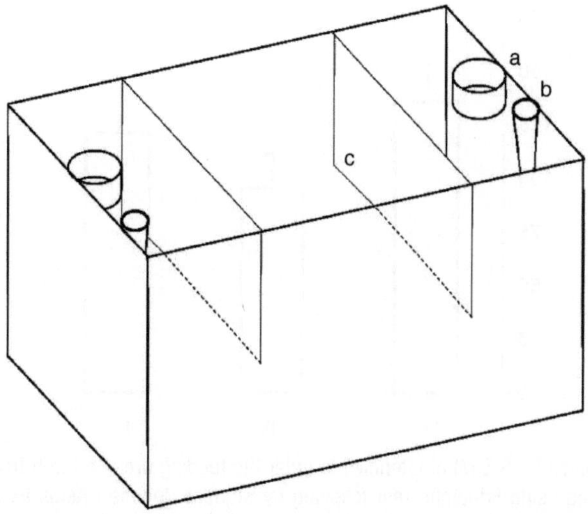

Fig. 10.4. Zebrafish color conditioning aquarium. **a** Feeder; **b** light; **c** barrier.

The animals had 5 min to enter the feeding area and those that did not enter the feeding area were excluded. Only one trial was carried out during each of the 5 days of experiment (D1, D2, D3, D4, and D5), and time to approach the feeder indicated by the green light was observed on each day.

4.2. Statistical Analysis

Results are reported as the mean ± standard error of the mean (S.E.M.). Data were not homogeneously distributed, therefore were analyzed using the non-parametric test of Friedman followed by Student Newman-Keuls multiple comparison test.

4.3. Results

Acquisition of the task can be inferred from a gradual decrease in the latencies to enter the feeding area indicated by the green light. This was observed in this experimental group with statistically significant difference between day 1 and 5 (Friedman $p = 0.0211$, $DF = 4$, $\chi^2 = 45994.75$; Student Newman-Keuls, $p < 0.05$). Data is presented in **Fig. 10.5**.

5. General Discussion

It has been shown that zebrafish have a natural preference for dark environments (12). Thus, in the first experiment of this study, the rise in time to enter the black compartment showed that the weight drop was an efficient aversive stimulus for the animals and caused a change in their natural preference. It was also observed that the animals were capable of remembering the experience of

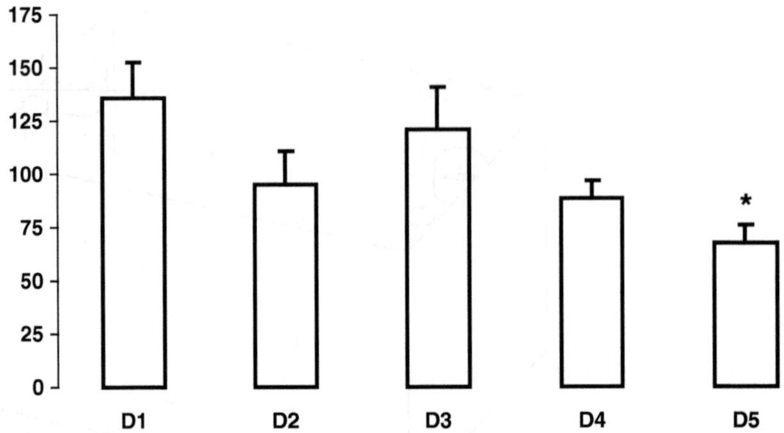

Fig. 10.5. Mean (± S.E.M) of latencies to enter the feeding area on the 5 training days (D1, D2, D3, D4, and D5). Data were analyzed using Friedman test followed by Student Newman-Keuls test when appropriate. Asterisk (*) indicates statistically significant differences in relation to D1 ($p < 0.05$).

the previous day, since the animals avoided the preferred compartment on test day. Thus, the experimental model of inhibitory avoidance seems to be an effective model to be used for studying learning and memory in zebrafish. However, some adaptations on the procedures could expand their use. For example, more days of testing could bring information on the extinction of this inhibitory response.

Other studies have supported the idea that zebrafish are capable of learning aversive events. Pradel et al. (13, 14) and Xu et al. (15) trained zebrafish in an active avoidance paradigm, in which animals had to learn to avoid a side of a shuttle box indicated by a light in which a mild electric shock would be delivered.

The inhibitory avoidance experimental procedure described here was first developed to be used in experiments using goldfish and presented many interesting results (16–18) because of which we investigated its utility for experiments using zebrafish. So, this experimental model was used to investigate the role of the histaminergic H1 receptor clorfeniramine, administered intraperitoneally, on learning in zebrafish (19).

Although inhibitory avoidance is a useful experimental model, it is not without its limitations or its confounding effects. The first one is that the animals are placed in the experimental aquarium only for the experimental procedure, and then are placed back into the maintenance aquarium. It is important to note that the transfer of animals between both aquariums is a source of stress reflected by the large number of animals that showed freezing behavior in the aquarium. Because of this freezing behavior, a high number of animals did not cross to the preferred compartment on the baseline trial, and were, therefore, excluded from the experiment. Since this was the most important limitation observed, we developed an experimental model where the animal would remain the entire duration of the experiment in the experimental aquarium. Therefore, in the second experiment, the animals were placed in the aquarium two days before the beginning of the procedures and were not moved until the end of the experiment. This reduced the number of fish excluded from the experiment because of freezing behavior on the first trial.

It is known that zebrafish are a shoaling (social) species (20), and therefore a large number of fish can be placed in a single aquarium for maintenance. Thus, placing a single animal per aquarium during the experiment could be stressful for the animals. To overcome this problem, the 20 experimental aquariums that were used in the experiment were placed side by side on two shelves, allowing visual contact between animals.

Learning and memory have been studied by many different approaches in zebrafish. In this second experiment, we used an aquarium with feeders placed at both ends, signaled each by a green or a red light. To reach food that was offered only by the feeder indicated by the green light, the animal had to swim below the plastic barrier.

It was observed that the animals were able to learn the task, since the time the animals took to enter the correct feeding area diminished after the days of training. Since the position of the lights was randomly changed every training day, we believe that the animals learned to discriminate the colors instead of associating the food offer to a particular side of the aquarium. In this experiment only one pellet of floating food was offered inside a feeder that was fixed to the aquarium wall in such a manner that its inferior extremity remained a few millimeters below water level. Therefore, once the food was dropped, it remained inside the feeder and the animals were not able to see it, indicating that the animals were able to associate the food with the light.

Our results are in agreement with previous studies that have shown that zebrafish could learn to discriminate between different pairs of colors (green and purple; red and blue) when one of the colors was paired with food reward (21). It has also been shown that zebrafish are able to discriminate between two achromatic patterns (21).

In the second experiment, the animals were able to learn the task on the 5th day of training, with one trial per day. This is a relatively short period of training for fish. In comparison, the olfactory conditioning procedure for zebrafish described by Braubach et al. (22) is a robust and simple model for the study of classically conditioned appetitive response. However, training to single odorants consisted of a total of 60 trials, conducted over the course of 5 days. New experimental approaches are being developed in order to overcome this problem. Recently a one-trial inhibitory avoidance task was developed for zebrafish, in which the animals learned to refrain from swimming to the preferred aquarium compartment in order to avoid an electric shock during a single-trial training session (23).

Different procedures using positive rewards were also effective for the learning and memory in this species. Visual access to a group of conspecifics has rewarding properties and this reinforcer can support associative learning (24). These and other authors (25) have successfully used computer images of zebrafish and elicited shoaling or reproductive behavioral responses to the images in experimental zebrafish. Pather and Gerlai (26) also used computer-animated images of zebrafish to support learning in zebrafish. In their experiment, the animated zebrafish was presented on a specific side of the aquarium (always on the same side for one group, randomly or on alternating sides for the other two groups). Learning of the task was expressed by increase of permanence of animals close to the side where the images were presented on the same side group and a decrease in this time on the alternating side group, since animals anticipated that the shoal would be presented on the opposite side of the aquarium after the intertrial interval.

Great effort has been put into practice by researchers from different laboratories to raise the number of experimental model options available for the study of learning and memory in zebrafish. The procedures described here, developed by our and other laboratories, contribute to a growing effort to establish different approaches to study learning and memory in zebrafish, in order to support investigations of learning, memory and behavioral plasticity, using behavioral or genetics approach.

References

1. Salas, C., Broglio, C., Duran, E., Gomez, A., Ocana, F. M., Jimenez-Moya, F. & Rodriguez, F. (2006) Neuropsychology of learning and memory in teleost fish. *Zebrafish* 3, 157–171.
2. Overmier, J. & Hollis, K. (1990) Fish in the think tank: learning, memory and integrated behavior. In Kesner, R. & Olton, D. (Eds.) *Neurobiology of Comparative Cognition*. Hillsdale, Lawrence Erlbaum Associates.
3. Panula, P., Sallinen, V., Sundvik, M., Kolehmainen, J., Torkko, V., Tiittula, A., Moshnyakov, M. & Podlasz, P. (2006) Modulatory neurotransmitter systems and behavior: towards zebrafish models of neurodegenerative diseases. *Zebrafish* 3, 235–247.
4. Bally-Cuif, L. (2006) Teleosts: simple organisms? Complex behavior. *Zebrafish* 3, 127–130.
5. Guo, S. (2004) Linking genes to brain, behavior and neurological diseases: what can we learn from zebrafish? *Genes Brain Behav* 3, 63–74.
6. Grunwald, D. J. & Eisen, J. S. (2002) Headwaters of the zebrafish – emergence of a new model vertebrate. *Nat Rev Genet* 3, 717–724.
7. Ingham, P. W. (2009) The power of the zebrafish for disease analysis. *Hum Mol Genet* 18, R107–R112.
8. Campbell, W. A., Yang, H., Zetterberg, H., Baulac, S., Sears, J. A., Liu, T., Wong, S. T., Zhong, T. P. & Xia, W. (2006) Zebrafish lacking Alzheimer

presenilin enhancer 2 (Pen-2) demonstrate excessive p53-dependent apoptosis and neuronal loss. *J Neurochem* 96, 1423–1440.
9. Flinn, L., Bretaud, S., Lo, C., Ingham, P. W. & Bandmann, O. (2008) Zebrafish as a new animal model for movement disorders. *J Neurochem* 106, 1991–1997.
10. Dooley, K. & Zon, L. I. (2000) Zebrafish: a model system for the study of human disease. *Curr Opin Genet Dev* 10, 252–256.
11. Shin, J. T. & Fishman, M. C. (2002) From zebrafish to human: modular medical models. *Annu Rev Genomics Hum Genet* 3, 311–340.
12. Serra, E. L., Medalha, C. C. & Mattioli, R. (1999) Natural preference of zebrafish (*Danio rerio*) for a dark environment. *Braz J Med Biol Res* 32, 1551–1553.
13. Pradel, G., Schachner, M. & Schmidt, R. (1999) Inhibition of memory consolidation by antibodies against cell adhesion molecules after active avoidance conditioning in zebrafish. *J Neurobiol* 39, 197–206.
14. Pradel, G., Schmidt, R. & Schachner, M. (2000) Involvement of L1.1 in memory consolidation after active avoidance conditioning in zebrafish. *J Neurobiol* 43, 389–403.
15. Xu, X., Scott-Scheiern, T., Kempker, L. & Simons, K. (2007) Active avoidance conditioning in zebrafish (*Danio rerio*). *Neurobiol Learn Mem* 87, 72–77.
16. Cofiel, L. P. & Mattioli, R. (2006) Involvement of histamine receptors in the acquisition of inhibitory avoidance in *Carassius auratus*. *Prog Neuropsychopharmacol Biol Psychiatry* 30, 1246–1250.
17. Faganello, F. R. & Mattioli, R. (2007) Anxiolytic-like effect of chlorpheniramine in inhibitory avoidance in goldfish submitted to telencephalic ablation. *Prog Neuropsychopharmacol Biol Psychiatry* 31, 269–274.
18. Medalha, C. C., Coelho, J. L. & Mattioli, R. (2000) Analysis of the role of histamine in inhibitory avoidance in goldfish. *Prog Neuropsychopharmacol Biol Psychiatry* 24, 295–305.
19. Serra, E. L., Medalha, C. C., Bueno, R. O. & Mattioli, R. (2002) The role of chlorpheniramine in inhibitory avoidance in *Danio rerio*. *J Fish Biol* 60, 1342–1345.
20. Spence, R., Gerlach, G., Lawrence, C. & Smith, C. (2008) The behaviour and ecology of the zebrafish, *Danio rerio*. *Biol Rev Camb Philos Soc* 83, 13–34.
21. Colwill, R. M., Raymond, M. P., Ferreira, L. & Escudero, H. (2005) Visual discrimination learning in zebrafish (*Danio rerio*). *Behav Processes* 70, 19–31.
22. Braubach, O. R., Wood, H. D., Gadbois, S., Fine, A. & Croll, R. P. (2009) Olfactory conditioning in the zebrafish (*Danio rerio*). *Behav Brain Res* 198, 190–198.
23. Blank, M., Guerim, L. D., Cordeiro, R. F. & Vianna, M. R. (2009) A one-trial inhibitory avoidance task to zebrafish: rapid acquisition of an NMDA-dependent long-term memory. *Neurobiol Learn Mem* 92(4), 529–534.
24. Al-Imari, L. & Gerlai, R. (2008) Sight of conspecifics as reward in associative learning in zebrafish (*Danio rerio*). *Behav Brain Res* 189, 216–219.
25. Turnell, E. R., Mann, K. D., Rosenthal, G. G. & Gerlach, G. (2003) Mate choice in zebrafish (*Danio rerio*) analyzed with video-stimulus techniques. *Biol Bull* 205, 225–226.
26. Pather, S. & Gerlai, R. (2009) Shuttle box learning in zebrafish (*Danio rerio*). *Behav Brain Res* 196, 323–327.

Chapter 11

Spatial Cognition in Zebrafish

Joshua L. Haight and Joseph A. Schroeder

Abstract

Studies of teleost spatial cognition have revealed that fish possess an impressive array of navigational abilities and are capable of spatial memory based tasks utilizing both egocentric and allocentric cues. The emergence of zebrafish as an optimal animal model for developmental, genetic, and chemical screening investigations necessitates a better understanding of this species behavior including spatial cognition. Investigations of zebrafish spatial cognition described here reveal that zebrafish quickly learn to execute spatial tasks based on visual cues to avoid simulated predator attacks and to obtain food reward. They are also capable of memorizing spatial alternation sequences for navigational tasks and memory of these tasks is retained for several weeks. Two additional protocols designed to evaluate complex navigational behavior in zebrafish are also described. Results from preliminary studies indicate that zebrafish can learn to navigate mazes comprised of multiple directional turns with minimal aid from allocentric visual cues. The growing collection of zebrafish spatial cognition protocols and the accumulation of data from carefully designed behavioral studies when combined with what is known about the molecular neurobiology of the species will ultimately lead to a better understanding of the neurological basis of spatial cognition.

Key words: Spatial memory, spatial alternation, cognitive maps, navigation, allocentric strategies, egocentric strategies, neurological basis of cognition, conditioned place preference, conditioned place aversion, associative learning, three-axis maze, multiple t-maze.

1. Introduction

Despite the growing popularity of zebrafish as models in genetics and molecular biology laboratories and especially given the increasing interest in understanding the genetic basis of behavior, relatively little is known about the spatial cognitive abilities of this versatile laboratory model. Other teleost species have been the subjects of spatial navigation experiments ranging from

A.V. Kalueff, J.M. Cachat (eds.), *Zebrafish Models in Neurobehavioral Research*, Neuromethods 52,
DOI 10.1007/978-1-60761-922-2_11, © Springer Science+Business Media, LLC 2011

simple landmark recognition (1) to ordered navigation (2) to construction of cognitive maps (3) in both open water or controlled laboratory designs. This chapter begins with a brief, broad perspective review of spatial cognition experiments employing a range of teleost species. Emphasis is placed on literature describing studies conducted in a controlled environment many of which could be adapted to evaluate the ability of zebrafish. This is followed by a discussion of the available literature on zebrafish. Finally, the review is supplemented with data from the authors' lab on zebrafish spatial memory and a discussion of the functional neuroanatomy of the zebrafish brain related to spatial cognition. For extensive reviews of spatial memory and navigation/orientation in a range of teleost species, see Braithwaite (4) or Braithwaite and de Perera (5).

A discussion of spatial navigation in fish must at the outset highlight several important defining aspects of the animals' interaction with their environment that are important when considering a spatial navigation experiment for comparison with land-dwelling species. In addition to the horizontal axes, orientation to the vertical axis is profoundly more relevant to navigation in an aquatic environment. Unique changes in the vertical axis such as light penetration, temperature, and water pressure are undoubtedly used by fish for navigation. Besides these cues and local visual and olfactory cues, teleosts have been shown to use the sun for compass navigation (6) and evidence suggests that they may also use the earth's magnetic fields (7). The complex interaction of these aspects of fish spatial navigation must not be overlooked when designing protocols to be used to evaluate the cognitive spatial abilities of zebrafish.

One of the simplest forms of spatial memory that has been studied in teleosts is the use of landmarks, or beacons to identify goals ranging from a food patch to an area of the environment that can be used for protection. The animal must recognize a landmark and associate it with a specific objective. Goldfish are able to use landmarks to identify the presence or absence of food patches hidden under the gravel of their tank (1). The Atlantic salmon (*Salmo salar*) has been the subject of several studies because of its lengthy migrations. As a testament to their extensive travel, Atlantic salmon who have been tagged in North America have been found off the coasts of the UK, New Foundland, and Norway (8) suggesting a superb navigational ability based on acute sensory discriminations. In the laboratory, salmon quickly learn to discriminate between channels identified with visual landmarks to obtain food (9). In this study, it was demonstrated that this discriminative ability is retained even when every effort is made to make the channels identical. Thus either the channels were not visually identical to the salmon eye, but had small imperfections that allowed the salmon to tell the difference between them, or "that the salmon parr may have deposited some form of olfactory cue on the rewarded patch to mark it" ((4), p. 92). Regardless of whether the fish were using a visual or olfactory cue, this study suggests that the ability of salmon to discriminate between navigational cues is sophisticated and adaptive, helping the salmon to navigate during extremely long migrations.

Other studies have also shown that fish can use both local and global landmarks to help them avoid a threat. One study employed 3-spined sticklebacks (*Gasterosteus aculeatus*) taken from either high or low predation environments and evaluated their ability to avoid a dangerous feeding patch (10). In general, fish originally captured from a high predation environment identified an unsafe feeding patch using global cues outside of the tank whereas fish from a low predation environment used internal tank landmarks to identify where to feed. This suggests that preference for a type of navigational landmark cue can vary within a species and is dependent on the type of environment the fish come from.

One method for evaluating the complex spatial navigation problems encountered by wild fish utilizes chains of landmarks. Instead of using one landmark to signal a food patch or shelter, fish can apparently memorize sequences of landmarks for use in piloting and navigation (2, 11). Reese (11) studied butterflyfish (family *Chaetodontidae*) in their wild habitat on coral reefs and demonstrated that the fish used landmarks to follow the same, distinct path everyday. When the landmark sequence was disrupted, the fish engaged in searching behavior and were able to resume their path when a path landmark was recognized instead of starting at the beginning of the route.

Another study by Girvan and Braithwaite (12) examined the difference in ability of wild 3-spined sticklebacks taken from either a river or a pond to navigate a spatial maze. Fish were trained to use either an egocentric, turn sequence learning strategy or an allocentric, landmark navigation strategy to complete the maze. Fish from the river population learned to solve the maze equally well in each test condition; fish from the pond population however were much faster at solving the maze in the condition containing local landmarks. This shows, much like the Huntingford and Wright (10) study, that some 3-spined sticklebacks preferred to "memorize" the algorithm of the maze (right-left-right), whereas others preferred to learn the association between the local landmarks and the correct path. It was suggested that a river environment is not conducive to visual landmarks, since currents are always shifting the surroundings, while the pond environment has relatively stable visual stimuli. This could potentially lead to the river sticklebacks developing alternative methods for navigation, other than visual recognition, while the pond sticklebacks can rely more heavily on visual cues.

Another interesting species that has been utilized in spatial navigation studies is the blind Mexican cave fish (*Astyanax fasciatus*). Teyke (13) observed that the blind fish learn about their surroundings by using their lateral line organ. When the fish were placed in an environment that was unfamiliar, they increased their swimming velocity, promoting lateral line stimulation. The fish localized their swimming to the perimeter of the tank and around novel stimuli, suggesting that they were learning about the size, dimensions, and characteristics of their environment. Once the fish became familiar with their surroundings, they decreased their swimming velocity. Data showed that swimming velocity would drop significantly from 2 to 4 h after being exposed to the new environment, and then steadily decline from 4 to 12 h, where it would begin to become relatively constant indicating that swimming velocity is a reliable indicator of familiarity with an environment. In addition, it has been demonstrated that blind Mexican cave fish can use the lateral line system to encode order of landmarks in their spatial maps and demonstrate possession of an "internal representation of order" ((2), p. 2133), suggesting that fish are not just simply associating one landmark with another, but have the cognitive ability to recognize an order in the landmarks of their environments. Zebrafish have a well-developed lateral line system (14). The lateral line system is used by fish to detect movement and vibration in the water, which aids in schooling behavior and avoiding predation (15). The sensory cells of the lateral line system are similar to the hair cells of the mammalian cochlea and likely function in fish in a similar fashion to the auditory system in other vertebrates during spatial navigation tasks. The zebrafish lateral line system has been proposed as a model for pharmacologic studies of hair cell ototoxicity (16).

Perhaps the most complex aspect of spatial cognition that has been shown in fish is the ability to form cognitive maps. This was first investigated by Aronson (17, 18) using the gobiid fish (*Bathygobius soperator*). The gobiid fish dwells in rocky tide pools during low tides and is quite unique in its behavior. When prompted with a predator attack, the gobiid fish will jump from its home tide pool to another tide pool or the open ocean with incredible accuracy. Aronson investigated how previous experience with an environment influenced jumping accuracy by constructing an artificial environment containing three tide pools and simulated tides. At low tide, the fish were confined to one of the three tide pools. At high tide, the fish could freely swim around the entire environment. Fish that were not allowed to explore the entire environment were very inaccurate at jumping to other tide pools when threatened with a predator attack, whereas fish that were allowed one high tide exploration of their environment were several times more accurate at jumping to another pool compared to their inexperienced counterparts. This study demonstrates that the gobiid fish displays an efficient ability to build a cognitive map of its surrounding environment from one exposure, and can refer to this cognitive map when prompted with a predator attack. A second study demonstrating the cognitive mapping abilities of fish was performed by Rodriguez et al. (3) using goldfish. In this study, goldfish were trained to reach a food reward in a four-arm maze using either an egocentric or allocentric strategy or both. Fish in the allocentric group were trained to use global landmarks outside of the tank to determine the location of a food reward in the maze, whereas fish in the egocentric group were trained to always turn a certain direction out of the start arm of the maze to reach the food. A third group was trained in both strategies. Results showed that the goldfish could learn both types of navigational strategies. Also, those trained in the allocentric strategy were able to reach the food reward when released from any arm of the maze, indicating that the fish had developed an internal representation of the relationships between the global landmarks outside of the tank and where the food reward was located.

As these studies indicate, the spatial navigation and memory capacity of teleosts is complex. They are able to learn subtle, defining spatial relationships between objects and are able to build cognitive maps of their environment. These abilities are important for helping fish remember the location of food patches, places of shelter, navigate their home territories, orient themselves in space, and perform complex migratory behaviors.

2. Examples of Spatial Memory in Zebrafish

Zebrafish are quickly becoming the primary model for neurodevelopmental studies. The exposed chorion and quick development of this fish make it an ideal candidate for the study of genetics and the molecular processes of development (19). While the biological studies of the zebrafish have charged to the forefront of genetic and neurological sciences, those assessing the cognitive functions of this fish have begun to emerge. Some literature exists on the basic memory functions of zebrafish, as well as the effects and timing of nicotine on improving memory (20, 21) and the role melatonin plays in suppressing the nighttime formation of memories (22). However a few studies directly assess the spatial capacities of zebrafish, and these are similar to studies that have been performed with other teleosts.

Arthur and Levin (19) used an experimental model similar to Huntingford and Wright (10) to assess whether zebrafish could remember the specific site of a simulated predator attack. A 40-l tank was split into three chambers separated by sliding plexiglass doors. The back of the tank was colored black to help orient the fish and one of the side chambers was designated as the "safe" chamber. Fish were individually placed into the start chamber for an orientation period of 60 s. After this period, the doors to each side chamber were opened and 20 s later a fish net was lowered into the middle chamber and moved around in a stereotypic fashion. If the fish swam into the "safe" chamber, the sliding door was closed and the fish net was removed, leaving the fish to swim freely for 60 s. If the fish swam into the other chamber, the door was closed and the fish net was placed into the chosen chamber and moved around for continued stress for 60 s. This was continued for 10 sessions, three trials per session. The learning parameters were then reversed for 18 sessions, with the previous "safe" side now associated with the net punishment. Results showed that the zebrafish significantly improved in their escape choice over time, learning to avoid the chamber associated with the net punishment. The fish also quickly learned to avoid the net-associated chamber when the test parameters were reversed. This shows that the fish were cognitively making a chamber choice based on avoidance behavior, and they could remember using spatial cues which chamber was associated with a punishment. This finding is similar to that of the Huntingford and Wright (10) study in which 3-spined sticklebacks were using visual cues to discriminate between compartments based on avoidance behavior.

In a follow-up to this study, Arthur and Levin (19) used the same basic protocol to determine if zebrafish could make the same discrimination using visual instead of spatial cues. In this experiment, the back wall of one chamber was colored blue, while the other was colored red of an equal intensity. Instead of keeping spatial cues constant, the fish had to learn to avoid a specific color chamber, which changed sides. The fish successfully learned to avoid the punishment condition by choosing the appropriate color compartment, and could learn to reverse their behavior when the test conditions were reversed. Thus zebrafish are capable of using both visual and spatial cues to avoid punishment, much like previous studies have demonstrated with other fish.

A second protocol utilizing zebrafish as test subjects was designed to assess the memory of this fish in a simple spatial alternation task (23). A small 2-gallon (30 × 20 × 15 cm) container was used as the experimental tank, and a laminated piece of white poster board was used to divide the tank in half, each half measuring 15 × 20 × 15 cm. The board divided the top 10 cm of the tank, leaving about a third of the tank open underneath, so the fish could have easy access to either side of the tank. A red card was taped onto one end of the tank, serving as a visual cue so the fish could orient themselves spatially. Fish were either trained to receive food on alternating sides of the tank, or on a randomly chosen side. At the start of the experiment, both groups of animals had an average choice ratio that was no better than chance. Over the course of 28 trials, the performance of the alternating side group significantly improved, while the performance of the random administration group never rose above chance levels. This demonstrates that zebrafish have the cognitive ability to orient themselves spatially and learn to alternate between two distinct chambers for a food reward.

As a second part to this study, Williams et al. (23) determined if the zebrafish could remember the spatial alternation task following a 10-day break in trials. After the first 28 trials, the fish from the alternating experimental group were removed from the test tank for 10 days. When this break period was over, the fish were reintroduced to the test tank and another 14 trials spatial alternation task, accurately choosing the side of the tank at which the food reward was administered at levels significantly better than chance.

Recent evidence from our laboratory has suggested that zebrafish are capable of more complex spatial egocentric navigation. Described below are two mazes and protocols we have developed and preliminary data on the performance of zebrafish in each maze. Both mazes were designed and constructed to determine if zebrafish were capable of memorizing a navigational "route." For both protocols, landmarks and external cues that the animals could use as beacons or for orientation were minimized. The three-axis maze requires fish to navigate a route based on x (forward/backward), y (left/right), and z (depth) axes. The maze consists of a 20 × 20 × 60 cm plexiglass tank. Four plexiglass inserts divide the tank into five 12 × 20 × 20 chambers. There is a 7 × 7 cm window cut out of one corner or the center of each insert. With the inserts in place, fish must swim from chamber to chamber through windows that are in a different location in each insert in order to reach a food reward in a floating feeding ring in the chamber on the opposite side of the tank. The order of inserts (i.e., location of each window) remained constant for each fish across trials. See **Fig. 11.1** for a diagram of the maze. Without the inserts, the tank served as the home tank for five fish at a time. At the start of each trial, all of the fish were removed and placed in a 1-l opaque circular holding tank containing home tank water. The filter, heater, and aerator were removed from the tank and white cardboard walls were placed outside, around the perimeter of the tank so that the fish could not use cues internal or external to the tank for navigation. Fish were food deprived for 2 days prior to the start of the experiment. Fish were placed individually in the start chamber and the latency to reach the goal chamber containing the feeding ring was recorded. Training consisted of two back-to-back trials per day. The average learning curve for 10 fish is displayed in **Fig. 11.2**.

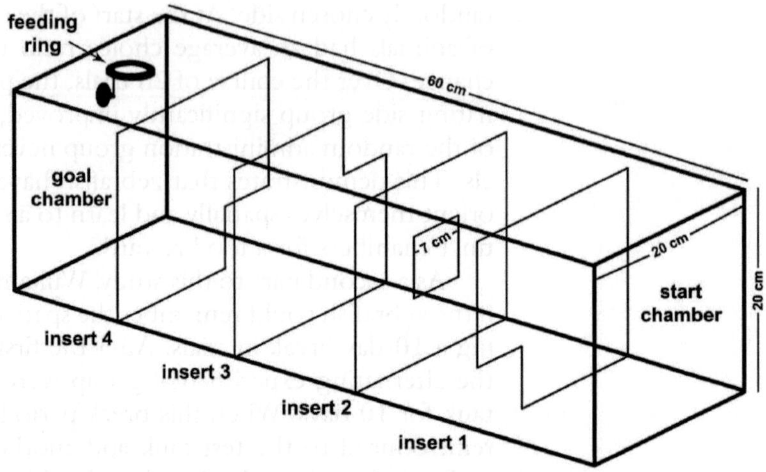

Fig. 11.1. Diagram of the zebrafish three-axis maze. Constructed of clear plexiglass, food-deprived fish are placed in the start chamber and must swim through the window in each insert to reach the feeding ring in the goal chamber (not drawn to scale).

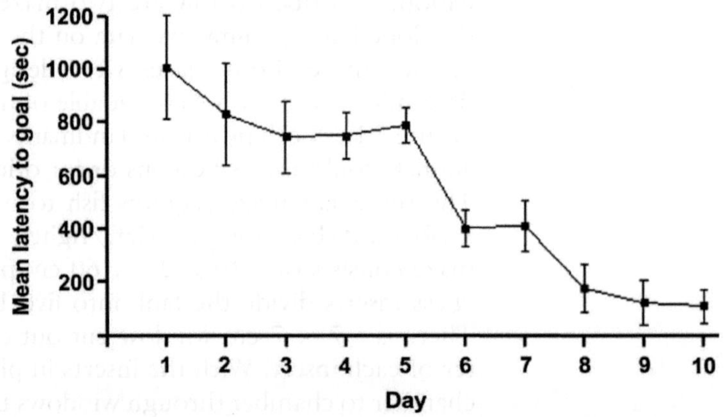

Fig. 11.2. Latency to goal as a function of maze exposure for the zebrafish three-axis maze. Each data point represents the average and standard deviation of two trials per day for 10 fish.

The multiple T-maze requires fish to memorize a series of left/right turns to navigate through a series of shallow brightly lit channels to their home tank which is deeper, covered, and contains plastic aquarium plants and two cohorts. The maze channels are 10 cm in height and 8 cm wide and are constructed of white plexiglass. The base of the maze is a large octagon platform (each side is 20 cm) constructed of white plexiglass. The platform rests in a large circular plastic pool that is 130 cm in diameter and filled to a depth of 20 cm. The home tank is a 20 × 20 × 20 cm plexiglass chamber inserted through the maze platform. The entry from the maze to the home tank can be closed with a sliding door. See **Fig. 11.3** for a diagram of the multiple t–maze. During the training period, the home tank housed three fish at a time. A 1-day pretraining period (straight-arm training) preceded maze training during which all sections of the maze except for the channel leading directly to the home tank entry were sealed off with plexiglass partitions. Fish were food deprived for 2 days prior to training. Each fish received five straight-arm, pretraining trials. All fish were transferred from the home tank to a 1-l circular holding tank before training. Fish were individually placed at the end of the training arm and learned to escape the maze channel by swimming to the home tank where they received a small amount of food. During maze training, individual fish were placed at the end of the start channel and the latency to enter the

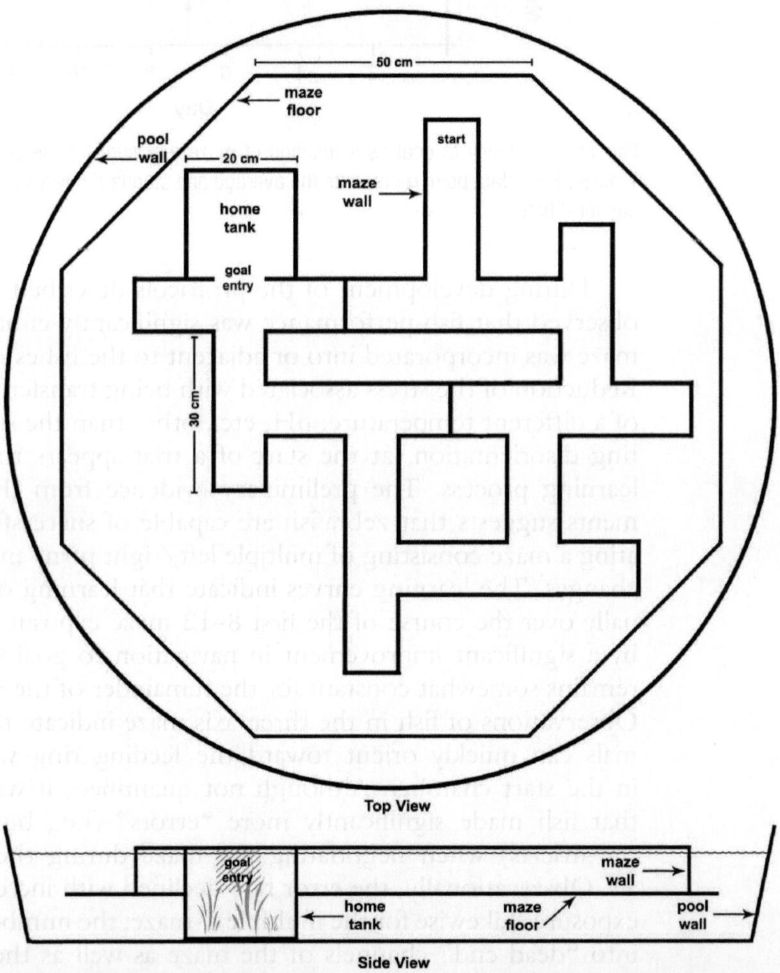

Fig. 11.3. Top and side view diagrams of the zebrafish multiple T-maze. The maze and home tank are constructed of white plexiglass on a plexiglass base inserted into a plastic pool. Fish are individually placed in the start arm and must navigate the brightly lit maze channels (depth = 10 cm) to reach the darker home tank (depth = 20 cm) containing plastic aquarium plants (not drawn to scale).

home tank was recorded. Each fish received two training trials per day with the second trial starting 1 h after the first trial. Fish were rewarded with a small amount of food following each trial. The average performance of 6 fish over the course of 12 days of training is displayed in **Fig. 11.4**.

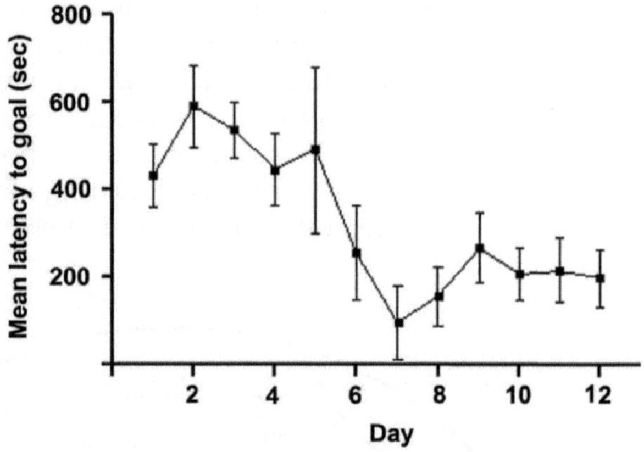

Fig. 11.4. Latency to goal as a function of maze exposure for the zebrafish multiple T-maze. Each data point represents the average and standard deviation of two trials per day for 6 fish.

During development of the protocols described here, it was observed that fish performance was significantly enhanced if the maze was incorporated into or adjacent to the fishes' home tank. Reduction of the stress associated with being transferred to water of a different temperature, pH, etc., other than the stress of netting disorientation, at the start of a trial appears to aid in the learning process. The preliminary evidence from these experiments suggests that zebrafish are capable of successfully negotiating a maze consisting of multiple left/right turns and/or depth changes. The learning curves indicate that learning occurs gradually over the course of the first 8–12 maze exposures, followed by a significant improvement in navigation to goal latency that remains somewhat constant for the remainder of the experiment. Observations of fish in the three-axis maze indicate that the animals can quickly orient toward the feeding ring when placed in the start chamber. Although not quantified, it was observed that fish made significantly more "errors" (i.e., bumping into the inserts) when negotiating the maze during the initial trials. Observationally, the error rate declined with increasing maze exposure. Likewise for the multiple T-maze, the number of entries into "dead-end" channels of the maze as well as the frequency of "back-tracing" through the maze was observationally more frequent during the early maze exposures. Criteria for defining and quantifying navigational errors in these spatial navigation protocols will be included in future studies. Potential navigational cues internal and external to the maze were minimized for these protocols suggesting that zebrafish are capable of egocentric spatial memory and navigation. Future experiments will determine how performance is affected by the inclusion of cues that enhance allocentric navigation. A study by Brown et al. (24) using zebra cichlids suggests that fish use the geometric shape of their home environment to develop navigational strategies. Cichlids raised in a circular tank adopted navigational strategies that were based on navigational cues rather than the geometry of the tank. This interesting finding could easily be adapted to zebrafish given the speed and ease of breeding this species in captivity.

3. What Areas of the Teleost Brain Have Been Implicated in Spatial Cognition?

As the above studies have shown, teleost fish have the ability to use multiple modes of spatial cognition to negotiate their environment (3). They can utilize allocentric strategies to navigate, which are viewpoint-centered, simple associations. These associations are directly cued, e.g., swimming toward a specific landmark. They can also navigate by using more complex egocentric strategies, which involve building a cognitive map and recognizing the 3-dimensional relationships between objects in the environment. Much like reptiles, birds, and mammals teleost fish can use these different strategies either alone or in conjunction. In mammals it is widely accepted that the hippocampus and related brain structures are responsible for the use of spatial memory in building these complex spatial maps (25). Identification of the brain areas responsible for spatial navigation in other vertebrates has yielded some interesting results.

The first studies to investigate whether nervous system structures analogous to the vertebrate hippocampus are responsible for allocentric spatial memory in the teleost fish brain were performed by Salas et al. (26, 27). The goal of these studies was to assess whether a lesion of the telencephalon in goldfish, containing neural structures similar to the hippocampus in vertebrates, would result in a loss of allocentric navigational ability. In one study (26), an opaque gray, diamond-shaped tank was used, only allowing for visual cues that were placed inside the tank. Two white and black striped panels that could be attached to the walls of the tank were used as visual cues. The fish were split into two different experimental groups, a "spatial constancy" group and a "directly cued" group. In the spatial constancy group, the correct exit to the tank was always in constant relationship with the visual cues, regardless of starting location. To correctly navigate to the exit, fish had to learn the spatial relationship between the visual cues and the exit. In the directly cued task, the visual cues were always adjacent to the correct exit, requiring the animals to form a simple association. After group assignment, animals were trained in their respective test condition for 3 days, until reaching performance levels with accuracy greater than 80%. Each test group was then divided into three operational groups: telencephalic ablation, sham operation, and controls. After a 4-day recovery period, fish were tested in their respective experimental conditions. Results show that animals that received the telencephalic ablation were significantly impaired in the spatial constancy condition, while no other group suffered significant deficits in performance. Removal of the telencephalon, which is analogous to the hippocampus in other animals, resulted in an inability to perform spatial task requiring the use of a cognitive map.

In another study by Salas et al. (27) an experimental procedure similar to that used by Rodriguez et al. (3) was employed with the addition of telencephalic ablation to each experimental group. Goldfish were trained for 20 sessions in four different conditions within a four-arm maze: place (allocentric), turn (egocentric), place-turn (both), and controls. Each group then received either sham operations or lesions of the telencephalon. Results showed that animals in the "place" experimental group suffered significant deficits in performance after lesion of the telencephalon. Along with the findings from the above, these studies show that lesion of the telencephalon in goldfish results in the inability to use allocentric, or cognitive-mapping, spatial navigation strategies.

Both of these studies implicated the involvement of the telencephalic region of the teleost fish brain in solving spatial problems where more complex, egocentric (cognitive mapping) strategies were necessary for successful navigation. Vargas et al. (28) expanded upon these studies to determine which parts of the telencephalon were most active during these types of cognitive activities. By using a silver nucleolar organizing region (AgNOR) neurohistochemical staining procedure, the experimenters examined morphological changes in the argyrophilic NOR-associated proteins, which are indicative of rRNA gene transcription, and thus cellular activity. Using goldfish as test subjects, one experimental group was trained in a spatial learning task similar to that in Salas et al. (26), while the other group was trained in a nonspatial task. Upon analysis of the telencephalon with the AgNOR stain, results showed increased AgNOR density in the dorsolateral telencephalic pallium of the fish trained in the spatial task over those trained in the control task. This indicates that the dorsolateral pallium of teleost fish is specially developed for complex spatial memory tasks, much like the hippocampus in humans.

Several additional studies have examined what types of spatial memory functions are inhibited by ablation of the lateral telencephalic pallium in teleosts. Summarized by Rodriguez et al. (29), lesions of the lateral pallium result in a total disruption of place learning and memory. The impairments observed are as severe as those observed when the entire telencephalic region is lesioned (26, 27). Also, these findings are only observed in lesions of the lateral pallium, and not the medial or dorsal pallium sections. These studies offer strong evidence that the lateral telencephalic pallium in teleost fish is the seat of complex spatial cognition and memory.

4. Conclusions

As zebrafish become the norm for genetic and neurodevelopmental research, it will be of significant interest to develop a better understanding of this species' cognitive abilities. Experiments employing a variety of other teleost species have yielded evidence that suggests that the spatial memory and navigation abilities of these animals have been underestimated. It has been demonstrated that fish are capable of using and combining multiple modes of spatial navigation. These include egocentric strategies suggestive of cognitive map formation as well as allocentric strategies that incorporate subtle environmental cues. The lateral telencephalic pallium has been identified as the brain region responsible for complex allocentric strategies using neurohistochemistry, and behavioral observations have verified this through ablation studies. Using these studies as a foundation, systematic and well-controlled behavioral protocols for evaluating the spatial cognition of zebrafish can be developed that when combined with genetic and molecular neurodevelopmental studies will have a profound impact on our understanding of the vertebrate brain.

References

1. Warburton, K. (1990) The use of local landmarks by foraging goldfish. *Anim Behav* 40, 500–505.
2. de Perera, T. B. (2004) Fish can encode order in their spatial map. *Proc R Soc Lond B Biol Sci* 271, 2131–2134.
3. Rodriguez, F., Duran, E., Vargas, J. P., Torres, B. & Salas, C. (1994) Performance of goldfish trained in allocentric and egocentric maze procedures suggests the presence of a cognitive mapping system in fishes. *Anim Learn Behav* 22, 409–420.
4. Braithwaite, V. A. (1998) Spatial memory, landmark use and orientation in fish. In Healy, S. (Ed.) *Spatial Representation in Animals.* New York, NY, Oxford University Press.
5. Braithwaite, V. A. & de Perera, T. B. (2006) Short-range orientation in fish: how fish map space. *Mar Freshw Behav Physiol* 39, 37–47.
6. Levin, L. E., Belmonte, P. & Gonzalez, O. (1992) Sun-compass orientation in the characid cheirodon-pulcher. *Environ Biol Fishes* 35, 321–325.

7. Walker, M. M. (1984) Learned Magnetic-Field Discrimination in yellowfin tuna, Thunnus-albacares. *J Comp Physiol* 155, 673–679.
8. Mills, D. (1989) *Ecology and Management of Atlantic Salmon*. New York, NY, Chapman and Hall.
9. Braithwaite, V. A., Armstrong, J. D., Mcadam, H. M. & Huntingford, F. A. (1996) Can juvenile Atlantic salmon use multiple cue systems in spatial learning? *Anim Behav* 51, 1409–1415.
10. Huntingford, F. A. & Wright, P. J. (1989) How sticklebacks learn to avoid dangerous feeding patches. *Behav Processes* 19, 181–189.
11. Reese, E. S. (1989) Orientation behavior of butterflyfishes (family chaetodontidae) on coral reefs – spatial-learning of route specific landmarks and cognitive maps. *Environ Biol Fishes* 25, 79–86.
12. Girvan, J. R. & Braithwaite, V. A. (1997) Orientation mechanisms in different populations of the three spined stickleback. *Orientation and Navigation-Bird, Humans and Other Animals*. United Kingdom Oxford, Royal Institute of Navigation.
13. Teyke, T. (1989) Learning and remembering the environment in the blind cave fish Anoptichthys-Jordani. *J Comp Physiol A Neuroethol Sens Neural Behav Physiol* 164, 655–662.
14. Metcalfe, W. K., Kimmel, C. B. & Schabtach, E. (1985) Anatomy of the posterior lateral line system in young larvae of the zebrafish. *J Comp Neurol* 233, 377–389.
15. Schellart, N. A. M. & Wubbels, R. J. (1998) The auditory and mechanosensory lateral line system. In Evans, D. H. (Ed.) *The Physiology of Fishes*. 2nd ed. Boca Raton, FL, CRC Press.
16. Whitfield, T. T. (2002) Zebrafish as a model for hearing and deafness. *J Neurobiol* 53, 157–171.
17. Aronson, L. R. (1951) Orientation and jumping behavior in the gobiid fish Bathygobius soporator. *Am Mus Novit* 1486, 1–22.
18. Aronson, L. R. (1971) Further studies on orientation and jumping behavior in the gobiid fish Bathygobius soporator. *Ann N Y Acad Sci* 188, 378–392.
19. Arthur, D. & Levin, E. D. (2001) Spatial and non-spatial visual discrimination learning in zebrafish (*Danio rerio*). *Anim Cogn* 4, 125–131.
20. Levin, E. D. & Chen, E. (2004) Nicotinic involvement in memory function in zebrafish. *Neurotoxicol Teratol* 26, 731–735.
21. Levin, E. D., Limpuangthip, J., Rachakonda, T. & Peterson, M. (2006) Timing of nicotine effects on learning in zebrafish. *Psychopharmacology (Berl)* 184, 547–552.
22. Rawashdeh, O., de Borsetti, N. H., Roman, G. & Cahill, G. M. (2007) Melatonin suppresses nighttime memory formation in zebrafish. *Science* 318, 1144–1146.
23. Williams, F. E., White, D. & Messer, W. S. (2002) A simple spatial alternation task for assessing memory function in zebrafish. *Behav Processes* 58, 125–132.
24. Brown, A. A., Spetch, M. L. & Hurd, P. L. (2007) Growing in circles: rearing environment alters spatial navigation in fish. *Psychol Sci* 18, 569–573.
25. Salas, C., Broglio, C., Duran, E., Gomez, A., Ocana, F. M., Jimenez-Moya, F. & Rodriguez, F. (2006) Neuropsychology of learning and memory in teleost fish. *Zebrafish* 3, 157–171.
26. Salas, C., Broglio, C., Rodriguez, F., Lopez, J. C., Portavella, M. & Torres, B. (1996a) Telencephalic ablation in goldfish impairs performance in a 'spatial constancy' problem but not in a cued one. *Behav Brain Res* 79, 193–200.
27. Salas, C., Rodriguez, F., Vargas, J. P., Duran, E. & Torres, B. (1996b) Spatial learning and memory deficits after telencephalic ablation in goldfish trained in place and turn maze procedures. *Behav Neurosci* 110, 965–980.
28. Vargas, J. P., Rodriguez, F., Lopez, J. C., Arias, J. L. & Salas, C. (2000) Spatial learning-induced increase in the argyrophilic nucleolar organizer region of dorsolateral telencephalic neurons in goldfish. *Brain Res* 865, 77–84.
29. Rodriguez, F., Lopez, J. C., Vargas, J. P., Broglio, C., Gomez, Y. & Salas, C. (2002) Spatial memory and hippocampal pallium through vertebrate evolution: insights from reptiles and teleost fish. *Brain Res Bull* 57, 499–503.

Chapter 12

The Behavioral Repertoire of Larval Zebrafish

Kandice Fero, Tohei Yokogawa, and Harold A. Burgess

Abstract

Shortly after larval zebrafish become free swimming their behavior is modulated by both autochthonous signals and external stimuli. Larvae show rapid responses to a range of sensory cues but are also capable of executing extended behavioral programs in response to changes in the environment. At this early stage, larvae have a small repertoire of discrete stereotyped movements which are deployed in different contexts to generate appropriate behavior. We outline the range of behaviors defined in zebrafish larvae to date and discuss insights into neural function revealed by behavioral assays. A growing body of work demonstrates that tractability of behavior and neural connectivity in larval zebrafish facilitate the analysis of neural pathways underlying vertebrate motor control and sensory processing.

Key words: Developing zebrafish, larvae, acoustic/vibrational stimuli, vestibular stimuli, lateral line stimuli, visual stimuli, chemical stimuli, locomotion, sensory cues, environmental adaptation, stereotypic behavior, neuroanatomy, neuronal pathways, motor control, sensory processing.

1. Scope

A major objective of the study of behavior is to reveal the functional anatomy of the nervous system – to define how patterns of neuronal connectivity and processing enable animals to generate appropriate motor responses. Zebrafish are unique among popular laboratory model organisms in that they were expressly chosen for their advantages in applying genetic analysis to the problem of neural development (1). Today, work by many laboratories has demonstrated that neurogenetic analysis in developing zebrafish is indeed a powerful tool for revealing neuronal pathways underlying behavior. In large part, this is due to the relative simplicity of the larval nervous system – it has been

A.V. Kalueff, J.M. Cachat (eds.), *Zebrafish Models in Neurobehavioral Research*, Neuromethods 52,
DOI 10.1007/978-1-60761-922-2_12, © Springer Science+Business Media, LLC 2011

estimated that the 5-day-old zebrafish brain is comprised of just 100,000 neurons, including an ever-growing inventory of identified neurons (2).

A second goal of behavioral research is to understand how animals use their behavioral repertoire in an ecological context. Laboratory strains of zebrafish show classic signs of domestication, differing significantly from wild populations in a manner consistent with loss of selection of antipredation behaviors including reduced fearfulness and shoaling (3, 4). Only recently have serious attempts been made to describe the natural environment in which zebrafish exist (5–7). This chapter therefore focuses on the repertoire of behaviors which can be elicited in zebrafish larvae under laboratory conditions.

Zebrafish are regarded as "larvae" from the time of hatching at around 3 days post fertilization (dpf) until sex differentiation at 19–23 dpf (8). However the first signs of motility in zebrafish occur much earlier, with spontaneous coiling movements observed by 17 h post fertilization (9). Several factors complicate analysis of behavior in very young fish: first, until 3 days post fertilization, the behavioral repertoire is very limited, a point which is underscored by the fact that the spinal cord alone is sufficient to mediate touch-evoked coiling and swimming responses in embryos (10). Second, from days 3 to 5, significant changes occur in movement kinematics, as animals transition from embryonic modes of motility to more stable larval patterns (11). Third, during the transition period the larva inflates its swim bladder and transitions from lying motionless on its side on the substrate to actively swimming about in the water (**Fig. 12.1**). At

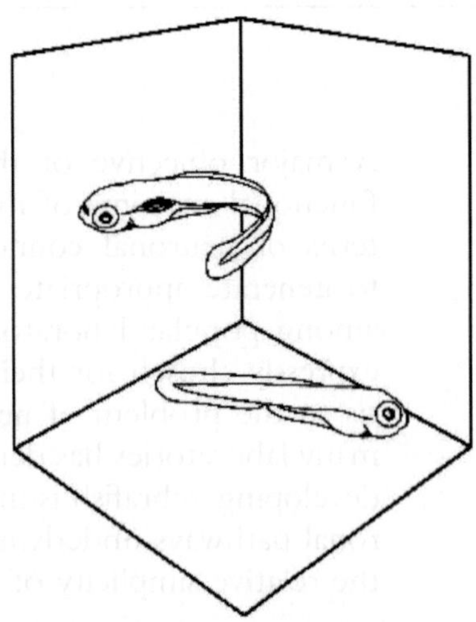

Fig. 12.1. A key transition in larval behavior occurs at 4 dpf. Shortly after hatching at 3–4 dpf, larval zebrafish inflate their swim bladder, gaining positive buoyancy. This transition marks the beginning of spontaneous swimming and increased responsiveness to stimuli.

the same time, it acquires greater responsiveness to acoustic and visual stimuli. This is a crucial distinction, as it has been directly shown that certain behavioral tests, for example the optomotor response, only elicit a response in free-swimming larvae (12). Fourth, the massive neuronal proliferation that has marked brain development becomes restricted to just a few areas by 5 dpf, with most regions of the brain comprised of post-mitotic neurons with well-elaborated neuronal arbors (13). Thus by 5 dpf, while the brain is still immature, it has at least reached a stage where development proceeds at a less explosive pace.

Additional factors confound analysis of behavior in larvae older than 7 dpf. Larvae from days 5 to 7 are relatively homogeneous within a clutch, with all healthy larvae having inflated their swim bladder and showing little variation in size or body weight. This is mainly because these animals still rely primarily on their yoke for nourishment. In contrast, older larvae must be fed in order to maintain normal levels of activity. After feeding commences, even when a large surplus of food is provided, greater variability in size is observed, presumably as a result of uneven success at predation (14). From a practical point of view, this necessitates sorting animals according to size as a proxy for developmental stage. Moreover, in the second week of development, significant changes occur that complicate behavioral analysis, including the first hints of social behavior (15). In this chapter we therefore focus almost entirely on behaviors manifest by larvae at ages between 5 and 7 days post fertilization.

As the optimal time window for testing zebrafish larvae is relatively narrow, it is particularly important that the fish be raised under strictly controlled conditions. The standard temperature for raising zebrafish is 28.5°C – at lower temperatures, fish develop more slowly (16). Low oxygen conditions can easily occur in high density cultures or when unfertilized eggs remain in the dish and become fodder for microorganisms. Under such hypoxic conditions, larval development is impeded (17) and adult behavior is modified (18). Even when efforts are made to maintain strict control over raising conditions, minor differences in other factors can confound comparison between results obtained in different laboratories. Unsurprisingly, the genetic background of the fish can exert a dramatic influence over behavior (19) but subtler differences, for instance osmolarity of the tank water, have been shown to alter neuroanatomical structure and therefore may very well affect behavior (20).

A final emphasis of this review will be on behaviors that can be elicited in free-swimming larvae. Many experiments by necessity are performed on larvae which have been immobilized. While restraint procedures are often required in order to analyze neuronal activity, they may also significantly disrupt behavioral responses (21, 22). Thus results obtained in immobilized larvae should ideally be followed up by experiments in free-swimming fish.

2. The Movement Repertoire

Larval zebrafish are amenable to a powerful approach for behavioral analysis which focuses on measuring the frequency of initiation of stereotyped movements. The pioneering ethologists Niko Tinbergen and Konrad Lorenz had the penetrating insight that even complex behaviors can be deconstructed into a linked series of simple, innate behavioral units, which they termed "fixed action patterns." In a classic application of this approach Tinbergen described hunting behavior in digger wasps as constituted by a series of movements in which the wasp first hovers down-wind of a honey bee, then lunges toward it, curls its abdomen to string the bee, clasps it, and finally initiates flight back to its burrow (23). Direct measurement of the initiation frequency of such distinct locomotor patterns has proven to be a powerful approach for the neurobiological analysis of behavior in invertebrate systems. A graduate student in Tinbergen's laboratory, Margaret Bastock, first described the sequence of motor acts employed by Drosophila during courtship and mating (24). Analysis of courtship behavior in Drosophila remains one of the most productive paradigms for the genetic analysis of neuronal networks underlying behavior (25). A similar approach has also been used to reverse engineer the neuronal architecture underlying other complex behaviors in invertebrates including navigation in *C. elegans* (26, 27) and locomotor choice in the leech *H. medicinalis* (28). Progress in describing the neuronal basis of behavior at a cellular level has thus benefited enormously from breaking behavior down into its constituent acts in invertebrates.

In contrast, behavioral analysis in vertebrate animals usually employs oblique measurements of neural function. In mice, the pattern and degree of locomotor activity in an open field test is considered to reflect levels of anxiety. Similarly, in tests of conditioned place preference, reward learning is quantified by the proportion of time a mouse lingers in a target compartment. While this approach has been productive, analysis of stereotyped motor patterns in vertebrates may be required to achieve a cellular level description of behavior. Such a strategy is certainly possible: Tinbergen famously described the complex series of stereotyped movements employed by male sticklebacks that together constitute courtship behavior (29). A major obstacle is that in mammals, stereotyped movements are difficult to recognize within the continuously modulated and highly plastic stream of movement. Over the course of the first year of life in humans, the cortex assumes responsibility for controlling movement and in consequence, motor acts become plastic and highly individualistic (30). This is not the case for larval zebrafish, where locomotion occurs as a series of distinct events, punctuated by longer bouts of inactivity. A typical locomotor sequence by a zebrafish larva is presented in **Fig. 12.2**. This trace demonstrates the curvature of the larva every 2 ms over a 28-s video recording while the fish engaged in spontaneous locomotion and responded to acoustic and visual stimuli. This demonstrates that movement episodes are clearly demarcated by long periods of inactivity. Bouts of activity are distinct acts, easily isolated within a larger behavioral sequence. Even casual inspection of the trace reveals that distinct curvature functions can be recognized in different episodes of movement. For example, the first two bouts of movement, marked "S" are simple, reasonably symmetrical sinusoids typical of slow swim movements. The third movement, marked "RT,"

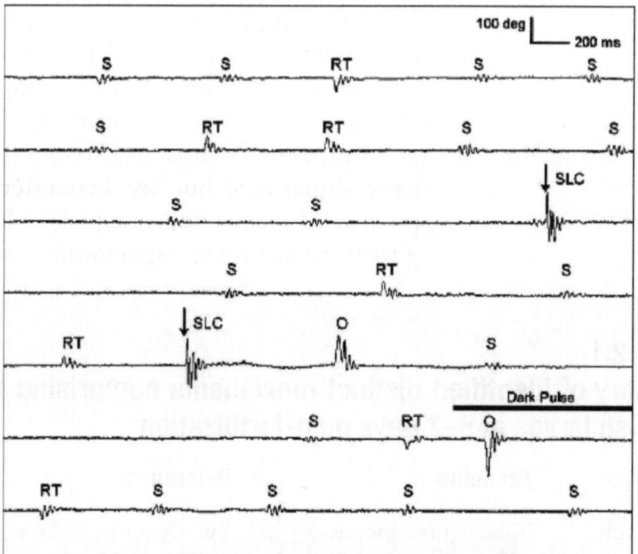

Fig. 12.2. Locomotor acts in zebrafish larvae occur as discrete movement episodes. In this example, a single larva was filmed for 28 s at 500 frames per second. The curvature of the larva was calculated in each frame and all 14,000 curvature points plotted, with time = 0 at the top left. Most spontaneously generated movements are either scoots (S) or routine turns (RT). The larva was also subjected to two acoustic stimuli (marked with arrows), which elicit a distinct type of maneuver, the short latency C-bend response (SLC). In addition a visual stimulus was used, consisting of a 1-s long dark pulse in which the overhead illumination was extinguished. This provoked a characteristic O-bend (O) response.

begins with a larger amplitude body flexion, followed by a few oscillations of the tail, characteristic of a routine turn movement. Thus larval zebrafish are a vertebrate model system in which it is possible to analyze behavior by measuring the frequency of initiation of discrete motor acts, similar to the approach that has been so profitable in invertebrate systems.

Kinematic analysis of bouts of movement produced under a large variety of stimulus conditions reveals that larvae have a small repertoire of stereotyped motor acts. These will be referred to in this chapter as "movement patterns" or "maneuvers." The maneuver repertoire of zebrafish larvae changes over the course of development. In practice, as many manipulations introduce a slight developmental delay into larval maturation, it is advantageous to avoid studying the behavior of newly hatched larvae, and analyze behavior in the window of day 5 through day 7 when the movement repertoire stabilizes at a set of 9 distinctive maneuvers (summarized in **Table 12.1** and illustrated in **Fig. 12.3**). By 5 dpf, forward propulsion is mainly achieved by deploying *slow swim* movements (which will also be referred to as *scoots*), characterized by a low tail beat amplitude and a short travel distance (31, 32). Despite the predominance of scoot maneuvers for forward swimming, 5-day-old larvae remain capable of much faster *burst swims*, sharing the large amplitude tail beats and sustained duration of the "cyclic swimming" pattern of newly hatched larvae (11). Burst swims and scoots not only have distinctive kinematic signatures, but are generated by different populations of spinal cord motor neurons (33). The *capture swim* is a third type of forward swim used specifically for striking at prey objects (34).

Table 12.1
Summary of identified distinct movements comprising the swimming repertoire of zebrafish larvae at 5–7 days post-fertilization

Maneuver	Stimulus	References
Slow swim	None (spontaneous)	Budick and O'Malley (31), Muller and van Leeuwen (11), and Burgess and Granato (32)
Burst swim	Looming predator escape	Budick and O'Malley (31) and Thorsen et al. (194)
Capture swim	Predation sequence	Borla et al. (34)
J-turn	Predation sequence	McElligott and O'Malley (38)
O-bend	Dark flash	Burgess and Granato (32)
Routine turn	Spontaneous/orienting	McElligott and O'Malley (38), Budick and O'Malley (31), and Burgess and Granato (32)
SLC	Acoustic/tactile startle	Kimmel et al. (37), Eaton et al. (36), and Burgess and Granato (35)
LLC	Acoustic startle	Burgess and Granato (35)
Struggle	Embedding	Liao and Fetcho (41)

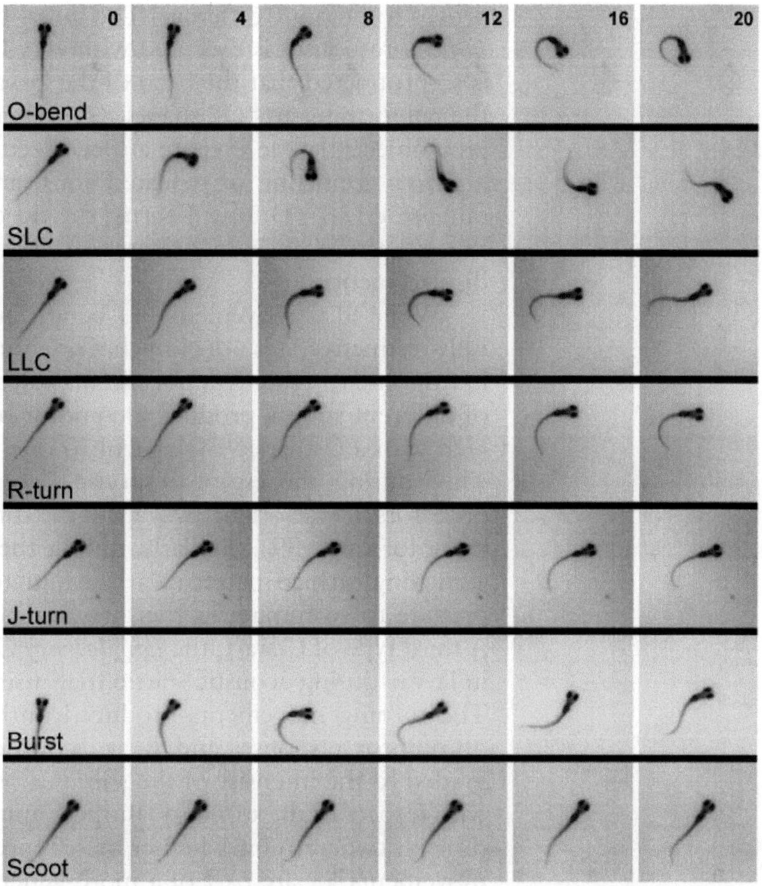

Fig. 12.3. Maneuvers are distinguished by distinctive kinematic features. Still frames at 4-ms intervals from the first 20 ms of different maneuvers illustrate the differences between stereotyped movement patterns.

These swims last less than 50 ms and are terminated with the bilateral extension of the pectoral fins for braking once the prey has been captured.

Larvae are capable of executing several different types of movements which result in a change of orientation. Under uniform conditions, with no overt stimulus supplied, larvae frequently initiate *routine turns* (R-turns), which result in a change of orientation of around 40°. Different types of external stimuli also produce movement responses which cause a change in orientation of the larva. In some cases, the kinematic properties of the response are sufficiently distinctive that it is classified as a distinct type of maneuver. Intense acoustic or tactile stimuli elicit a response at latencies of less than 15 ms, initiated with a very high angular velocity C-bend to one side, here referred to as a *short latency C-bend (SLC)* (35–37). Sudden decrements in light intensity produce a response which looks similar to the eye, involving a large C-bend, but when measured with high speed videography, proves to be quite different. The initial bend movement of these *O-bend* responses is executed with very low angular velocity, but is so protracted that the head of the larva swings around to meet the tail forming an "O" shape (32). When orienting toward small prey objects, larvae execute a specialized type of turn designated a *J-turn* consisting of repeated small amplitude flexions to the same side (38). During J-turns, the tip of the tail hooks around causing the change in orientation, but also a slight backward displacement.

While SLC movements, O-bends, and J-turns are kinematically distinctive, it is not clear that orienting movements produced in response to other types of stimulation are also unique. A variety of different stimuli produce locomotor responses which are initiated with a C-shaped movement to one side, much like R-turns. Thus during the escape response from simulated approaching predators, larvae orient away from the threat before rapidly swimming forward (39). Similarly, during the optomotor response to a moving striped pattern, larvae initiate turn movements as a prelude to swimming in the direction of the moving bars (40). A third type of C-start, the *long latency C-bend (LLC)* is observed in larvae during acoustic startle trials using a weak stimulus (35). The turning movements produced during predator escape, the optomotor response, and weak acoustic startle trials are often graded to the intensity of the stimulus or orientation of the larva with respect to the stimulus (Burgess, unpublished data); thus it is difficult to determine whether these represent unique stereotyped movements, or are part of a more general mechanism employed by larvae for steering in general. This is likely to be resolved only by a detailed understanding of the neural machinery required for the execution of each of these movements.

Finally, when trapped, larvae can initiate a *struggle* maneuver, consisting of a large amplitude body wave traveling from the tail toward the head (41). Surprisingly, while struggling movements are relatively easy to elicit in larvae (for example by embedding in agarose), the behavioral context in which they are generated has not yet been strictly defined.

The different patterned movements described above are produced by discrete modes of activity of spinal cord pattern generators. The spinal cord contains an array of identifiable classes of neurons which are recruited in unique patterns to generate stereotyped movements. The identification of spinal motor circuits which generate patterned movements is an active field of study (reviewed in (42)). Reticulospinal inputs to the spinal cord trigger the production of different maneuvers (40, 43), thus measurement of movement frequency reveals the output of brainstem centers that activate spinal circuits. In subsequent sections, we describe how different stimuli trigger the production of distinct subsets of the larval movement repertoire.

In larval zebrafish, kinematic analysis reliably distinguishes between different elements of the movement repertoire (31, 32, 34, 44). Kinematic analysis can be automated through computational methods, enabling very high throughput analysis of behavior (32, 35). Large numbers of trials can be quickly analyzed, allowing rigorous statistical analysis to dissect the effect of genetic mutations, pharmacological exposure, and anatomical manipulations. It is worth noting that a similar type of analysis is possible at least in principle for adult zebrafish, in which several distinctive movement patterns have been described (45, 46). However, as adult zebrafish swim continuously, movement patterns are not conveniently separated by periods of inactivity, making it computationally more difficult to perform kinematic analysis.

3. The Behavioral Repertoire

A longstanding controversy in behavioral analysis concerns how behavioral responses should be interpreted and described. The celebrated physiologist Jacques Loeb, who pioneered the study of animal tropisms, warned against too easily imputing purpose or intention from behavioral observations, writing (47)

> The history of science has taught us that confusion always reigns when anthropomorphic motives are brought into scientific research. Before the time of Galileo a body sinking in fluid "sought its place." Galileo and his followers put an end to the sovereignty of this psychology, at least in inanimate nature.

Similar sentiments have been expressed by many researchers, notably those from the behaviorist school. Attempts to assign meaning to a behavioral act fall afoul of the temptations of anthropomorphism. Understanding the adaptive value of a behavior in its ecological niche is a worthwhile goal, but it should be kept in mind that experiments under laboratory conditions may reveal responses that are mere epiphenomena. Worse, describing behaviors by assumed motivational state or purpose can imply an internal state that may simply not exist. There is therefore a strong case for confining behavioral analysis to documenting the response of animals under a carefully defined stimulus environment. On the other hand, unlike Galileo's cannon balls, animals are governed by a nervous system capable of sustaining a variety of internal states. Even in larval zebrafish, internal processes like circadian oscillations exert a significant effect on behavior. From a practical point of view, it is essential to recognize the existence of internal states in animals, and where possible, reduce response variability by establishing conditions where internal states are controlled.

We take a pragmatic approach here in describing the behavioral repertoire of zebrafish larvae. It would be cumbersome to merely enumerate the pattern of locomotor responses of larvae to stimuli without using terms like "startle response" or "predation." Moreover, teleological terms are frequently used in the existing literature on larval behavior and consistency is at least as valuable as philosophical precision. Thus, in this chapter, we frame our outline of the behavioral repertoire using the Behaviorist approach, and describe zebrafish behaviors first and foremost in terms of the stimulus that elicits them. However, we do not avoid using commonly employed terminology to refer to a behavior and where its function is reasonably clear, we do not shy away from offering a tentative interpretation. Certain behaviors, like the startle response, can be elicited by stimulation of diverse sensory modalities. However in these cases the sensory pathways transmitting the stimulus differ, thus this approach has the virtue of focusing on the neurobiological basis of the response.

While zebrafish larvae are able to generate a relatively circumscribed set of basic locomotor patterns, as outlined above, their behavioral repertoire is clearly far more extensive, even given the incomplete state of our knowledge. Many assays measuring locomotor behavior in zebrafish have been described (**Fig. 12.4**) and it is likely that careful measurement will reveal a far more extensive set of behavioral skills. Larval behaviors are generally modulated by specific aspects of the testing environment. Where possible, we thus also outline how behavioral responses vary with the testing regime, recognizing that modulation of behavior draws upon distinctive neural circuits.

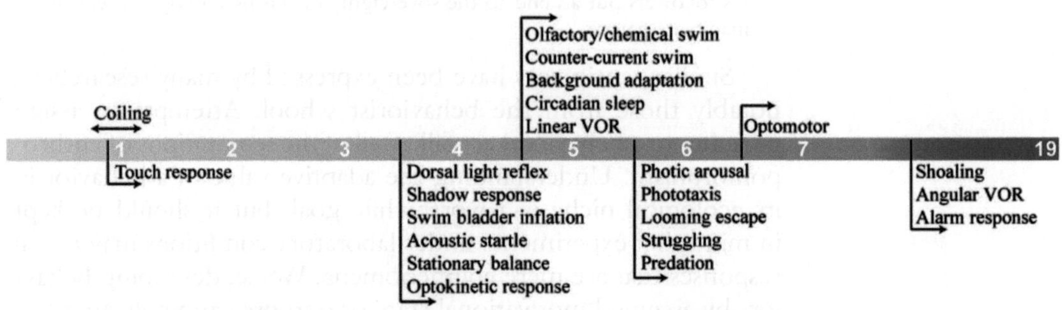

Fig. 12.4. Developmental timeline of the emergence of behaviors. The timeline shows the earliest stage at which behaviors are manifest. See text for details and references.

4. Acoustic/ Vibrational Stimuli

Larvae respond to abrupt acoustic/vibrational stimuli with a fast C-bend followed by a bout of high-amplitude tail beat swimming. In a seminal early paper by Kimmel, four types of response to an abrupt acoustic/vibrational stimulus (a ball dropping into water) were described (37). However, when larvae are stimulated with a brief, calibrated whole-body vibration, only two modes of response are observed (35). Short latency C-bend responses (SLC) occur within 15 ms of the stimulus, while a second wave of long latency C-bend (LLC) responses are initiated 20–60 ms after the stimulus. SLC and LLC responses have distinct kinematic features, suggesting that they are generated by distinct neural circuits (35, 48). Indeed SLC responses require the Mauthner cells, a bilateral pair of giant reticulospinal neurons in the hindbrain (35, 43, 48–50), whereas LLC responses are Mauthner cell independent. Consistent with this, SLC responses have an all-or-nothing character: stimuli of increasing intensity increase the probability of eliciting an SLC response, but do not alter the kinematics of the response. In contrast, LLC responses are graded with stimulus intensity – more intense stimuli increase both the probability and angular velocity of LLC responses. In addition, electrophysiological recordings from hindbrain during stimulation of the body with a vibrating probe have directly revealed that Mauthner cell responses occur around 15 ms faster than non-Mauthner responses, consistent with the timing of SLC and LLC responses (51).

SLC responses are "true" startle responses not only because the short latency and explosive speed of the movement recall startle responses in higher vertebrates. These responses also fulfill a key behavioral criterion for startle movements in that they override ongoing movements (35). Indeed, larvae already engaged in swim bouts at the moment that a sudden acoustic stimulus is delivered show increased responsiveness to the stimulus. Thus the Mauthner cell mediated response is a defensive behavior enabling larvae to rapidly escape from predator strikes. In the wild, such predators might include Dragonfly larvae, known to lurk in the shallow waters occupied by zebrafish juveniles and notorious for the lightning fast labial strikes with which they capture larval fish (5). Interestingly, in young embryos before hatching, the Mauthner cell also drives powerful tail flips. As embryos entrapped in the chorion can not escape from predators this is unlikely to function as a defensive reflex. During the hatching period, many spontaneous Mauthner cell spikes are observed, leading to the suggestion that the early function of Mauthner-mediated axial contractions is to facilitate hatching by rupturing the chorionic membrane (36).

The function of LLC responses is less clear. If a stimulus is potentially important enough to respond to at all, then why don't larvae always execute a rapid SLC response rather than wait an extra 20 ms and execute a slower response? An intriguing possibility is that this second response mode somehow serves a navigational role, allowing larvae additional time to process directional information about the stimulus. Alternatively, larvae may respond to less threatening stimuli with a more calculated response, taking into account the position of obstacles in the environment. As yet, there is no evidence to support either of these propositions. However, startle responses in adult fish are both highly directional with respect to the position of the stimulus and executed so as to avoid running into obstacles in the environment (21, 52).

Acoustic startle responses are usually elicited in zebrafish larvae by stimuli which are intrinsically multimodal – by tapping the dish with a solenoid, or using a shaker to rapidly vibrate the whole dish. Such stimuli are multimodal, exposing the fish to whole body acceleration, bulk water movement, and waves of compression and rarefaction of water molecules. Composed of calcium carbonate, the otoliths are denser than the body of the fish and therefore move more slowly during whole body acceleration. Both SLC and LLC responses are likely triggered by differential acceleration of the otoliths of the inner ear compared to the body. While it was demonstrated that LLC responses were selectively lost in *keinstein* mutants, which lack all four otoliths (35), *keinstein* mutants show a severe balance defect and tend to lie in contact with the bottom of the testing arena. These larvae may have effectively received a mechanosensory stimulus triggering SLC responses. In fact, laser ablation of inner-ear otoliths prevented Mauthner-mediated responses to direct otic vesicle stimulation (48), suggesting that SLC responses are triggered by otolith movement. It remains possible that "tap" stimuli also trigger SLC responses through the lateral line.

Detailed measurements of the kinematic performance of SLC movements can yield insights into neural function. In particular the first C-bend and counterbend of SLC movements are extremely stereotyped – the coefficient of variation for the magnitude of the first C-bend in response to an acoustic startle stimulus is just 14.7% ($n = 77$ TL strain larvae). By comparison, the mean intrastrain coefficient of variation for the acoustic startle response in rats has been estimated at 40% (53). As the magnitude of the C-bend is a function of both the angular velocity of the bend, and the interval before the counterbend begins (the "duration"), separate analysis of these parameters is often informative. When analyzing genetic mutant strains, specific errors in the timing of the activation of the counterbend suggest defects in neural wiring of pattern generators which generate startle responses. Where the timing of the response is normal but the magnitude and angular velocity of the C-bend are reduced, it is likely that the defect lies with motor neurons or muscle cells. An interesting possibility arises where the initial C-bend of a movement is "long and slow," with an extended C1 duration accompanied by reduced angular velocity – this may suggest that a movement is being simultaneously initiated on both sides of the spinal cord due to a failure of reciprocal inhibition (49). Finally, quantification of the degree of tilting or rolling during startle responses can reveal whether balance is normal.

The degree of responsiveness of larvae to acoustic startle stimuli is a function of sensory acuity and sensorimotor integration. In adult zebrafish acoustic startle responses are initiated in response to activation of the saccular organ of the inner ear. The swim bladder acts as a hydrophone, with sound pressure transduced to the inner ear via the Weberian ossicles. Deflation of the swim bladder eliminates startle responses in adults (54), but not in larval zebrafish, indicating that acoustic stimuli are detected differently (55). Yet surprisingly adult and larval fish have a very similar response threshold to acoustic stimuli suggesting that Mauthner cell excitability is continuously adjusted over the course of development to maintain a set response threshold (55). Moreover, though the kinematics of the execution of SLC responses in larval zebrafish are extremely stereotyped, the degree of responsiveness, or sensitivity to the stimulus, is susceptible to modulation by several behavioral paradigms. Thus startle modulation is a fertile field in larval zebrafish for studying both ecological modulation of behavior, and the neural mechanisms that subserve these effects. Several inhibitory mechanisms serving to modulate the excitability of the Mauthner cell have been described, providing ample neural substrate for this modulation to occur (reviewed in (56)).

4.1. Habituation of the Startle Response

Behavioral responsiveness to an acoustic startle stimulus decreases when the stimulus is repeatedly presented at intervals of less than 15–20 s (36, 57). At short interstimulus intervals, the Mauthner cell responds only to the first stimulus in a train, with subsequent stimuli failing to elicit a Mauthner spike (36). In keeping with this finding, kinematic analysis shows that SLC responsiveness diminishes extremely rapidly, while LLC responses are still generated even after many trials (35). This is consistent with the notion that Mauthner-mediated startle responses are reserved for occasional situations requiring extremely fast and powerful responses.

Loss of startle responsiveness may be due to true habituation of the response, fatigue, or sensory adaptation. Certainly many neurons receiving Mauthner cell output show reduced responsiveness with repetitive stimulation of the Mauthner cell indicative of fatigue (reviewed in (58)). However, as Mauthner cell responsiveness itself declines with repeated stimulation, fatigue is unlikely to account for the whole reduction in response (36). Reduced responsiveness is not due to sensory adaptation because the VIII cranial nerve is known to follow sensory stimulation even at 1 Hz (59). In contrast, consistent with an effect of habituation, responsiveness recovers spontaneously after several minutes with no stimulus presentation, and also when a light flash stimulus is presented in the middle of the acoustic stimulus train (36, 57).

While light flash stimuli do not trigger Mauthner-mediated startle responses (32), after the first few trials of a repeated acoustic startle stimulus, very few Mauthner-mediated SLC responses are produced. Dishabituation is therefore likely to be selectively affecting LLC responses and this suggests that neural elements mediating light flash responses are shared with circuits mediating LLC responses.

Glycinergic interneurons are likely to mediate habituation of the startle response, as strychnine injections can abrogate the loss of responsiveness to repeated stimulus presentation (36). Intriguingly, serotonin can enhance glycinergic inhibition of the Mauthner cell suggesting that habituation may itself be regulated, possibly by the state of arousal of the fish (60).

4.2. Prepulse Inhibition of the Startle Response

Acoustic startle responses are modulated in larval zebrafish by a phenomenon resembling prepulse inhibition (PPI) of startle in higher vertebrates (35). In this paradigm, the larva is exposed to a sequence of brief vibrations. The first stimulus is weak and by itself elicits only rare responses. The second stimulus is intense, calibrated so that when used by itself, it triggers SLC responses in a majority of larvae. When the two stimuli are combined into a stimulus train, with the weak stimulus occurring between 10 and 1,000 ms before the intense stimulus, larvae produce significantly fewer SLC movements in response to the intense stimulus. The inhibitory effect is confined to SLC responsiveness – when SLC responses are produced, their kinematics are essentially identical to larvae responding to the intense stimulus alone. Moreover, PPI in larval zebrafish does not affect either LLC responsiveness or kinematics. The observation that LLC kinematics are not altered in PPI suggests that PPI is not due to attenuation of sensory reception: under normal conditions weaker acoustic stimuli produce LLC responses with slower kinematic performance. PPI in larval zebrafish therefore likely constitutes a true phenomenon of sensorimotor gating, where sensory signals are perceived normally, but prevented from activating normal motor responses.

In support of this, PPI has also been described in adult zebrafish where electrophysiological recordings demonstrate that Mauthner cell responsiveness is reduced due to a combination of at least three inhibitory processes (61). Interestingly, it is thought that PPI in mammals is also mediated by a combination of inhibitory processes, with overlapping temporal functions (62). Similarities between PPI in fish and mammals are extensive. Perhaps most importantly, in both groups, the interval between the weak prestimulus and the startle-inducing intense stimulus is optimal at interstimulus intervals of 50–300 ms. In addition, PPI in zebrafish and mammals is suppressed by dopamine agonists and glutamatergic antagonists. In fact, the amount of inhibition produced by the prepulse is greater when larvae are bathed in the dopamine receptor antagonist haloperidol, indicating that endogenous dopamine modulates PPI in fish (35).

This mode of startle regulation is likely common to the entire vertebrate lineage, as a process similar to PPI has also been documented in birds (63) and in tadpoles (Burgess, unpublished). Indeed, the key neuronal elements which generate PPI in mammals are thought to be confined to the brainstem (62), the most conserved part of the vertebrate brain. One would imagine that a behavior which is conserved over the entire vertebrate lineage should confer some critical behavioral advantage. It is therefore surprising that the behavioral role of PPI remains poorly understood. One hypothesis is that PPI is part of the normal mechanism for protection of sensory processing. According to this hypothesis, when a salient stimulus is recognized, subsequent incoming stimuli are temporarily gated so that a continuous flow of sensory information does not flood the brain, allowing the earlier stimulus to be fully processed. This model is intriguing because schizophrenic patients show both reduced PPI and a tendency for confused thought due to the brain being constantly flooded with sensory information (64). The link to schizophrenia goes even deeper, as a leading hypothesis concerning the etiology of schizophrenia postulates that patients suffer from overactivity in the mesolimbic dopaminergic projection (reviewed in (65)). Dopaminergic antagonists which increase PPI in zebrafish and some mammals are classic antipsychotic agents. Thus, it is hoped that the relatively simple nervous system of zebrafish may permit the identification of neuronal cell pathways mediating PPI, and provide clues to the pathogenesis of schizophrenia.

A key distinction between PPI in zebrafish and mammals is that in zebrafish, the prepulse reduces the probability of eliciting an SLC response, whereas in mammals, the prepulse reduces the magnitude of startle responses. However this anomaly is easily explained by anatomical differences. In mammals, startle responses are initiated by the thousands of giant reticulospinal neurons in the pontine nucleus caudalis (PnC). The magnitude of a startle response is determined by the number of these neurons that are activated by a stimulus (66). In mammalian PPI, the effect of a weak prepulse is to reduce both the cohort size and spike frequency of PnC neurons activated by a subsequent startle stimulus, thus reducing the magnitude of the response (67, 68). In contrast, SLC responses to acoustic stimuli are completely eliminated in larval zebrafish by ablation of the bilateral pair of Mauthner neurons. Startle responses are all-or-nothing events initiated by the firing of a single Mauthner cell. Thus by inhibiting Mauthner cell activation, a weak prepulse effectively alters the probability of generating a startle response, rather than its magnitude.

A second difference between PPI in zebrafish and mammals concerns the effect of a weak prepulse when the intense stimulus follows after an interval of greater than 1s. In zebrafish larvae, the amount of inhibition simply diminishes, and larvae return to baseline levels of startle responsiveness. In mammals, long lead intervals generate facilitation rather than inhibition of the startle response (69). As there is no hint of this process in zebrafish, presumably long lead interval facilitation of startle is a separate phenomenon which is simply not implemented in the zebrafish brain.

There are many open questions regarding PPI in zebrafish. It is not known whether startle responses elicited by touch stimuli are also susceptible to inhibition by a weak prepulse. In higher vertebrates, PPI can not only be elicited in several sensory modalities, but it also operates cross-modally – for example a weak tactile prepulse can inhibit an acoustic startle response (70, 71). It would be fascinating to uncover a similar process in zebrafish, but the experiment is technically difficult due to the problem of producing reliable tactile stimuli of defined intensity. In mammals, PPI can be elicited when the prepulse is the cessation of a background tone, however this process requires cortical processing. The anatomy and function of the homologous region, the dorsal forebrain, in fish remain largely mysterious. It would be extremely valuable to establish a robust behavioral assay dependent on this part of the brain in larval zebrafish. However to date there are no reports on whether a cessation stimulus can generate PPI in zebrafish.

5. Vestibular Stimuli

In addition to detecting acoustic/vibrational cues, the otoliths of the inner ear play a critical role in sensing the direction of gravity in order to maintain an upright posture. A third role for the otoliths is in driving the vestibular-ocular reflex, to be discussed below in relation to the related behavior, the optokinetic response. By 4 dpf, not only do larvae maintain a dorsal-up orientation when stationary, but they also show near perfect upright stability while engaged in vigorous locomotor activity (49, 72). Balance is not simply a property of the morphology of the larvae, but requires active maintenance – after anesthetic treatment with tricaine methanesulfonate, larvae float upside-down or at a vertically oblique angle.

Genetic mutations that prevent otolith formation impair both stationary and locomotor balance in zebrafish larvae (73). However only the anterior otolith is required for balance, as manipulation of the site of otolith formation in the *monolith* mutant shows that individuals lacking both anterior otoliths have defective balance, while fish lacking only the posterior otoliths have normal equilibrium (72). A collection of genetic "circler" mutants with morphologically normal inner ears, but with balance defects has been isolated (74). Most of these mutants show greatly impaired startle responses, a failure to engage hindbrain reticulospinal neurons and/or abnormal hair cell morphology, indicating a defect in acceleration detection rather than in neuronal pathways underlying balance performance (75). In contrast, the *twitch twice* mutant has a stationary and locomotor balance phenotype with close to normal responsiveness to vibrational startle stimuli suggesting that the balance defect lies within central pathways for maintaining equilibrium (49).

Body position is also determined by visual cues in most fish. Like many other aquatic animals, fish orient their dorsal side toward the brightest part of the visual field (reviewed in (76)). As many fish have dark dorsal surfaces, this "dorsal light reaction" is generally thought to aid in camouflage. Generally of course, the strongest incident light will be from above, reinforcing the effect of gravity. But by positioning a strong light source laterally, fish can be induced to tilt sideways in the direction of the light (77). While this behavior is fairly subtle in normal fish, manipulations that disrupt vestibular control of equilibrium can produce a much more overt response. Zebrafish are no exception – in mutant strains that effect balance, a strong dorsal light reaction can be induced (75). This behavior can be elicited as early as 3 dpf in larvae raised in simulated microgravity to impair development of vestibular function (78).

6. Touch Stimuli

Experimentally, touch stimuli are among the most difficult to deliver in a reproducible fashion. Zebrafish are remarkably sensitive to the approach of a probe in the water, which likely activates both lateral line and acoustic/vibrational modalities. Thus fish often escape from a probe well before physical contact is made. To overcome this difficulty, tactile stimuli are generally delivered by firing dye pellets at the fish (43). The moment of contact is easy to determine so long as a dye with sufficient contrast is used, making it easy to verify that responses are selectively activated by contact with dye pellets. A clever solution to the problem of delivering reproducible touch stimuli is to convert mechanosensory neurons into light-sensitive neurons in transgenic fish. Expression of the light-gated cation channel, channelrhodopsin-2, in Rohon Beard or trigeminal neurons makes them responsive to light, allowing intense pulses of blue light to trigger "touch-" mediated escape responses (79).

Larvae respond to tactile stimuli with highly directional escape movements (36). Touches directed to the head result in very large angle C-bends that orient larvae away from the stimulus prior to the execution of an escape swim. Tail touches elicit small initial C-bends so that the animals rapidly move away from the stimulus. Touches to the torso trigger bends away from the stimulus. At least part of the neuronal basis for the difference between escape responses elicited by head or tail touches has been elucidated. Application of water pulses to the head triggers activation of both the Mauthner cell as well as two other reticulospinal neurons, MiD2 and MiD3, located in adjacent caudal rhombomeres (80). However, such pulses may also activate the inner ear, and indeed a careful comparison of behavioral response latencies shows that selective head touch stimuli trigger MiD3 responses, while otic vesicle stimulation drives Mauthner cells responses (48). Consistent with this finding, laser ablation of the MiD3 neuron (together with the Mauthner cell and the MiD2 neuron) abolishes responses to head touch stimuli (43). Tactile stimuli directed at the tail drive Mauthner cell responses (48, 80) and after selective laser ablation of the Mauthner cell, fast responses to tail-directed touches are abolished (43). The picture that emerges is that the Mauthner cell mediates acoustic and tail-touch responses, while the MiD3 cell governs head-touch responses and possibly modulates bend amplitude during Mauthner-mediated escapes (48), although selective ablation of MiD3 will be required to confirm this model.

7. Lateral Line Stimuli

The lateral line is a specialized organ for detecting changes in water motion that has been likened to a sense of "distant touch". The mechanosensory hair cells of the lateral line are located in neuromasts, either superficially located on the surface of the skin, or subdermally in canals. Canal neuromasts of adult fish detect water acceleration indicative of potential prey objects (81); however there are several arguments against the lateral line playing such a role in zebrafish larvae. First, canals do not begin to form until several weeks of development in zebrafish (82). Second, it has been argued that the wide variability in superficial neuromast sensitivity in zebrafish larvae makes the lateral line unsuitable for localizing spatial cues (83). Finally, experiments on prey capture have failed to show significant residual prey capturing ability for larvae tested in the dark (38, 44). While little work has directly addressed the behavioral function of the lateral line in zebrafish, in adult fish of other species, the lateral line is also used for schooling (84), localization of stationary objects (85) and counter-current swimming (86). In larval zebrafish, the lateral line may be involved in predator escape, as has been shown for herring larvae (87). This is probably also true for zebrafish larvae as abrupt initiation of water movement in an impulse flow chamber has been shown to elicit escape responses, dependent on lateral line neuromasts (88). Direct synaptic connections between the posterior lateral line nerve and the Mauthner cell have been described (89). Thus it is very likely that these responses are Mauthner-mediated rapid escape movements, similar to acoustic SLC responses, but this is yet to be directly shown. Intriguingly, the lateral line receives excitatory innervation from hypothalamic neurons, leading to the suggestion that lateral line sensitivity is increased at night or when light levels are low (90).

If lateral line stimulation can truly trigger startle responses in larvae, it is obviously important to ensure that this does not occur as a result of self-generated movement. The lateral line receives inhibitory input from nuclei in the hindbrain which could well play an important role in reducing sensory sensitivity or gating sensory feedback during movement (89, 90). However, discordant results on lateral line inhibition have been obtained in different fish species (81, 91, 92), so this remains to be tested in zebrafish larvae.

8. Visual Stimuli

Zebrafish larvae have at least two light-sensitive tissues, the retina and the pineal. While photic stimulation of the pineal can drive behavior in other fish species (93), this remains to be demonstrated in larval zebrafish. On the other hand, many behavioral assays have explored retinal responses to visual stimuli. By 3 dpf, the output neurons of the retina, the ganglion cells, project to 10 identified arborization fields in the larval brain (94). It has been suggested that these 10 areas may represent functional segregation of pathways driving distinct behaviors (44). To date, only a behavioral role for the largest and most accessible area, the optic tectum, has been described. However as outlined below, the richness of the visual behavioral repertoire makes this an appealing hypothesis.

8.1. Optomotor Response

Fish species native to moving water typically orient and swim in the direction of movement of a complex visual stimulus (95, 96).

Though they are generally found in slow moving or still waters (5, 6), zebrafish show a robust optomotor response; in fact this behavior was one of the first to be quantitatively described in zebrafish (12). The optomotor response is thought to be one mechanism by which the fish minimizes the "slip" of the visual world on the retina. A second such mechanism, the optokinetic response, is discussed below. The optomotor response is generally elicited using a striped pattern of moving bars, which either drift in a straight line underneath the fish, in which case the fish will swim in the same direction as the drift, or move in a circular pattern, as on a rotating cylindrical drum, in which case larvae will swim around the perimeter of the drum. This assay lends itself to analysis of visual processing, as many parameters, including spatial frequency, contrast, and color can be easily adjusted. Indeed, the optomotor response was first used to estimate that acuity is limited to objects occupying greater than 8° of the visual field in 6 dpf larvae (12).

The optomotor response reveals that visual processing in zebrafish larvae is surprisingly sophisticated. Similar to human motion perception, the optomotor response in zebrafish larvae is predominantly driven by long and middle wavelength cone photoreceptors (97). Moreover, the optomotor response can be elicited by second-order motion, such as the drifting of a region in the visual field where objects are made to flicker (98). On the other hand, the anatomical pathways by which zebrafish larvae process motion remain poorly defined. In many species, including goldfish, the optomotor response requires the function of the optic tectum (99). Surprisingly, this is not the case in zebrafish, where optomotor swimming is not disrupted after laser ablation of the optic tectum, or in mutants with widespread disruption of retinotectal projections (100, 101). Mutants with defects in the optomotor behavior have been isolated by genetic screening, and should provide insight into the neural circuits which mediate this behavior (100, 102).

Most fish will swim against a current ("rheotaxis"), generally by orienting to face upstream and swimming forward. There has been a longstanding controversy over the nature of the stimulus that evokes rheotaxis. In a classic paper, EP Lyon argued that rheotaxis is a visually evoked response to the apparent motion of the environment (96). In this view, counter-current swimming is no more than the optomotor response, elicited when the current displaces the fish such that the visual world appears to move (103). However, as blind fish successfully oriented against the current when they touched the bottom substrate of the stream, Lyon accepted that rheotaxis may also be evoked by mechanical stimulation. More recent work in a variety of adult fish has demonstrated that at low flow rates, the superficial neuromasts of the lateral line mediate counter-current swimming (86). Zebrafish larvae as young as 4 dpf show robust counter-current swimming (12, 104). Interestingly, when young larvae are subjected to ultra-marathon-like episodes of counter-current swimming, yolk utilization is significantly increased and larvae become resistant to hypoxia (104). Very little else is known about rheotaxis in zebrafish larvae, including whether counter-current swimming is in fact an optomotor response and how motor patterns are used to swim at a suitable rate against the current.

8.2. Optokinetic Reflex

The optokinetic reflex (OKR) is triggered by visual stimuli similar to those that operate the optomotor reflex. In immobilized fish, moving stripes of light and dark cause the eyes to smoothly track in the direction of movement of the visual world, with regular rapid saccades in the reverse direction. Optokinetic movements serve to stabilize retinal "slip" when the visual field moves. The response is present in larvae as early as 3 dpf, corresponding to the period when the lens focuses light onto photoreceptor outer segments and extraocular muscles are fully differentiated (105, 106). By 6 dpf, the OKR has a high degree of contrast sensitivity and larvae are able to track slowly moving stimuli with adult-like precision (107). The robustness of the OKR has made it an attractive behavioral assay for zebrafish research and the assay of choice in several screens for genetic mutations affecting behavior (100, 102, 105, 108). In zebrafish, the OKR can be elicited by stimulating one eye only (monocular OKR) causing robust movement of the stimulated eye, and weaker, but coordinated same-direction movement of the shielded eye. In many species, the monocular OKR has a peculiar intrinsic asymmetry, with the response being relatively insensitive to motion of the visual field in a nasal to temporal (anterior to posterior) direction (reviewed in (109)). This property is recapitulated in larval zebrafish (49, 101, 110), although the effect was not found in at least one study (111). In mammals, the OKR is processed both by brainstem and cortical pathways, with cortical control predominating except in neonates (112, 113). The neuroanatomical basis of the OKR in fish remains poorly characterized, in part because many of the existing mutants with defects in the OKR show disrupted formation of the retina itself.

A related behavior, the vestibular ocular reflex (VOR) is less well characterized. In the VOR, linear acceleration or angular rotation of the head triggers eye movement to counteract visual flow during head movement. Linear acceleration of the head is detected by the otoliths not only during locomotion when there is direct displacement of the position of the head, but also when the head tilts with respect to gravity. Linear VOR has been directly demonstrated in zebrafish by confining larvae to a capillary tube and tilting the tail up and down (78). By 4 dpf, larvae respond by rotating the eyes to compensate for the tilting of the head.

Analysis of the *monolith* mutant, in which a single otolith forms in each otic vesicle, either in an anterior or posterior position, has shown that linear VOR is mediated by the anterior otolith (72). The linear VOR also drives compensatory eye movements to maintain a vertical corneal orientation when the larva tilts slightly to one side (107).

In other species, the VOR triggered by pure angular rotation requires flow through the semicircular canals. Early reports that the angular VOR could be elicited by 74 hpf in zebrafish larvae (107) were not replicated when larvae were tested in the dark, removing any possibility of an additional optokinetic stimulus (111). Indeed, when tested in the dark larvae do not show an angular VOR until 14 dpf. The absence of a VOR in younger larvae has been attributed to the narrowness of the semicircular canal lumen, which would restrict endolymph flow (111). Another possibility is that because the purpose of the angular VOR is to reduce visual flow, it is not performed under conditions of darkness. It may be relevant to note that retinal rod cells, which greatly improve visual sensitivity in dim lighting, become active at around 15 dpf in larval zebrafish, coincident with the emergence of angular VOR under dark conditions (114).

Intriguingly eye movements in larval zebrafish also occur spontaneously. Spontaneous eye movements occur every few seconds and have adult-like kinematic features as early as 4 dpf, however their function remains mysterious (107, 111). A fascinating possibility is that larvae use spontaneous eye movements to survey different regions of the visual world without initiating a body movement that might attract the attention of a predator.

8.3. Looming Escape Response

Responses to looming stimuli are common among animals including humans (115, 116). Rapidly approaching objects have a unique optical signature, that of a dark patch rapidly growing on the retina. Such "looming" stimuli alert animals to impending collisions or the approach of a predator. Of particular importance to many fish species is the threat of predation from above. This is certainly the case for zebrafish which tend to occupy relatively shallow and clear waters allowing them to be snatched by avian predators (5–7). In a pair of classic studies, Dill demonstrated that adult zebrafish flee before a mock bird suspended from a wire allowed to swoop down toward the tank (117, 118). Escape responses occurred when the rate of increase in the angle subtended on the retina exceeded a threshold, however more sophisticated analyses have shown that the response threshold is dynamically scaled so that the looming stimulus is more sensitive to objects further away (22). The visual escape response is sufficiently robust to have been used in genetic screens by the Dowling lab, where in place of a mechanical bird, a dark vertical bar on a rotating cylinder triggered escape responses in adult zebrafish as it swept toward them (119). Zebrafish larvae respond in a similar way to a dark stimulus sweeping across their field of vision (39). In this case, an approaching object is simulated by illuminating the testing arena with a projector then suddenly sweeping a dark box across the illuminated region. Larvae respond to this stimulus by first executing a turn movement that orients them in the same direction as the moving box, then deploying a series of "burst" swim maneuvers in order to rapidly move in the same direction as the shadow (Burgess, unpublished). While the looming response in adult goldfish is mediated by the Mauthner cell and has similar kinematic properties to the acoustic startle response (22), this is probably not the case for larval zebrafish, where C-bends away from the stimulus do not show the explosive angular velocity of Mauthner cell responses (Burgess, unpublished). Adult zebrafish become sensitized by repeated presentation of a simulated predator, initiating escape responses more rapidly (118). It is not known whether this is also the case for larvae or whether the psychophysical parameters required to generate responses to looming stimuli in larvae are similar to those in adults. In pigeons, looming stimuli are detected by thalamic neurons (120) raising the possibility that this assay may be a useful tool for probing the function of visual areas outside the optic tectum in zebrafish (94).

8.4. Shadow and Light Flash Responses

Sudden increments or decrements in light intensity elicit acute locomotor responses in zebrafish larvae (32). Sharp decrements in light intensity trigger dramatic responses that have been described as a "visual startle response" or "shadow evoked startle" (37, 106, 121, 122). Certainly, acute reductions in light intensity elicit a very large amplitude C-bend response that to the eye resembles a tactile or acoustic startle response. However, kinematic analysis of high-speed video recordings of these movements shows that they have distinct kinematic features (32). The initial C-bend during Mauthner-mediated tactile or acoustic startle responses has extremely high angular velocity, but is also short, being followed within 8 ms by a counterbend in the opposite direction. In contrast, larvae respond to dark flashes with slow bends of very long duration before the counterbend begins. Larvae assume an O-shape in these movements, with the head touching the tip of the tail, thus these movements are referred to as O-bends. In adult fish, visual stimuli are capable of exciting the Mauthner cells (123), although it is not clear that abrupt changes in light intensity stimuli bring the Mauthner cell to firing threshold (22). However, in larvae O-bends are executed independent of the Mauthner cell, as after ablation of both Mauthner cells, O-bend responsiveness and kinematics are unaffected (32).

Are O-bend responses to dark flash stimuli nevertheless a type of "startle" response? Dark flash responses have one curious feature that suggests that these movements are not startle responses – the initial bend in the movement tends to be directed toward the "shadow." Thus if the stimulus light is placed asymmetrically in the environment, for example, at one end of the testing arena, when it is turned off, the initial bend will be in the direction of the extinguished light (32). If O-bend responses were a way to avoid predators passing overhead, it seems unlikely that they would serve to displace the fish toward the predator. Instead, it is plausible to suggest that the O-bend response is a navigational movement, rather than a startle response, serving to orient larvae toward the most recent source of light in their environment.

Dark flash responses are mediated by the lateral eyes, and in fact full responsiveness to dark flash stimuli emerges close to the time when the lens begins to focus light onto photoreceptor outer segments (106). Photoreceptor neurons separately transmit information about increments and decrements in light intensity through ON and OFF streams in the retina and by 5 dpf, the larval retina contains dedicated OFF ganglion cells which relay light decrement events into the brain (124). Only larvae which are fully light adapted respond to dark flashes with a stereotyped O-bend response. To achieve full dark flash responsiveness requires 20 min of light exposure in larvae which have been completely dark adapted (32). This is an important point, as it means that dark flash responsiveness is not a reliable way to assess sensory sensitivity "sleeping" larvae (125). Even in larvae which are fully light adapted, after a step change to a higher light intensity, O-bend responsiveness takes around 1 min to fully recover, presumably reflecting the diverse processes involved in retinal light adaptation (32).

O-bend movements are not generic responses to abrupt changes in light intensity. Sudden increases in light intensity, which do not activate retinal OFF circuitry, do not trigger O-bends (32). Rather, increases in light intensity elicit only routine turn movements in free-swimming larvae. Very little is known about the behavioral function of turn responses to full field increases in light intensity. One possibility is that larvae are attempting to turn away from bright regions in order to prevent retinal saturation. Experiments using asymmetrically placed light sources might resolve this question, but to date no studies have been reported which employ such a test.

It is not known how abrupt the decrement in light intensity needs to be in order to trigger a stereotyped dark flash response, or how zebrafish larvae respond to gradual dimming of the visual field. In the natural environment, a gradual reduction in light intensity occurs each day at twilight, and it would not be surprising if the nervous system was hardwired with a suitable behavioral program for the approach of night. Indeed "dimming" receptors have been described in the zebrafish retina but no behavioral studies have explicitly addressed such "crepuscular" behavior (124).

8.5. Phototaxis

Many fish species show a pronounced tendency to swim toward regions of strong illumination. George Romanes, the founder of the field of comparative psychology and a close friend of Charles Darwin, attributed this response to a sense of curiosity, approvingly citing a poem by Shelley describing its fatal consequences (126):

> And the fisher, with his lamp
> And speak, about the low rocks damp
> Crept, and struck the fish which came
> To worship the delusive flame.

Other fish species, presumably less inquisitive, tend to shy away from strongly lit regions. Positive and negative phototaxis have been reported in zebrafish larvae, and both dark and light preference in adults (97, 105, 122, 127–129, 130). The conditions used in each study have been markedly different, making it difficult to assess whether zebrafish undergo a developmental switch in their affinity for illumination (as has been reported in other species (131) and suggested for zebrafish (132)). It is also possible that the differing results are due to differences in circadian state at the time of testing, or color preference. Color preference is a distinct behavior from flux preference in frogs, being mediated by different central pathways (133). Like frogs, adult zebrafish show a bias for short-wavelength stimuli (134), however it is not clear that this is truly color preference rather than flux preference. In larval zebrafish, a careful comparison of the effectiveness of light of different wavelengths has shown that blue and red cone photoreceptors are most effective at driving positive phototaxis (97). In frogs, evidence suggests that phototaxis is mediated by the thalamus rather than the optic tectum (133), thus phototaxis behavior in larval zebrafish may provide another route for exploring the function of extra-tectal visual areas.

8.6. Predation

Although larvae can be sustained by the yolk alone until around 10 dpf (135), larval survival is greatly reduced without access to food by 6 dpf. A small number of studies have looked at larval predation behavior. Paramecia are typically used as prey objects, as it appears that objects of this size most effectively elicit the predation sequence. By using the optomotor response to assess spatial acuity, it is clear that 6 dpf zebrafish larvae are easily able to resolve an object occupying 30 min of an arc, similar to the angle subtended by a paramecium at the distance at which larvae initiate the capture sequence (12). During predation, larvae orient toward the prey object using a specialized maneuver termed a J-turn. J-turns are distinct from other modes of turning, in that bending occurs near the tip of the tail, with the animal repeatedly bending its caudal tail segment to the same side in order to slowly orient toward the prey (38). This movement allows the larva to reorient, generating a minimum of turbulence that could cause the prey object to be passively displaced. Once larvae are positioned about 0.5 mm from the prey, they deploy a specialized capture swim to launch their attack strike (38). Larvae do not always have to give chase for successful feeding, using "suction-feeding" to capture paramecia that venture too close (31, 34).

Mauthner cell ablation does not disrupt predation in larvae, making it unlikely that these neurons are directly involved in prey capture (34) but the possibility remains that larvae, like adult fish, may initiate a "voluntary" Mauthner-mediated startle response after capturing prey at the surface (136). Ablation of the optic tectum almost completely disrupts predation behavior, as would be expected of a behavior requiring an orienting movement (44). Laser ablation has also identified two bilateral pairs of reticulospinal neurons downstream of the optic tectum required for predation behavior. Ablation of the MeLr and MeLc neurons produces specific defects in the ability of zebrafish larvae to orient toward prey (44). Neurons which trigger the performance of capture swims have not yet been identified.

Larvae surrounded by a large excess of food show markedly improved survival suggesting that 5–7 dpf larvae are not highly efficient hunters (14) and there is speculation that predation behavior may improve with practice (34). Several groups have demonstrated that predation behavior is dramatically diminished when fish are tested in the dark or when blind fish are used (38, 44). Thus larval zebrafish are almost entirely visual hunters, with the possibility that passive suction feeding may occasionally allow capture of close-swimming prey objects in the dark.

8.7. Visual Background Adaptation

Starting on the fourth day of development, zebrafish larvae employ a camouflage response to adjust their coloration to the environment (137). When larvae are placed on a dark background, melanin granules in melanophores disperse, making the larvae look heavily pigmented, while in the light, granules aggregate so that larvae appear pale. The response is relatively slow, requiring around 20 min for the color change to occur (138). Blind fish, unable to sense light in the environment, appear heavily pigmented, making this a useful phenotype for genetic screening aimed at isolating mutants with visual defects (100). Ethanol exposure mimics the effect of darkness in this assay, causing melanin granules to disperse (139), a phenotype sufficiently robust that genetic screening has recovered mutants with reduced sensitivity to ethanol (138). As for other fish species, visual background adaptation in zebrafish larvae does not require an intact optic tectum (101), rather information about illumination is thought to be relayed to the hypothalamus for triggering a neuroendocrine response.

8.8. Lateralized Visual Behavior

A curious feature of the larval zebrafish brain is the notable asymmetry of the habenula nucleus of the diencephalic roof (140). The left habenula is larger than the right and projects to a distinct region of the interpeduncular nucleus (141). In many vertebrates, such lateralized brain regions assume specialized functions (reviewed in (142)). The lateral position of the eyes on the head, and complete crossing of the retinotectal tract means that each side of the larval zebrafish brain receives a different view of the visual world. Thus, cerebral lateralization in zebrafish may mean that the eyes are selectively used for inspecting distinct types of visual stimuli. Consistent with this notion, adult zebrafish show a small but significant tendency to use the right eye for preparing motor responses to visual cues, but prefer to use the left eye for inspecting novel stimuli (reviewed in (132)). Unfortunately, investigations examining behavioral lateralization in zebrafish larvae yielded inconsistent results. Thus, visual startle responses have been shown to exhibit a rightward bias (121, 122) while others have failed to find any such directionality (39). There is also conflicting evidence for preferential eye use during self-inspection using a mirror (39, 121, 143, 144). Several reasons may explain the difficulty in reconciling results obtained by different groups. First, effect sizes are small and even without adjusting for multiple hypothesis testing, significance values are frequently only barely significant or "suggestive" (122, 143). Second, different strains of zebrafish show wildly discordant patterns of eye use (143, 144). Third, the mirror test is both acutely susceptible to duration of viewing effects, and not very robust. Thus it has been reported that larvae prefer to use the left eye after 5 min of self-viewing (144), shift from left to right preference after 4 min of self-viewing (121), shift from no bias to left preference after 4 min of self-viewing (143), or show no population level eye preference at any time point tested (39). Valuable tools for analysis of lateralized brain function are genetic manipulations which produce reversals in brain asymmetry and the availability of transgenic reporter lines for identifying fish with brain reversals (39, 121). Fish with reversed brain laterality show a profound and persistent reduction in swimming behavior when placed in a novel testing chamber, suggesting that brain asymmetry in larvae is related to exploratory behavior or anxiety (39). The robust effect obtained in this assay holds promise for dissecting the behavioral role of anatomically lateralized brain circuits.

9. Chemical Stimuli

Zebrafish acquire chemical information about the environment through olfaction, gustation, and through the use of specialized skin cells called "club cells." Although the developmental timing of when these systems become functional is still debated (145), morphological and behavioral evidence support early emergence of the olfactory system. The olfactory system becomes morphologically established within 3 dpf. By this time, the olfactory epithelium is developed, beating cilia facilitate water flow within the nares, and there is active neural connectivity to the olfactory bulb (145–147). Concomitantly, zebrafish exhibit various behaviors in response to chemical cues at this stage.

9.1. Locomotor Responses

Of the four classes of chemical stimuli to which fish are sensitive (amino acids, bile salts, prostoglandins and steroid hormones: (148)), studies examining chemosensory-based behavior in larval zebrafish have focused primarily on amino acids, to which both appetitive and aversive responses occur. Zebrafish produce aversive responses to certain amino acids such as L-cysteine at 3 dpf (147, 149). Aversive responses in these cases are characterized by spatial displacement of larvae away from the odor source (149). When olfactory epithelia are damaged, by cadmium exposure, aversion to L-cysteine is absent supporting that the behavior is mediated by olfaction (150). Additionally, fish with abnormal neuronal connections between the olfactory epithelia and the olfactory bulb, as found in the *laure* mutant, do not produce the aversive responses seen in wild-type fish (149). Other authors have found negligible responses to amino acids at 3 dpf, instead observing responses at 4 dpf, defined by marked increases in the occurrence and velocity of swimming (151). Noxious chemical stimuli are detected by a distinct mechanism. By 5 dpf, trigeminal neurons, Rohon Beard cells, and cranial sensory ganglia express the TRPA1 channel, responsible for detecting noxious chemicals in a variety of organisms (152). Larvae respond to such chemicals with increased locomotor activity, an effect which is lost in *TRPA1b* mutants. Thus, larvae respond to a variety of chemical and olfactory stimuli with increases in locomotor activity.

Missing from existing analyses of locomotor responses to chemical stimuli is information that would show whether larvae can use spatially located chemical cues for navigating (76). While a simple increase in locomotor activity, such as has been demonstrated in existing studies, can enable an animal to move away from an aversive stimulus, it would be interesting to know whether larvae show orienting responses that would enable chemotaxis, either toward an appetizing stimulus or away from an aversive source. Studies conducted on fish larvae in other species such as cod and herring have observed positive chemokinesis in response to food odors (153, 154). In cod larvae, swimming decreases in the presence of increasing amino acid concentrations; the decrease in activity may reflect an attempt to localize the odor source (154). Similar studies may prove fruitful in zebrafish larvae. Another approach to analyzing chemotaxis in larvae may be to make use of olfactory imprinting. Olfactory imprinting has been implicated as one mechanism by which adult fish orient back to natal home ranges (155). While olfactory imprinting is involved in establishing kin preference in zebrafish (Gerlach et al. (156), discussed below), it is not known whether it also effects place preference in zebrafish. Tantalizingly though, it has been shown that zebrafish reared in phenylethyl alcohol (PEA) exhibit a preference for the odor in adulthood, occurring as a result of changes in gene expression at 2 or 3 dpf in the olfactory epithelium upon exposure to the odor (157). The odor from decomposing fish has been proposed as another relevant olfactory cue for natal habitat, however, adult zebrafish prefer the odor regardless of whether they are exposed during rearing (147). Thus there are several potential avenues for establishing an odor preference assay in zebrafish larvae.

9.2. Alarm Responses

First identified by Karl von Frisch in 1938, alarm or "fright" response occurs when fish perceive an odor that is released from the injured skin of conspecifics. The odor may function as a cue that fish use for predator avoidance. The response is species specific and in zebrafish, it is characterized by rapid swimming upon detection of the odor, followed by aggregation when multiple individuals are present (158–160). The response has been reported in zebrafish larvae that are 2–3 weeks old, which follows the beginning of alarm odor production in the skin by club cells (147). Juveniles display ontogenetic differences in the response where younger individuals (42 dpf) are attracted to the stimulus before adopting an adult-like fright response (160). Examinations of this behavior in zebrafish larvae are limited to date but the stereotyped and context-dependent nature of this response in zebrafish makes it a promising candidate for neurogenetic investigation (158).

10. Complex Behavioral Programs

While most behavioral work in larvae has analyzed acute responses to abrupt, or continuing sensory triggers, there is also a substantial literature on complex behaviors that are generated over a longer time period. While it is relatively clear that some of these (for example, circadian sleep) are generated by internally timed programs, others (shoaling, swim bladder inflation) may ultimately be shown to be driven by a series of sensory releasing stimuli.

10.1. Swim Bladder Inflation

Among the most complex behavioral programs that larvae execute is inflation of the swim bladder. Swim bladder inflation typically occurs on 3 or 4 dpf, once the embryo has hatched from the chorion and the pharynx has opened at 74 hpf (161). Once inflated, the swim bladder is actively maintained by transfer of gas from the blood, although it appears that unlike other teleosts, zebrafish do not have a dedicated gas gland for this purpose (162). In contrast, the initial act of inflation requires that larvae reach the surface of the water, where they gulp a small bubble of air. Peristalsis then forces the air down the esophagus and through a specialized pneumatic duct into the swim bladder. Zebrafish are considered physostomous fish, in that the pneumatic duct is not lost. Nevertheless, larvae that are obstructed from reaching the surface to inflate their swim bladder by 7 dpf are no longer able to do so when released (Burgess, unpublished observations). This is likely because muscle forms around the ostium of the pneumatic duct early in development, effectively closing it off and preventing air from leaking back out of the swim bladder (163).

Failure to inflate the swim bladder impairs feeding, presumably as larva struggle to maintain buoyancy in the water column. In turn, lack of nourishment results in skeletal malformations and eventually death by starvation (164). Thus, swim bladder inflation is critical for survival into the juvenile stage. Swim bladder inflation must rely on a sophisticated behavioral program, as prior to inflation, larvae are negatively buoyant and therefore must actively ascend to the air-water interface. It has been suggested that this is accomplished simply by a combination of swimming and use of the attachment gland to gradually ascend along the sides of the tank (165), however surprisingly, this behavior has not yet been described. It is noteworthy that at least 90% of mutants recovered in large-scale genetic screens fail to inflate their swim bladder (166), underscoring the degree of coordinated use of sensory and motor functions required for this behavior.

10.2. Sleep

Locomotor activity in larval zebrafish shows circadian rhythmicity, with larvae showing more swimming activity during daytime and immobility at night. Two processes align activity levels with the diurnal cycle. First, by 4 dpf, activity is driven by an endogenous, entrainable circadian clock that even in constant dark conditions stimulates swimming during the period corresponding to daylight (167, 168). In constant darkness, the intrinsic oscillator drives motor activity with a period of around 25 h, which under normal conditions remains synchronized with subjective day through entrainment. Second, levels of activity in both zebrafish larvae and adults are acutely regulated by light itself (32, 169). This regulation is similar to "locomotor masking" in higher vertebrates, where absolute irradiance levels override the circadian clock to drive states of activity or immobility (170, 171). When larval zebrafish are subjected to repeated cycles of 1 h of light and 1 h of darkness, the periodicity of their spontaneous locomotor activity exactly matches the rhythm of the light cycle (32). In larvae which are robustly light adapted, extinguishing illumination of the testing arena triggers an extended behavioral program consisting of four steps (32). First, larvae respond to the sudden decrement in light with an O-bend response with a latency of several hundred milliseconds, as outlined above. Second, over the next 5 s, larvae execute routine turn movements at high frequency. Third, by approximately 1 min, larvae show increased movement activity. Fourth, levels of activity slowly decline over the next 20 min until larvae show a very low baseline level of activity corresponding to the rate of movement in their dark phase. Thus, even during subjective day, zebrafish larvae will show greatly reduced locomotor activity in the absence of illumination.

A significant question is whether the night phase immobility of larval zebrafish constitutes "sleep" in any meaningful way. Consensus about the definition of human sleep was reached after discovery of electrical currents generated by the brain and measured from scalp called the electroencephalogram (EEG). EEG patterns characteristic of different modes of sleep have been observed during behavioral quiescence in most mammalian species tested (172). Debatable cases remain (173, 174), but methodological issues for identification of sleep in mammals is largely settled. In contrast, since EEG is measured on the cerebral cortex of mammals, variation in brain architecture makes it difficult or impossible to use the electroencephalogram to characterize sleep in non-mammalian animals. Thus, the other approach to define and measure sleep is to use behavioral criteria. Behavioral sleep has been observed in a variety of species including worms, insects, fish, amphibians, reptiles, birds, and mammals (175). There are four key accepted behavioral criteria: behavioral quiescence, increased arousal threshold, rapid reversibility, and homeostatic regulation. In addition to these core four criteria, the presence of a characteristic posture and the existence of circadian influences are closely associated with sleep.

Applying behavioral criteria, two groups have reported a sleep-like state in larval zebrafish (125, 176). Prolonged periods of immobility are observed predominantly at night and are associated with characteristic postures in which larvae either float with the head pointing downwards or remain in a preferred position close to the bottom of the testing chamber. The arousal threshold for mechanical and visual stimuli is significantly increased during subjective night compared to subjective daytime. Rapid reversibility from behavioral quiescence has not been directly reported in larval zebrafish, however locomotor activity returns to baseline levels after several minutes during the transition from the quiescent state to the active state. Decreased locomotor activity suggestive of rebound sleep is observed after 6 h of rest deprivation at the end of the night (176) and in fact similar conditions cause sleep deprivation in adult zebrafish (169). Taken together, this evidence provides strong support for the existence of behavioral sleep in zebrafish larvae.

Neurotransmitter systems which regulate sleep in mammals are also present in zebrafish. These include dopaminergic, histaminergic, noradrenergic, and serotonergic systems. The distribution of enzymes involved in neurotransmitter synthesis and neurotransmitter receptors has homology to mammals (177, 178) and in fact hypnotic drugs for mammals also affect larval zebrafish, indicating that there is a high degree of conservation in GABAergic and histaminergic systems which modulate sleep in larval zebrafish (179, 180). There may also be unique features of sleep regulation in zebrafish. The hypocretin/orexin system plays a key role in wake maintenance in mammals (reviewed in (181)). However, there are conflicting reports as to whether hypocretin receptor neurons in larval zebrafish coexpress monoaminergic markers as in mammals (125, 169, 178). Moreover, manipulation of the hypocretin/orexin system has yielded inconsistent results. A hypocretin receptor deletion mutant failed to exhibit a phenotype during wake periods, but showed severe fragmentation of nighttime sleep (169). Sleep fragmentation has also been reported after disruption of the hypocretin system in mammals (182, 183). In contrast, overexpression of hypocretin ligands in transgenic fish promoted locomotor activity (125). Although further work will be required to reconcile these results, it is clear that the zebrafish model offers significant opportunities to contribute to the understanding of mechanisms of sleep regulation. Considerable variation in sleep architecture exists even in mammalian species (184), but it seems likely that sleep has a fundamental function which has been maintained over the course of evolution. The strength of genetic and neuroanatomical approaches in zebrafish may ultimately provide crucial clues to answer the longstanding question as to why humans sleep.

10.3. Exploratory Behavior

A few studies have examined the pattern of activity of larvae placed in a novel testing environment. Upon being placed in an unfamiliar testing environment, larvae show a drop in locomotor activity which normalizes after around 3 min (32), however it is not clear whether this is due to the handling or the features of the environment. The familiarity of the testing chamber may influence behavior – when raised in tanks with conspicuous vertical stripes, larvae show a reduced tendency to steer away from similar features in a swimway (122). This also suggests that larvae retain at least a short-term memory of their previous environment. Intriguingly, as noted above, larvae with a reversal of habenula asymmetry show greatly reduced swimming in a novel testing chamber, although in most respects such larvae show normal locomotor activity (39). This suggests that the habenula may be involved in motivational aspects of behavior in zebrafish larvae, although it is not clear why brain reversal would generate such a robust phenotype. Many manipulations, in particular pharmacological treatments, alter the proportion of zebrafish larvae found in the center and edges of the testing arena (139; Burgess, unpublished). This has been interpreted as thigmotaxis, or wall-seeking behavior. It must be kept in mind that a reduced concentration of fish in the center of the testing chamber could simply result from a change in the pattern of spontaneous movement. Larvae normally deploy both routine turns and scoot maneuvers during unstimulated locomotion. Any manipulation which reduces the ratio of turn movements to scoots will tend to cluster larvae at edges of the testing arena. In *C. elegans*, a similar change in the ratio of turns to forward movements arises under conditions of food deprivation. Animals enter a dispersal state, moving away from their location in search of new food (26). In a confined space, this would lead to animals aggregating at the sides of the testing chamber, but clearly not represent thigmotaxis. To date, a clear demonstration of active wall-seeking behavior has yet to be presented in zebrafish larvae.

10.4. Social Behavior

Nascent signs of social behavior emerge in larval zebrafish at the early flexion stage (15, 128), corresponding to 9–10 dpf in larvae raised at 28°C (185). It is not yet known what sensory cues trigger aggregation in larvae. While adult zebrafish prefer to shoal with similarly pigmented conspecifics (186, 187), larval zebrafish do not show such a preference (15). Aggregation preference is acquired by learning the pigmentation pattern of conspecifics during rearing, and is not readily reprogrammed by subsequent exposure to other fish (188–190). Preference is imprinted by the juvenile stage, however the boundaries of the critical period for imprinting are yet to be defined (189). Though pigmentation does not influence choice of shoaling partners in larvae, it is possible that other visual cues, including motion and form, play a significant role in triggering aggregation.

Olfactory cues also influence choice of shoaling partners in juvenile zebrafish (191). Remarkably, 3–4-week-old fish can distinguish between water conditioned by familiar siblings, and water conditioned by unfamiliar siblings (192). Odor preferences in juvenile zebrafish are imprinted by exposure to an odor during larval stages. A critical period has been identified at 6 dpf where exposure to siblings predicts preference for sibling odor later in life (156). The imprinting window is remarkably narrow, restricted to a single day during larval development. Moreover, larvae only learn a preference for the odor of related individuals and do not imprint a preference for the odor of larvae derived from distinct parental crosses. The exquisite selectivity of this effect suggests the involvement of a genetic predisposition which becomes activated through experience.

A disadvantage to neurobiological analysis of shoaling behavior in zebrafish is that the shoaling phenotype is relatively weak compared to wild strains, almost certainly because captive zebrafish do not face the threat of predation (3, 4). Studies analyzing neural pathways involved in shoaling may benefit from the availability of strains which have been maintained in captivity for only a small number of generations.

11. Concluding Remarks

In this chapter we have sought to outline the basic behavioral repertoire of larval zebrafish, drawing attention to the many outstanding questions yet to be addressed. However, if one assumes that "what we know we don't know" is only a small subset of "what we don't know we don't know" then it is clear that our knowledge of the behavioral repertoire remains far from complete. For instance, as poikilotherms, body temperature and metabolism in fish critically depend on the temperature of the surrounding water. It is surprising that so little is known about the ability of larval zebrafish to detect and respond to changes in temperature. A single study has shown that larvae respond with locomotor activity when exposed to bath water less than 16°C, or greater than 37°C (152). These values broadly correspond to data gathered from ecological studies, which have found active and healthy zebrafish inhabiting water ranging from 16.5 to 38.6°C (5, 7). However it would be of great interest to know whether larvae have a preferred temperature, and whether they will actively navigate through a temperature gradient to reach it. Another poorly explored area is the role of associative memory in zebrafish larvae. Conditioned place preference tasks are routinely used in adult zebrafish and are sufficiently robust to allow screening for genetic mutants (193), however it is not known whether such a task can be successfully modified for use with larvae. Young juvenile larvae, 3–4 weeks of age, failed to acquire a spatial alternation task requiring associative learning (194), but it remains possible that the locomotor demands in this task were excessive. Finally, in zebrafish larvae, research has mainly focused on locomotor behaviors, eye movements, and neuroendocrine responses. Other types of motor response including fin and jaw movements are less well documented and remain a very fertile ground for future studies (34, 195). A wealth of information exists describing behavior in the larvae of other fish species and this should greatly assist in achieving a comprehensive account of the abilities of larval zebrafish.

In many ways, analysis of the neurobiological underpinnings of behavior in larvae is still in its infancy. The difficulty in applying the traditional toolbox of neuroscience to larvae, such as electrophysiological recordings and lesion analysis, has impeded progress. Recent breakthroughs in applying neurogenetic techniques to zebrafish will greatly accelerate the rate of discovery. Key advances include the availability of genetically encoded calcium indicators to monitor neural activity (196, 197), light-activated cation channels to drive activity in vivo (79), and a burgeoning collection of transgenic fish to facilitate visualization and manipulation of the larval brain (198–200). These techniques draw upon the unique advantages of the larval zebrafish model, its optical accessibility and relatively limited neuronal complexity, and vastly expand the opportunity to link the function of identified neurons to behavior.

Niko Tinbergen pointed out that "… one expects to find an innate base beneath the plastic behavior of mammals" (201). In higher vertebrates, cortical control of behavior dominates the function of brainstem circuits. Phylogenetically ancient but intact subcortical circuits nevertheless assume operational control of behavior under certain conditions in humans: during infancy (30), when attention is distracted (116), after cortical damage (202), or when stimuli are carefully designed to access only subcortical pathways (203). In contrast, teleost fish lack direct corticospinal connections and indeed the brainstem is held to be chiefly responsible for the organization of behavior (reviewed in (204)). Larval zebrafish share the basic architectural features common to all vertebrate brains and allow the neuronal basis of behavior to be decoded at a stage when genetically encoded neural circuits predominate. It is thus not unrealistic to hope that the neurobiological analysis of behavior in zebrafish larvae will provide profound insights into the operation of the human brain.

Acknowledgments

This work was supported by the Intramural Research Program of the Eunice Kennedy Shriver National Institute of Child Health and Human Development, National Institutes of Health.

References

1. Grunwald, D. J. & Eisen, J. S. (2002) Headwaters of the zebrafish – emergence of a new model vertebrate. *Nat Rev Genet* 3, 717–724.
2. Kimmel, C. B. (1982) Reticulospinal and vestibulospinal neurons in the young larva of a teleost fish, *Brachydanio rerio. Prog Brain Res* 57, 1–23.
3. Robison, B. D. & Rowland, W. (2005) A potential model system for studying the genetics of domestication: behavioral variation among wild and domesticated strains of zebra danio (*Danio rerio*). *Can J Fish Aquat Sci* 62, 2046–2054.
4. Wright, D., Nakamichi, R., Krause, J. & Butlin, R. K. (2006) QTL analysis of behavioral and morphological differentiation between wild and laboratory zebrafish (*Danio rerio*). *Behav Genet* 36, 271–284.
5. Engeszer, R. E., Patterson, L. B., Rao, A. A. & Parichy, D. M. (2007b) Zebrafish in the wild: a review of natural history and new notes from the field. *Zebrafish* 4, 21–40.
6. McClure, M. M., McIntyre, P. B. & McCune, A. R. (2006) Notes on the natural diet and habitat of eight danionin fishes, including the zebrafish *Danio rerio. J Fish Biol* 69, 553–570.
7. Spence, R., Fatema, M. K., Reichard, M., Huq, K. A., Wahab, M. A., Ahmed, Z. F. & Smith, C. (2006) The distribution and habitat preferences of the zebrafish in Bangladesh. *J Fish Biol* 69, 1435–1448.
8. Uchida, D., Yamashita, M., Kitano, T. & Iguchi, T. (2002) Oocyte apoptosis during the transition from ovary-like tissue to testes during sex differentiation of juvenile zebrafish. *J Exp Biol* 205, 711–718.
9. Saint-Amant, L. & Drapeau, P. (1998) Time course of the development of motor behaviors in the zebrafish embryo. *J Neurobiol* 37, 622–632.
10. Downes, G. & Granato, M. (2006) Supraspinal input is dispensable to generate glycine-mediated locomotive behaviors in the zebrafish embryo. *J Neurobiol* 66, 437–451.
11. Muller, U. & van Leeuwen, J. (2004) Swimming of larval zebrafish: ontogeny of body waves and implications for locomotory development. *J Exp Biol* 207, 853–868.
12. Clark, D. T. (1981) Visual Responses in Developing Zebrafish (*Brachydanio rerio*). Oregon, Eugene.
13. Mueller, T. & Wullimann, M. F. (2005) *Atlas of Early Zebrafish Brain Development. A Tool for Molecular Neurogenetics.* Amsterdam, Elsevier B.V.
14. Eaton, R. C. & Farley, R. D. (1974) Growth and the reduction of depensation of zebrafish, *Brachydanio rerio*, reared in the laboratory. *Copeia* 1, 204–209.
15. Engeszer, R. E., da Barbiano, L. A., Ryan, M. J. & Parichy, D. M. (2007a) Timing and plasticity of shoaling behaviour in the zebrafish, *Danio rerio. Anim Behav* 74, 1269–1275.
16. Kimmel, C. B., Ballard, W. W., Kimmel, S. R., Ullmann, B. & Schilling, T. F. (1995) Stages of embryonic development of the zebrafish. *Am J Anat* 203, 253–310.
17. Küster, E. & Altenburger, R. (2008) Oxygen decline in biotesting of environmental samples-Is there a need for consideration in the acute zebrafish embryo assay? *Environ Toxicol* 23, 745–750.
18. Marks, C., West, T. N., Bagatto, B., Moore, F. B. G. & Taylor, C. M. (2005) Developmental environment alters conditional aggression in zebrafish. *Copeia* 2005, 901–908.
19. Burgess, H. A. & Granato, M. (2008) The neurogenetic frontier–lessons from misbehaving zebrafish. *Brief Funct Genomic Proteomic* 7, 474–482.
20. Liu, N. A., Liu, Q., Wawrowsky, K., Yang, Z., Lin, S. & Melmed, S. (2006) Prolactin receptor signaling mediates the osmotic response of embryonic zebrafish lactotrophs. *Mol Endocrinol* 20, 871–880.
21. Preuss, T. & Faber, D. S. (2003) Central cellular mechanisms underlying temperature-dependent changes in the goldfish startle-escape behavior. *J Neurosci* 23, 5617–5626.
22. Preuss, T., Osei-Bonsu, P. E., Weiss, S. A., Wang, C. & Faber, D. S. (2006) Neural representation of object approach in a decision-making motor circuit. *J Neurosci* 26, 3454–3464.
23. Tinbergen, N. (1972) On the orientation of the Digger Wasp *Philanthus triangulum* Fabr: II. The hunting behavior (1935). *The Animal in Its World.* Cambridge, Harvard University Press.
24. Bastock, M. (1956) A gene mutation which alters a behavior pattern. *Evolution* 10, 421–439.
25. Villella, A., Hall, J. C. & Jeffrey, C. H. (2008) Chapter 3 Neurogenetics of courtship and mating in Drosophila. *Advances in Genetics.* San Diego, CA, Academic Press.
26. Gray, J. M., Hill, J. J. & Bargmann, C. I. (2005) A circuit for navigation in *Caenorhabditis elegans. Proc Natl Acad Sci USA* 102, 3184–3191.

27. Mori, I. & Ohshima, Y. (1995) Neural regulation of thermotaxis in *Caenorhabditis elegans. Nature* 376, 344–348.
28. Briggman, K., Abarbanel, H. & Kristan, W., Jr. (2005) Optical imaging of neuronal populations during decision-making. *Science* 307, 896–901.
29. Tinbergen, N. (1952b) The curious behavior of the stickleback. *Sci Am* 187, 22–26.
30. Sarnat, H. B. (2003) Functions of the corticospinal and corticobulbar tracts in the human newborn. Amsterdam, IOS Press.
31. Budick, S. & O'Malley, D. (2000) The behavioral repertoire of larval zebrafish: swimming, escaping and prey capture. *J Exp Biol* 203, 2565–2579.
32. Burgess, H. & Granato, M. (2007a) Modulation of locomotor activity in larval zebrafish during light adaptation. *J Exp Biol* 210, 2526–2539.
33. McLean, D. L., Fan, J., Higashijima, S., Hale, M. E. & Fetcho, J. R. (2007) A topographic map of recruitment in spinal cord. *Nature* 446, 71–75.
34. Borla, M. A., Palecek, B., Budick, S. & O'Malley, D. M. (2002) Prey capture by larval zebrafish: evidence for fine axial motor control. *Brain Behav Evol* 60, 207–229.
35. Burgess, H. & Granato, M. (2007b) Sensorimotor gating in larval zebrafish. *J Neurosci* 27, 4984.
36. Eaton, R., Farley, R., Kimmel, C. & Schabtach, E. (1977) Functional development in the Mauthner cell system of embryos and larvae of the zebra fish. *J Neurobiol* 8, 151–172.
37. Kimmel, C. B., Patterson, J. & Kimmel, R. O. (1974) The development and behavioral characteristics of the startle response in the zebra fish. *Dev Psychobiol* 7, 47–60.
38. McElligott, M. & O'Malley, D. (2005) Prey tracking by larval zebrafish: axial kinematics and visual control. *Brain Behav Evol* 66, 177–196.
39. Facchin, L., Burgess, H. A., Siddiqi, M., Granato, M. & Halpern, M. E. (2009) Determining the function of zebrafish epithalamic asymmetry. *Philos Trans R Soc Lond B Biol Sci* 364, 1021–1032.
40. Orger, M., Kampff, A., Severi, K., Bollmann, J. & Engert, F. (2008) Control of visually guided behavior by distinct populations of spinal projection neurons. *Nat Neurosci* 11, 327.
41. Liao, J. C. & Fetcho, J. R. (2008) Shared versus specialized glycinergic spinal interneurons in axial motor circuits of larval zebrafish. *J Neurosci* 28, 12982–12992.
42. McLean, D. L. & Fetcho, J. R. (2008) Using imaging and genetics in zebrafish to study developing spinal circuits in vivo. *Dev Neurobiol* 68, 817–834.
43. Liu, K. S. & Fetcho, J. R. (1999) Laser ablations reveal functional relationships of segmental hindbrain neurons in zebrafish. *Neuron* 23, 325–335.
44. Gahtan, E., Tanger, P. & Baier, H. (2005) Visual prey capture in larval zebrafish is controlled by identified reticulospinal neurons downstream of the tectum. *J Neurosci* 25, 9294–9303.
45. Blaser, R. & Gerlai, R. (2006) Behavioral phenotyping in zebrafish: comparison of three behavioral quantification methods. *Behav Res Methods* 38, 456–469.
46. Liu, D. W. & Westerfield, M. (1988) Function of identified motoneurones and co-ordination of primary and secondary motor systems during zebra fish swimming. *J Physiol* 403, 73–89.
47. Loeb, J. (1905) *Studies in General Physiology.* Chicago, University of Chicago Press.
48. Kohashi, T. & Oda, Y. (2008) Initiation of Mauthner- or non-Mauthner-mediated fast escape evoked by different modes of sensory input. *J Neurosci* 28, 10641–10653.
49. Burgess, H. A., Johnson, S. L. & Granato, M. (2009) Unidirectional startle responses and disrupted left-right co-ordination of motor behaviors in robo3 mutant zebrafish. *Genes Brain Behav* 8, 500–511.
50. Kimmel, C., Eaton, R. & Powell, S. (1980) Decreased fast-start performance of zebrafish larvae lacking Mauthner neurons. *J Comp Physiol A* 140, 343–350.
51. Eaton, R. C., Nissanov, J. & Wieland, C. M. (1984) Differential activation of Mauthner and non-Mauthner startle circuits in the zebrafish: implications for functional substitution. *J Comp Physiol A Sens Neural Behav Physiol* 155, 813–820.
52. Eaton, R. C. & Emberley, D. S. (1991) How stimulus direction determines the trajectory of the Mauthner-initiated escape response in a teleost fish. *J Exp Biol* 161, 469–487.
53. Glowa, J. R. & Hansen, C. T. (1994) Differences in response to an acoustic startle stimulus among forty-six rat strains. New York, Springer-Verlag.
54. Canfield, J. G. & Eaton, R. C. (1990) Swimbladder acoustic pressure transduction initiates Mauthner-mediated escape. *Nature* 347, 760–762.
55. Zeddies, D. G. & Fay, R. R. (2005) Development of the acoustically evoked behavioral response in zebrafish to pure tones. *J Exp Biol* 208, 1363–1372.

56. Faber, D. S. & Korn, H. (1978) Electrophysiology of the Mauthner cell: basic properties, synaptic mechanisms and associated networks. In Faber, D. S. & Korn, H. (Eds.) *Neurobiology of the Mauthner Cell*. New York, NY, Raven Press.
57. Best, J. D., Berghmans, S., Hunt, J. J., Clarke, S. C., Fleming, A., Goldsmith, P. & Roach, A. G. (2008) Non-associative learning in larval zebrafish. *Neuropsychopharmacology* 33, 1206–1215.
58. Diamond, J. (1971) The Mauthner cell. In Hoar, W. S. & Randall, D. J. (Eds.) *Fish Physiology*. New York, NY, Academic Press.
59. Eaton, R. & Farley, R. (1975) Mauthner neuron field potential in newly hatched larvae of the zebra fish. *J Neurophysiol* 38, 502–512.
60. Mintz, I., Gotow, T., Triller, A. & Korn, H. (1989) Effect of serotonergic afferents on quantal release at central inhibitory synapses. *Science* 245, 190–192.
61. Neumeister, H., Szabo, T. M. & Preuss, T. (2008) Behavioral and physiological characterization of sensorimotor gating in the goldfish startle response. *J Neurophysiol* 99, 1493–1502.
62. Bosch, D. & Schmid, S. (2008) Cholinergic mechanism underlying prepulse inhibition of the startle response in rats. *Neuroscience* 155, 326–335.
63. Schall, U., Keysers, C. & Kast, B. (1999) Pharmacology of sensory gating in the ascending auditory system of the pigeon (*Columba livia*). *Psychopharmacology (Berl)* 145, 273–282.
64. Braff, D., Stone, C., Callaway, E., Geyer, M., Glick, I. & Bali, L. (1978) Prestimulus effects on human startle reflex in normals and schizophrenics. *Psychophysiology* 15, 339–343.
65. Laruelle, M., Kegeles, L. S. & Abi-Dargham, A. (2003) Glutamate, dopamine, and schizophrenia: from pathophysiology to treatment. *Ann N Y Acad Sci* 1003, 138–158.
66. Koch, M., Lingenhöhl, K. & Pilz, P. K. D. (1992) Loss of the acoustic startle response following neurotoxic lesions of the caudal pontine reticular formation: possible role of giant neurons. *Neuroscience* 49, 617–625.
67. Carlson, S. & Willott, J. F. (1998) Caudal pontine reticular formation of C57BL/6 J mice: responses to startle stimuli, inhibition by tones, and plasticity. *J Neurophysiol* 79, 2603–2614.
68. Lingenhohl, K. & Friauf, E. (1994) Giant neurons in the rat reticular formation: a sensorimotor interface in the elementary acoustic startle circuit? *J Neurosci* 14, 1176–1194.
69. Putnam, L. E. & Vanman, E. J. (1999) Long lead interval startle modification. In Dawson, M. E., Schell, A. M. & Bohmelt, A. H. (Eds.) *Startle Modification, Implications for Neuroscience, Cognitive Science and Clinical Science*. Cambridge, Cambridge University Press.
70. Aubert, L., Reiss, D. & Ouagazzal, A. M. (2006) Auditory and visual prepulse inhibition in mice: parametric analysis and strain comparisons. *Genes Brain Behav* 5, 423–431.
71. Hoffman, H. S. & Ison, J. R. (1980) Reflex modification in the domain of startle: I. Some empirical findings and their implications for how the nervous system processes sensory input. *Psychol Rev* 87, 175–189.
72. Riley, B. B. & Moorman, S. J. (2000) Development of utricular otoliths, but not saccular otoliths, is necessary for vestibular function and survival in zebrafish. *J Neurobiol* 43, 329–337.
73. Whitfield, T. T., Granato, M., van Eeden, F. J., Schach, U., Brand, M., Furutani-Seiki, M., Haffter, P., Hammerschmidt, M., Heisenberg, C. P., Jiang, Y. J., Kane, D. A., Kelsh, R. N., Mullins, M. C., Odenthal, J. & Nusslein-Volhard, C. (1996) Mutations affecting development of the zebrafish inner ear and lateral line. *Development* 123, 241–254.
74. Granato, M., van Eeden, F. J., Schach, U., Trowe, T., Brand, M., Furutani-Seiki, M., Haffter, P., Hammerschmidt, M., Heisenberg, C. P., Jiang, Y. J., Kane, D. A., Kelsh, R. N., Mullins, M. C., Odenthal, J. & Nusslein-Volhard, C. (1996) Genes controlling and mediating locomotion behavior of the zebrafish embryo and larva. *Development* 123, 399–413.
75. Nicolson, T., Rusch, A., Friedrich, R. W., Granato, M., Ruppersberg, J. P. & Nusslein-Volhard, C. (1998) Genetic analysis of vertebrate sensory hair cell mechanosensation: the zebrafish circler mutants. *Neuron* 20, 271–283.
76. Fraenkel, G. & Gunn, D. (1961) *The Orientation of Animals Kineses, Taxes and Compass Reactions*. New York, Dover Publications.
77. Gibbs, M. A. & Northmore, D. P. (1996) The role of torus longitudinalis in equilibrium orientation measured with the dorsal light reflex. *Brain Behav Evol* 48, 115–120.
78. Moorman, S. J., Burress, C., Cordova, R. & Slater, J. (1999) Stimulus dependence of the development of the zebrafish (*Danio rerio*) vestibular system. *J Neurobiol* 38, 247–258.
79. Douglass, A. D., Kraves, S., Deisseroth, K., Schier, A. F. & Engert, F. (2008) Escape behavior elicited by single,

channelrhodopsin-2-evoked spikes in zebrafish somatosensory neurons. *Curr Biol* 18, 1133–1137.
80. O'Malley, D. M., Kao, Y. H. & Fetcho, J. R. (1996) Imaging the functional organization of zebrafish hindbrain segments during escape behaviors. *Neuron* 17, 1145–1155.
81. Palmer, L. M., Deffenbaugh, M. & Mensinger, A. F. (2005) Sensitivity of the anterior lateral line to natural stimuli in the oyster toadfish, Opsanus tau (Linnaeus). *J Exp Biol* 208, 3441–3450.
82. Webb, J. F. & Shirey, J. E. (2003) Postembryonic development of the cranial lateral line canals and neuromasts in zebrafish. *Dev Dyn* 228, 370–385.
83. van Trump, W. J. & McHenry, M. J. (2008) The morphology and mechanical sensitivity of lateral line receptors in zebrafish larvae (*Danio rerio*). *J Exp Biol* 211, 2105–2115.
84. Partridge, B. L. & Pitcher, T. J. (1980) The sensory basis of fish schools: relative roles of lateral line and vision. *J Comp Physiol A Sens Neural Behav Physiol* 135, 315–325.
85. Weissert, R. & Campenhausen, C. (1981) Discrimination between stationary objects by the blind cave fish Anoptichthys jordani (Characidae). *J Comp Physiol A Sens Neural Behav Physiol* 143, 375–381.
86. Montgomery, J. C., Baker, C. F. & Carton, A. G. (1997) The lateral line can mediate rheotaxis in fish. *Nature* 389, 960–963.
87. Blaxter, J. H. S. & Fuiman, L. A. (1990) The role of the sensory systems of herring larvae in evading predatory fishes. *J Mar Biol Assoc UK* 70, 413–427.
88. McHenry, M. J., Feitl, K. E., Strother, J. A. & van Trump, W. J. (2009) Larval zebrafish rapidly sense the water flow of a predator's strike. *Biol Lett* 5, 477–479.
89. Metcalfe, W. K., Kimmel, C. B. & Schabtach, E. (1985) Anatomy of the posterior lateral line system in young larvae of the zebrafish. *J Comp Neurol* 233, 377–389.
90. Bricaud, O., Chaar, V., Dambly-Chaudiere, C. & Ghysen, A. (2001) Early efferent innervation of the zebrafish lateral line. *J Comp Neurol* 434, 253–261.
91. Palmer, L. M., Giuffrida, B. A. & Mensinger, A. F. (2003) Neural recordings from the lateral line in free-swimming toadfish, Opsanus tau. *Biol Bull* 205, 216–218.
92. Russell, I. J. & Roberts, B. L. (1974) Active reduction of lateral-line sensitivity in swimming dogfish. *J Comp Physiol A Neuroethol Sens Neural Behav Physiol* 94, 7–15.
93. Yoshizawa, M. & Jeffery, W. R. (2008) Shadow response in the blind cavefish Astyanax reveals conservation of a functional pineal eye. *J Exp Biol* 211, 292–299.
94. Burrill, J. D. & Easter, S. S., Jr. (1994) Development of the retinofugal projections in the embryonic and larval zebrafish (*Brachydanio rerio*). *J Comp Neurol* 346, 583–600.
95. Clausen, R. G. (1931) Orientation in fresh water fishes. *Ecology* 3, 541–546.
96. Lyon, E. P. (1904) On rheotropism. I. Rheotropism in fishes. *Am J Physiol* 12, 149–161.
97. Orger, M. & Baier, H. (2005) Channeling of red and green cone inputs to the zebrafish optomotor response. *Vis Neurosci* 22, 275–281.
98. Orger, M. B., Smear, M. C., Anstis, S. M. & Baier, H. (2000) Perception of Fourier and non-Fourier motion by larval zebrafish. *Nat Neurosci* 3, 1128–1133.
99. Springer, A. D., Easter, S. S., Jr. & Agranoff, B. W. (1977) The role of the optic tectum in various visually mediated behaviors of goldfish. *Brain Res* 128, 393–404.
100. Neuhauss, S., Biehlmaier, O., Seeliger, M., Das, T., Kohler, K., Harris, W. & Baier, H. (1999) Genetic disorders of vision revealed by a behavioral screen of 400 essential loci in zebrafish. *J Neurosci* 19, 8603.
101. Roeser, T. & Baier, H. (2003) Visuomotor behaviors in larval zebrafish after GFP-guided laser ablation of the optic tectum. *J Neurosci* 23, 3726–3734.
102. Muto, A., Orger, M. B., Wehman, A. M., Smear, M. C., Kay, J. N., Page-McCaw, P. S., Gahtan, E., Xiao, T., Nevin, L. M., Gosse, N. J., Staub, W., Finger-Baier, K. & Baier, H. (2005) Forward genetic analysis of visual behavior in zebrafish. *PLoS Genet* 1, e66.
103. Easter, S. S., Jr. (1972) Pursuit eye movements in goldfish (*Carassius auratus*). *Vision Res* 12, 673–688.
104. Bagatto, B., Pelster, B. & Burggren, W. W. (2001) Growth and metabolism of larval zebrafish: effects of swim training. *J Exp Biol* 204, 4335–4343.
105. Brockerhoff, S., Hurley, J., Janssen-Bienhold, U., Neuhauss, S., Driever, W. & Dowling, J. (1995) A behavioral screen for isolating zebrafish mutants with visual system defects. *Proc Natl Acad Sci* 92, 10545–10549.
106. Easter, S. & Nicola, G. (1997a) The development of vision in the zebrafish (*Danio rerio*). *Dev Biol* 180, 646–663.
107. Easter, S. S., Jr. & Nicola, G. N. (1997b) The development of eye movements in the zebrafish (*Danio rerio*). *Dev Psychobiol* 31, 267–276.

108. Gross, J. M., Perkins, B. D., Amsterdam, A., Egana, A., Darland, T., Matsui, J. I., Sciascia, S., Hopkins, N. & Dowling, J. E. (2005) Identification of zebrafish insertional mutants with defects in visual system development and function. *Genetics* 170, 245–261.
109. Huang, Y. Y. & Neuhauss, S. C. F. (2008) The optokinetic response in zebrafish and its applications. *Front Biosci* 13, 1899–1916.
110. Qian, H., Zhu, Y., Ramsey, D. J., Chappell, R. L., Dowling, J. E. & Ripps, H. (2005) Directional asymmetries in the optokinetic response of larval zebrafish (*Danio rerio*). *Zebrafish* 2, 189–196.
111. Beck, J. C., Gilland, E., Tank, D. W. & Baker, R. (2004) Quantifying the ontogeny of optokinetic and vestibuloocular behaviors in zebrafish, medaka, and goldfish. *J Neurophysiol* 92, 3546–3561.
112. Harris, L. R., Lepore, F. & Guillemot, J. P. (1980) Abolition of optokinetic nystagmus in the cat. *Science* 210, 91–92.
113. Naegele, J. R. & Held, R. (1982) The postnatal development of monocular optokinetic nystagmus in infants. *Vision Res* 22, 341.
114. Bilotta, J., Saszik, S. & Sutherland, S. (2001) Rod contributions to the electroretinogram of the dark-adapted developing zebrafish. *Dev Dyn* 222, 564–570.
115. Ball, W. & Tronick, E. (1971) Infant responses to impending collision: optical and real. *Science* 171, 818–820.
116. King, S. M., Dykeman, C., Redgrave, P. & Dean, P. (1992) Use of a distracting task to obtain defensive head movements to looming visual stimuli by human adults in a laboratory setting. *Perception* 21, 245–259.
117. Dill, L. (1974a) The escape response of the zebra danio (*Brachydanio rerio*). I. The stimulus for escape. *Anim Behav* 22, 710–721.
118. Dill, L. M. (1974b) The escape response of the zebra danio (*Brachydanio rerio*). II. The effect of experience. *Anim Behav* 22, 723–730.
119. Li, L. & Dowling, J. E. (1997) A dominant form of inherited retinal degeneration caused by a non-photoreceptor cell-specific mutation. *Proc Natl Acad Sci USA* 94, 11645–11650.
120. Sun, H. & Frost, B. J. (1998) Computation of different optical variables of looming objects in pigeon nucleus rotundus neurons. *Nat Neurosci* 1, 296–303.
121. Barth, K. A., Miklosi, A., Watkins, J., Bianco, I. H., Wilson, S. W. & Andrew, R. J. (2005) fsi Zebrafish show concordant reversal of laterality of viscera, neuroanatomy, and a subset of behavioral responses. *Curr Biol* 15, 844–850.
122. Watkins, J., Miklosi, A. & Andrew, R. J. (2004) Early asymmetries in the behaviour of zebrafish larvae. *Behav Brain Res* 151, 177–183.
123. Zottoli, S., Hordes, A. & Faber, D. (1987) Localization of optic tectal input to the ventral dendrite of the goldfish Mauthner cell. *Brain Res* 401, 113–121.
124. Emran, F., Rihel, J., Adolph, A. R., Wong, K. Y., Kraves, S. & Dowling, J. E. (2007) OFF ganglion cells cannot drive the optokinetic reflex in zebrafish. *Proc Natl Acad Sci USA* 104, 19126–19131.
125. Prober, D. A., Rihel, J., Onah, A. A., Sung, R. -J. & Schier, A. F. (2006) Hypocretin/orexin overexpression induces an insomnia-like phenotype in zebrafish. *J Neurosci* 26, 13400–13410.
126. Romanes, G. (1883) *Animal Intelligence.* New York, NY, D. Appleton and Co.
127. Gerlai, R., Lahav, M., Guo, S. & Rosenthal, A. (2000) Drinks like a fish: zebra fish (*Danio rerio*) as a behavior genetic model to study alcohol effects. *Pharmacol Biochem Behav* 67, 773–782.
128. Maximino, C., Marques, T., Dias, F., Cortes, F. V., Taccolini, I. B., Pereira, P. M., Colmanetti, R., Gazolla, R. A., Tavares, R. I. & Rodrigues, S. T. K. (2007) A comparative analysis of the preference for dark environments in five teleosts. *Int J Comp Psychol* 20, 351–367.
129. Serra, E. L., Medalha, C. C. & Mattioli, R. (1999) Natural preference of zebrafish (*Danio rerio*) for a dark environment. *Braz J Med Biol Res* 32, 1551–1553.
130. Burgess, H. A., Schoch, H. & Granato, M. (2010) Distinct retinal pathways drive spatial orientation behaviors in zebrafish navigation. *Curr Biol* 20, 381–386.
131. Bulkowski, L. & Meade, J. W. (1983) Changes in phototaxis during early development of walleye. *Trans Am Fish Soc* 112, 445–447.
132. Miklosi, A. & Andrew, R. J. (2006) The zebrafish as a model for behavioral studies. *Zebrafish* 3, 227–234.
133. Kicliter, E. (1973) Flux, wavelength and movement discrimination in frogs: forebrain and midbrain contributions. *Brain Behav Evol* 8, 340–365.
134. Colwill, R. M., Raymond, M. P., Ferreira, L. & Escudero, H. (2005) Visual discrimination learning in zebrafish (*Danio rerio*). *Behav Processes* 70, 19–31.
135. Jardine, D. & Litvak, M. K. (2003) Direct yolk sac volume manipulation of zebrafish embryos and the relationship between off-

spring size and yolk sac volume. *J Fish Biol* 63, 388–397.
136. Canfield, J. G. & Rose, G. J. (1993) Activation of Mauthner neurons during prey capture. *J Comp Physiol A Sens Neural Behav Physiol* 172, 611–618.
137. Logan, D. W., Burn, S. F. & Jackson, I. J. (2006) Regulation of pigmentation in zebrafish melanophores. *Pigment Cell Res* 19, 206–213.
138. Peng, J., Wagle, M., Mueller, T., Mathur, P., Lockwood, B. L., Bretaud, S. & Guo, S. (2009) Ethanol-modulated camouflage response screen in zebrafish uncovers a novel role for cAMP and extracellular signal-regulated kinase signaling in behavioral sensitivity to ethanol. *J Neurosci* 29, 8408–8418.
139. Lockwood, B., Bjerke, S., Kobayashi, K. & Guo, S. (2004) Acute effects of alcohol on larval zebrafish: a genetic system for large-scale screening. *Pharmacol Biochem Behav* 77, 647–654.
140. Concha, M. L. & Wilson, S. W. (2001) Asymmetry in the epithalamus of vertebrates. *J Anat* 199, 63–84.
141. Gamse, J. T., Kuan, Y. S., Macurak, M., Brosamle, C., Thisse, B., Thisse, C. & Halpern, M. E. (2005) Directional asymmetry of the zebrafish epithalamus guides dorsoventral innervation of the midbrain target. *Development* 132, 4869–4881.
142. Vallortigara, G. & Rogers, L. J. (2005) Survival with an asymmetrical brain: advantages and disadvantages of cerebral lateralization. *Behav Brain Sci* 28, 575–589.
143. Andrew, R. J., Dharmaretnam, M., Gyori, B., Miklosi, A., Watkins, J. A. & Sovrano, V. A. (2009) Precise endogenous control of involvement of right and left visual structures in assessment by zebrafish. *Behav Brain Res* 196, 99–105.
144. Sovrano, V. A. & Andrew, R. J. (2006) Eye use during viewing a reflection: behavioural lateralisation in zebrafish larvae. *Behav Brain Res* 167, 226–231.
145. Hansen, A. & Zeiske, E. (1993) Development of the olfactory organ in the zebrafish, *Brachydanio rerio. J Comp Neurol* 333, 289–300.
146. Li, J., Mack, J. A., Souren, M., Yaksi, E., Higashijima, S., Mione, M., Fetcho, J. R. & Friedrich, R. W. (2005) Early development of functional spatial maps in the zebrafish olfactory bulb. *J Neurosci* 25, 5784–5795.
147. Whitlock, K. E. (2006) The sense of scents: olfactory behaviors in the zebrafish. *Zebrafish* 3, 203–213.
148. Hara, T. J. (1994) The diversity of chemical stimulation in fish olfaction and gustation. *Rev Fish Biol Fish* 4, 1–35.
149. Vitebsky, A., Reyes, R., Sanderson, M., Michel, W. & Whitlock, K. (2005) Isolation and characterization of the laure olfactory behavioral mutant in the zebrafish, *Danio rerio. Dev Dyn* 234, 229–242.
150. Matz, C. J. & Krone, P. H. (2007) Cell death, stress-responsive transgene activation, and deficits in the olfactory system of larval zebrafish following cadmium exposure. *Environ Sci Technol* 41, 5143–5148.
151. Lindsay, S. M. & Vogt, R. G. (2004) Behavioral responses of newly hatched zebrafish (*Danio rerio*) to amino acid chemostimulants. *Chem Senses* 29, 93–100.
152. Prober, D. A., Zimmerman, S., Myers, B. R., McDermott, B. M., Jr., Kim, S. H., Caron, S., Rihel, J., Solnica-Krezel, L., Julius, D., Hudspeth, A. J. & Schier, A. F. (2008) Zebrafish TRPA1 channels are required for chemosensation but not for thermosensation or mechanosensory hair cell function. *J Neurosci* 28, 10102–10110.
153. Dempsey, C. H. (1978) Chemical stimuli as a factor in feeding and intraspecific behaviour of herring larvae. *J Mar Biol Ass UK* 58, 739–747.
154. Døving, K. B., Mårstøl, M., Andersen, J. R. & Knutsen, J. A. (1994) Experimental evidence of chemokinesis in newly hatched cod larvae (*Gadus morhua* L.). *Mar Biol* 120, 351–358.
155. Atema, J., Kingsford, M. J. & Gerlach, G. (2002) Larval reef fish could use odour for detection, retention and orientation to reefs. *Mar Ecol Prog Ser* 241, 151–160.
156. Gerlach, G., Hodgins-DAVIS, A., Avolio, C. & Schunter, C. (2008) Kin recognition in zebrafish: a 24-hour window for olfactory imprinting. *Proc Biol Sci* 275, 2165–2170.
157. Harden, M. V., Newton, L. A., Lloyd, R. C. & Whitlock, K. E. (2006) Olfactory imprinting is correlated with changes in gene expression in the olfactory epithelia of the zebrafish. *J Neurobiol* 66, 1452–1466.
158. Jesuthasan, S. J. & Mathuru, A. S. (2008) The alarm response in zebrafish: innate fear in a vertebrate genetic model. *J Neurogenet* 22, 211–228.
159. Speedie, N. & Gerlai, R. (2008) Alarm substance induced behavioral responses in zebrafish (*Danio rerio*). *Behav Brain Res* 188, 168–177.
160. Waldman, B. (1982) Quantitative and developmental analysis of the alarm reaction in the Zebra Danio, *Brachydanio rerio. Copeia* 1, 1–9.

162. Wallace, K. & Pack, M. (2003) Unique and conserved aspects of gut development in zebrafish. *Dev Biol* 255, 12–29.
162. Finney, J. L., Robertson, G. N., McGee, C. A., Smith, F. M. & Croll, R. P. (2006) Structure and autonomic innervation of the swim bladder in the zebrafish (Danio rerio). *J Comp Neurol* 495, 587–606.
163. Robertson, G. N., McGee, C. A., Dumbarton, T. C., Croll, R. P. & Smith, F. M. (2007) Development of the swim-bladder and its innervation in the zebrafish, *Danio rerio*. *J Morphol* 268, 967.
164. Goolish, E. & Okutake, K. (1999) Lack of gas bladder inflation by the larvae of zebrafish in the absence of an air-water interface. *J Fish Biol* 55, 1054–1063.
165. Laale, H. W. (1977) A literature review. *J Fish Biol* 10, 121–173.
166. Amsterdam, A., Nissen, R. M., Sun, Z., Swindell, E. C., Farrington, S. & Hopkins, N. (2004) Identification of 315 genes essential for early zebrafish development. *Proc Natl Acad Sci* 101, 12792.
167. Cahill, G., Hurd, M. & Batchelor, M. (1998) Circadian rhythmicity in the locomotor activity of larval zebrafish. *Neuroreport* 9, 3445–3449.
168. Hurd, M. & Cahill, G. (2002) Entraining signals initiate behavioral circadian rhythmicity in larval zebrafish. *J Biol Rhythms* 17, 307–314.
169. Yokogawa, T., Marin, W., Faraco, J., Pezeron, G., Appelbaum, L., Zhang, J., Rosa, F., Mourrain, P. & Mignot, E. (2007) Characterization of sleep in zebrafish and insomnia in hypocretin receptor mutants. *PLoS Biol* 5, e277.
170. Aschoff, J. (1960) Exogenous and endogenous components in circadian rhythms. *Cold Spring Harb Symp Quant Biol* 25, 11–28.
171. Mrosovsky, N. & Hattar, S. (2003) Impaired masking responses to light in melanopsin-knockout mice. *Chronobiol Int* 20, 989–999.
172. Zepelin, H. (2005) Mammalian sleep. In Kryger, M. H., Roth, T. & Dement, W. C. (Eds.) *Principles and Practice of Sleep Medicine*. 4th ed. Philadelphia, PA, Saunders.
173. Lyamin, O., Pryaslova, J., Lance, V. & Siegel, J. (2005) Animal behaviour: continuous activity in cetaceans after birth. *Nature* 435, 1177.
174. Siegel, J. M., Manger, P. R., Nienhuis, R., Fahringer, H. M. & Pettigrew, J. D. (1996) The echidna Tachyglossus aculeatus combines REM and non-REM aspects in a single sleep state: implications for the evolution of sleep. *J Neurosci* 16, 3500–3506.
175. Zimmerman, J. E., Naidoo, N., Raizen, D. M. & Pack, A. I. (2008) Conservation of sleep: insights from non-mammalian model systems. *Trends Neurosci* 31, 371–376.
176. Zhdanova, I. V., Wang, S. Y., Leclair, O. U. & Danilova, N. P. (2001) Melatonin promotes sleep-like state in zebrafish. *Brain Res* 903, 263–268.
177. Faraco, J. H., Appelbaum, L., Marin, W., Gaus, S. E., Mourrain, P. & Mignot, E. (2006) Regulation of hypocretin (orexin) expression in embryonic zebrafish. *J Biol Chem* 281, 29753–29761.
178. Kaslin, J. & Panula, P. (2001) Comparative anatomy of the histaminergic and other aminergic systems in zebrafish (*Danio rerio*). *J Comp Neurol* 440, 342–377.
179. Renier, C., Faraco, J. H., Bourgin, P., Motley, T., Bonaventure, P., Rosa, F. & Mignot, E. (2007) Genomic and functional conservation of sedative-hypnotic targets in the zebrafish. *Pharmacogenet Genomics* 17, 237–253.
180. Zhdanova, I. V. (2006) Sleep in zebrafish. *Zebrafish* 3, 215–226.
181. Taheri, S., Zeitzer, J. M. & Mignot, E. (2002) The role of hypocretins (orexins) in sleep regulation and narcolepsy. *Annu Rev Neurosci* 25, 283–313.
182. Lin, L., Faraco, J., Li, R., Kadotani, H., Rogers, W., Lin, X., Qiu, X., de Jong, P. J., Nishino, S. & Mignot, E. (1999) The sleep disorder canine narcolepsy is caused by a mutation in the hypocretin (orexin) receptor 2 gene. *Cell* 98, 365–376.
183. Mochizuki, T., Crocker, A., McCormack, S., Yanagisawa, M., Sakurai, T. & Scammell, T. E. (2004) Behavioral state instability in orexin knock-out mice. *J Neurosci* 24, 6291–6300.
184. Allada, R. & Siegel, J. M. (2008) Unearthing the phylogenetic roots of sleep. *Curr Biol* 18, R670–R679.
185. Cubbage, C. C. & Mabee, P. M. (1996) Development of the cranium and paired fins in the zebrafish *Danio rerio* (Ostariophysi, Cyprinidae). *J Morphol* 229, 121–160.
186. Rosenthal, G. G. & Ryan, M. J. (2005) Assortative preferences for stripes in danios. *Anim Behav* 70, 1063–1066.
187. Saverino, C. & Gerlai, R. (2008) The social zebrafish: behavioral responses to conspecific, heterospecific, and computer animated fish. *Behav Brain Res* 191, 77–87.
188. Engeszer, R., Ryan, M. & Parichy, D. (2004) Learned social preference in zebrafish. *Curr Biol* 14, 881–884.
189. Engeszer, R. E., Wang, G., Ryan, M. J. & Parichy, D. M. (2008) Sex-specific perceptual

spaces for a vertebrate basal social aggregative behavior. *Proc Natl Acad Sci USA* 105, 929–933.
190. McCann, L. I. & Carlson, C. C. (1982) Effect of cross-rearing on species identification in zebra fish and pearl danios. *Dev Psychobiol* 15, 71–74.
191. Mann, K. D., Turnell, E. R., Atema, J. & Gerlach, G. (2003) Kin recognition in juvenile zebrafish (*Danio rerio*) based on olfactory cues. *Biol Bull* 205, 224–225.
192. Gerlach, G. & Lysiak, N. (2006) Kin recognition and inbreeding avoidance in zebrafish, *Danio rerio*, is based on phenotype matching. *Anim Behav* 71, 1371–1377.
193. Darland, T. & Dowling, J. E. (2001) Behavioral screening for cocaine sensitivity in mutagenized zebrafish. *Proc Natl Acad Sci USA* 98, 11691–11696.
194. Williams, F. E., White, D. & Messer, W. S. (2002) A simple spatial alternation task for assessing memory function in zebrafish. *Behav Processes* 58, 125–132.
195. Thorsen, D., Cassidy, J. & Hale, M. (2004) Swimming of larval zebrafish: fin-axis coordination and implications for function and neural control. *J Exp Biol* 207, 4175–4183.
196. Higashijima, S., Masino, M. A., Mandel, G. & Fetcho, J. R. (2003) Imaging neuronal activity during zebrafish behavior with a genetically encoded calcium indicator. *J Neurophysiol* 90, 3986–3997.
197. Yaksi, E., Judkewitz, B. & Friedrich, R. W. (2007) Topological reorganization of odor representations in the olfactory bulb. *PLoS Biol* 5, e178.
198. Asakawa, K., Suster, M. L., Mizusawa, K., Nagayoshi, S., Kotani, T., Urasaki, A., Kishimoto, Y., Hibi, M. & Kawakami, K. (2008) Genetic dissection of neural circuits by Tol2 transposon-mediated Gal4 gene and enhancer trapping in zebrafish. *Proc Natl Acad Sci USA* 105, 1255–1260.
199. Davison, J. M., Akitake, C. M., Goll, M. G., Rhee, J. M., Gosse, N., Baier, H., Halpern, M. E., Leach, S. D. & Parsons, M. J. (2007) Transactivation from Gal4-VP16 transgenic insertions for tissue-specific cell labeling and ablation in zebrafish. *Dev Biol* 304, 811–824.
200. Scott, E. K., Mason, L., Arrenberg, A. B., Ziv, L., Gosse, N. J., Xiao, T., Chi, N. C., Asakawa, K., Kawakami, K. & Baier, H. (2007) Targeting neural circuitry in zebrafish using GAL4 enhancer trapping. *Nat Methods* 4, 323–326.
201. Tinbergen, N. (1952a) The behavior of the stickleback. *Sci Am* 187, 22–26.
202. Rafal, R., Smith, J., Krantz, J., Cohen, A. & Brennan, C. (1990) Extrageniculate vision in hemianopic humans: saccade inhibition by signals in the blind field. *Science* 250, 118–121.
203. Howard, I. P. & Ohmi, M. (1984) The efficiency of the central and peripheral retina in driving human optokinetic nystagmus. *Vision Res* 24, 969–976.
204. Aronson, L. R. (1970) Functional evolution of the forebrain in lower vertebrates. In Aronso, N. L. R., Tobac, H. E., Lehrma, N. D. S. & Rosenblat, T. J. S. (Eds.) *Development and Evolution of Behavior*. San Francisco, CA, WH. Freeman and Company.

Subject Index

A.V. Kalueff, J.M. Cachat (eds.), *Zebrafish Models in Neurobehavioral Research*, Neuromethods 52,
DOI 10.1007/978-1-60761-922-2, © Springer Science+Business Media, LLC 2011

MIX
Papier aus verantwortungsvollen Quellen
Paper from responsible sources
FSC® C105338

If you have any concerns about our products,
you can contact us on
ProductSafety@springernature.com

In case Publisher is established outside the EU,
the EU authorized representative is:
Springer Nature Customer Service Center GmbH
Europaplatz 3, 69115 Heidelberg, Germany

Printed by Libri Plureos GmbH
in Hamburg, Germany